Essays on Geometry

A. Muhammed Uludağ • Ayberk Zeytin
Editors

Essays on Geometry

Celebrating the 65th Birthday of Athanase Papadopoulos

Springer

Editors
A. Muhammed Uludağ
Department of Mathematics
Galatasaray University
Beşiktaş, Istanbul, Türkiye

Ayberk Zeytin
Department of Mathematics
Galatasaray University
Beşiktaş, Istanbul, Türkiye

ISBN 978-3-031-76259-8 ISBN 978-3-031-76257-4 (eBook)
https://doi.org/10.1007/978-3-031-76257-4

This Springer imprint is published by the registered company Springer Nature Switzerland AG
The registered company address is: Gewerbestrasse 11, 6330 Cham, Switzerland

Preface

We are delighted to present a selection of papers, some by colleagues who were unable to attend the "Conference in Honor of the 65th Birthday of Athanase Papadopoulos," that were written following the event. This event, held to celebrate the contributions and achievements of our dear friend Athanase, was a landmark occasion for our team and our department. It marked the first conference we organized since 2019, following the "Journées Arithmetiques," a series with a rich and storied history.

This event also underlines the long-lasting collaboration between our team and Athanase. Over the years, this partnership has fostered numerous fruitful discussions, research initiatives, and academic advancements. Indeed, together we have organized many conferences at Galatasaray University, exchanged many students and research visits, and collaborated in joint research projects among others.

The onset of the COVID-19 pandemic had a profound impact on academic gatherings worldwide, and our conferences were no exception. The pandemic not only disrupted our usual schedule but also posed significant logistical and financial challenges. Travel restrictions, health concerns, and social distancing measures made it nearly impossible to host in-person events. As a result, many conferences were either canceled or transitioned to virtual formats, which, while effective, could not fully replicate the dynamic interactions and networking opportunities of traditional face-to-face meetings.

We have managed to secure generous financial support of the personal grant of Sumio Yamada, the Turkish Mathematical Society, and the TÜBİTAK grant 119F405, for which we are thankful. Without their contributions, this conference would not have been possible. We thank our colleagues at the Department of Mathematics of Galatasaray University for their unwavering support and collaboration throughout the planning and execution of this event.

We are also grateful to the attendees for their active participation and engagement, which made the conference a truly enriching experience for everyone involved. We also appreciate the efforts of the organizing committee, namely İbrahim Emir Çiçekli, Eliz Gündüz, İsmail Sağlam, İlayda Serbes, Zeliha Sevgi, and

Özge Ülkem. The conference featured an impressive list of speakers, each bringing their expertise to the event. We were honored to host the following list of speakers in the conference: Gérard Besson, Zehra Bilgin, Charalampos Charitos, Gilles Courtois, Viveka Erlandsson, Krishnendu Gongopadhyay, Sachiko Hamano, Xavier Hascher, Daniel Massart, Dmitry Millionshchikov, Vassili Nestoridis, Ken'ichi Ohshika, Ioannis Papadoperakis, Athanase Papadopoulos, Massimo Picardello, Ahmad Rafiqi, İsmail Sağlam, Andrea Seppi, Hiroshige Shiga, Muhammed Uludağ, and Sumio Yamada.

Each presentation contributed to the vibrant and dynamic discussions. The breadth of the topics covered and the depth of the insights shared are a proof of the intellectual legacy of Athanase. Last but not least, we are grateful to the referees of each chapter in this volume for their meticulous work and insightful comments, which have contributed to improving the quality and clarity of the content. Their expertise have been invaluable in shaping the final version of this book.

We hope that the collection of articles contained within this volume will serve as a valuable resource for both researchers and scholars. It is our sincere wish that the knowledge and ideas exchanged during this conference will inspire continued innovation and collaboration. We look forward to organizing many more such events in the future, continuing the tradition of excellence and friendship that defines our academic community.

Each article has undergone a meticulous review process by leading researchers, ensuring the quality and relevance. The book offers a blend of original research and insightful survey articles explaining recent advancements and open problems. Figures 1, 2, 3, 4, 5, and 6 present a series of photographs captured at various moments throughout the conference.

Fig. 1 Athanase Papadopoulos delivering his talk during the conference

Fig. 2 During the traditional Bosphorus tour (left to right: Sachiko Hamano, Susmu Tanabé, and Athanase Papadopoulos)

Fig. 3 During the traditional Bosphorus tour (left to right: Krishnendu Gongopadhyay, A. Muhammed Uludağ, and Athanase Papadopoulos)

Fig. 4 A group picture with some of the participants (left to right: Ken'ichi Ohshika, Sumio Yamada, Xavier Hascher, Athanase Papadopoulos, Vassilis Nestoridis, Hiroshige Shiga, and Sachiko Hamano)

Fig. 5 A group picture with some of the participants (left to right: Krishnendu Gongopadhyay, İlayda Serbes, Viveka Erlandsson, A. Muhammed Uludağ, Daniel Massart, İsmail Sağlam, İ. Emir Çiçekli, Daniel Massart, Massimo Picardello, Sofiane Souaifi, Gilles Courtois, Özge Ülkem, Dmitry Millionshchikov, Eliz Gündüz, Gérard Besson, Athanase Papadopoulos, Teoman Yalçınkaya, Zehra Bilgin, K. Salih Bilgin, and Öznur Turhan)

Fig. 6 During the birthday dinner (left to right: Ken'ichi Ohshika and Athanase Papadopoulos)

The authors acknowledge financial support by TÜBİTAK grants 119F405 and 221N171 as well as a GSU research grant FIR-2022-1089 during the preparation of this book.

Istanbul, Türkiye
August 2024

A. Muhammed Uludağ
Ayberk Zeytin

Contents

Contributors

Vassiliki Farmaki Department of Mathematics, National and Kapodistrian University of Athens, Panepistemiopolis, Athens, Greece

Krishnendu Gongopadhyay Indian Institute of Science Education and Research (IISER) Mohali, S.A.S. Nagar, Punjab, India

Sachiko Hamano Department of Mathematics, Faculty of Science, Kyoto Sangyo University, Kyoto, Japan

Nariya Kawazumi Department of Mathematical Sciences, University of Tokyo, Komaba Meguro-ku, Tokyo, Japan

François Laudenbach Laboratoire de Mathématiques Jean Leray, UMR 6629 du CNRS, Faculté des Sciences et Techniques, Université de Nantes, Nantes cedex 3, France

Chandan Maity Indian Institute of Science Education and Research (IISER) Berhampur, Berhampur, Odisha, India

Hideki Miyachi School of Mathematics and Physics, College of Science and Engineering, Kanazawa University, Kakuma-machi, Kanazawa, Ishikawa, Japan

Stelios Negrepontis Department of Mathematics, National and Kapodistrian University of Athens, Panepistemiopolis, Athens, Greece

Ken'ichi Ohshika Department of Mathematics, Gakushuin University, Mejiro, Toshima-ku, Tokyo, Japan

Robert Penner Institut des Hautes Études Scientifiques, Bures-sur-Yvette, France

Massimo A. Picardello Dipartimento di Matematica, Università di Roma "Tor Vergata", Roma, Italy

Arkady Plotnitsky Literature, Theory, and Cultural Studies Program, Philosophy and Literature Program, Purdue University, W. Lafayette, IN, USA

A. Muhammed Uludağ Department of Mathematics, Galatasaray University, Beşiktaş, Istanbul, Türkiye

Ayberk Zeytin Department of Mathematics, Galatasaray University, Beşiktaş, Istanbul, Türkiye

Chapter 1
Introduction

A. Muhammed Uludağ and Ayberk Zeytin

Abstract This volume presents a collection of diverse and insightful articles authored by renowned experts. In this introduction, we give an overview of these contributions.

The book begins with an interview with Athanase Papadopoulos, conducted by A. Muhammed Uludağ. The interview delves into both the personal and professional journey of Athanase. A wide array of topics is covered, from Papadopoulos's multicultural upbringing in a Levantine society to his educational path in France, his decision to pursue mathematics over engineering, and his subsequent academic career. Papadopoulos shares insights into his research contributions. The interview also highlights his interest in the history of mathematics, detailing his work on Arabic manuscripts and his admiration for historical figures like Euler and Thurston. Additionally, Papadopoulos reflects on the intersection of mathematics and music, a subject close to his heart given his family's musical background. The dialogue provides a comprehensive look at Papadopoulos's multifaceted career, his views on mathematical research, education, and the broader cultural and intellectual contexts that have shaped his work.

In Chap. 3, entitled "A glance at Novikov's theory of multivalued Morse functions and related harmonic bifurcations", François Laudenbach discusses a generalization of Morse theory in the context of dynamical systems. Specifically, a Morse complex is defined over a smooth, closed manifold M, and is associated with a Morse function $f : M \to \mathbb{R}$ together with a gradient vector field X. The dynamics of the gradient of a Morse function is well understood. More generally, a Morse-Novikov complex is defined by analogy, where the Morse function f is replaced by a non-exact closed differential form of degree one on M. The author first introduces the twisted Morse complex and then the Morse-Novikov complex. This complex turns out to be a graded differential complex whose homology depends

A. M. Uludağ · A. Zeytin (✉)
Department of Mathematics, Galatasaray University, Beşiktaş, Istanbul, Türkiye
e-mail: muludag@gsu.edu.tr; azeytin@gsu.edu.tr

A. M. Uludağ, A. Zeytin (eds.), *Essays on Geometry*,
https://doi.org/10.1007/978-3-031-76257-4_1

on the chosen one-form. After introducing a fixed one-form, the author explains how the Morse-Novikov complex changes when X is deformed by a homoclinic bifurcation.

In Chap. 4, entitled "A twisted invariant of a compact Riemann surface" Nariya Kawazumi discusses the study of functions on the moduli space of compact Riemann surfaces of genus g, which are regarded as analytic invariants of a compact Riemann surface. Abelian integrals and arithmetic geometry, which includes Arakelov geometry, serve as effective tools for understanding compact Riemann surfaces which are also relevant in arithmetic geometry. Faltings introduced an analytic invariant of compact Riemann surfaces based in the context of Arakelov geometry, while Hain and Reed introduced another invariant based on the Hodge line bundle and intermediate Jacobian. De Jong proved that the Faltings invariant, the Hain–Reed invariant, and the Kawazumi–Zhang invariant are linearly dependent. In this work, the authors introduce a twisted version of the Kawazumi–Zhang invariant of a compact Riemann surface, which is a smooth tensor field on the moduli space $\mathcal{M}_g$ of Riemann surfaces of genus g and discuss its relation to the first Mumford–Morita–Miller class and the original Kawazumi–Zhang invariant. This is the first main result of the chapter. The second main result is that when $g = 3$, the Kawazumi–Zhang invariant on $\mathcal{M}_3$ (endowed with the metric coming from the moduli space $\mathcal{A}_3$ of principally polarized Abelian varieties of dimension 3) is not an eigenfunction for the Laplacian.

In Chap. 5, entitled "Directional moduli and pseudoconvexity", Sachiko Hamano focuses on understanding how the conformal invariants associated with the open Riemann surface vary when deformed over a pseudoconvex domain. This problem is relevant when given an open Riemann surface S one asks how many other compact Riemann surfaces there are in which S embeds preserving the conformal structure. Formulating this problem in terms of Schiffer spans in $g = 0$ and hyperbolic spans in $g = 1$ leads to interesting results. For compact hyperbolic or elliptic Riemann surfaces the concept of **a**-span and **a**-directional moduli disk is introduced as a way to measure the rigidity of a pseudoconvex domain. The main result of the chapter is that when deforming over a pseudoconvex domain, the diameter of the **a**-directional moduli disk is subharmonic.

In Chap. 6, entitled "Angle defect for super triangles", Robert Penner continues his walk on the path of Lobachevsky[1] and discusses the angle defect theorem in the context of super-hyperbolic geometry. Recall that total area of a hyperbolic triangle is $\pi - \alpha - \beta - \gamma$; where α, β and γ are the interior angles of a hyperbolic triangle. The angle defect theorem states that for any hyperbolic triangle the defect (that is the difference π − sum of interior angles) is equal to the area of the triangle. In this contribution, the author asks the analogous question in the context of $\mathcal{N} = 1$ super-hyperbolic geometry, i.e. the geometry of the usual two-real dimensional hyperbolic plane enriched with two real anti-commuting variables. Standard arguments, such as

[1] "In order to understand a geometry, both locally and globally, it is enough to understand its triangles." as rephrased by Athanase Papadopoulos.

the additivity of the area of triangles do not hold true in super hyperbolic geometry. The proof is obtained through lengthy computations.

Chapter 7 is concerns the Lipschitz constant and quasiconformal dilatation of geographical maps. More precisely, the authors Hideki Miyachi and Ken'ichi Ohshika focus on the Delisle map—a projection of a region of the sphere onto a cone. This map was historically used by Euler for mapping the Russian Empire, and the article provides numerical evidence that the Delisle map is optimal among several similar maps. The paper compares several projections from the sphere to a cone, study their Lipschitz and quasiconformal properties to assess whether they are suitable for cartography. The chapter also discusses historical contributions to the field, including work by Euler, Ptolemy, and Tissot, and explores how these concepts have evolved within mathematical geography. The chapter's ultimate aim is to determine whether Euler's preference for the Delisle map was justified, particularly for representing the Russian Empire, and concludes that it was indeed a reasonable choice based on the metrics considered.

Chapter 8, entitled "Positive definite functions on semi-homogeneous trees and spherical representations" is a work by Massimo A. Picardello where the group of isometries, say G, of a semi-homogeneous tree (that is, a bipartite tree having vertices of type $+$ and $-$ so that every vertex of type $+$ is of degree q_+ and every vertex of type $-$ is of degree q_-) is studied. An application of such a study is a sequence of unitary spherical representation of G. To this end, the author considers the action of G on the set of vertices of type $+$ (resp. $-$), denoted V_+ (resp. V_-). This action allows one to define positive definite functions via a convolution product. This convolution product turns out to be closely related to the Laplacian. Then the study focuses on multiplicative functionals on the of spherical functions. It is proven that the spherical functions are positive definite if and only if they are real-valued and bounded. Additionally, the spherical function corresponding to a real eigenvalue in the ℓ^1-spectrum of the Laplacian gives rise to two unitary representations of G, one for each parity.

In Chap. 9, entitled "Trees of Fractions" A. Muhammed Uludağ explores various submonoids of the monoid $GL_2(\mathbb{N})$ and related trees such as the Calkin–Wilf, Stern–Brocot, and others. The chapter begins by defining $SL_2(\mathbb{N})$ as the monoid of unimodular 2×2 matrices with non-negative integral entries and establishing that $SL_2(\mathbb{N})$ is freely generated by the matrices T and T^t. The chapter further investigates the structure and properties of $GL_2(\mathbb{N})$, showing it is generated by the matrices T, T^t, and U, and explores the relationships and actions of these matrices on the set of positive rational numbers, $\mathbb{Q}^+$. The chapter discusses the construction and properties of the Calkin-Wilf tree, a tree structure representing positive rationals, showing how every rational number can be uniquely represented within the tree through the matrices T and T^t. It also covers the Stern-Brocot tree, which provides another enumeration of rationals through mediant operations, and discusses its relationship with the Calkin-Wilf tree. Further sections introduce other trees like the Drib and Bird trees, each with unique generation rules and properties. The chapter is rich in exercises and propositions, inviting the reader to explore the details and connections between these mathematical structures.

Chapter 10, entitled “Binary quadratic forms: modern developments” by Ayberk Zeytin, provides a comprehensive historical overview and contemporary algebraic developments of binary quadratic forms. It traces the origins of binary quadratic forms back to ancient times, starting with Pythagorean triples documented in the Babylonian tablet Plimpton 322 around 1800 BCE. The chapter highlights significant contributions from Euclid, Diophantus, and Brahmagupta, and it notes a gap until Fermat, who made advances in the study of equations of Pell type. The eighteenth century saw a resurgence in the study of binary quadratic forms with contributions from Euler, Lagrange, Legendre, and particularly Gauss, who introduced fundamental concepts like the Gauss product on the set of classes of primitive binary quadratic forms, leading to the birth of algebraic number theory. In modern developments, the chapter covers the work of Bhargava, who generalized the Gauss product using Bhargava cubes, providing a new composition law for binary quadratic forms. It also discusses Zagier’s alternative reduction theory and Penner’s geometric interpretation of the Gauss product for definite binary quadratic forms. The author introduces the concept of Çarks, a combinatorial viewpoint interpreting the theory of indefinite binary quadratic forms as part of the theory of dessins d’enfants, emphasizing the interplay between combinatorics and number theory. The chapter concludes by exploring algorithmic solutions to the representation problem and the impact of these modern advancements on the broader field of algebraic number theory.

In Chap. 11, entitled “Reversibility of unipotent matrices” by Krishnendu Gondopadhyay and Chandan Maity real elements in the group of unipotent upper-triangular matrices and also in related groups are studied. More explicitly, for a lie group G, $\mathfrak{g}$ can be endowed with a notion of reality. This notion can be used to define an element as reversible (i.e., conjugate to its own inverse) and strongly reversible (i.e. expressed as a product of two involutions). Four questions related to such elements are treated. The author not only discusses shorter proofs of some of the existing results about these questions, but also present some new results.

The penultimate chapter entitled “‘Le complément supérieur’: On the Poetics of Mathematics” by Arkady Plotnitsky explores the asymmetrical relationship between modern mathematics and modern physics, with a focus on the significant impact of mathematics on physics. The author argues that while modern mathematics, claimed to have emerged around 1800, has separated itself from natural objects and physics, modern physics, especially in the twentieth century, has heavily relied on these abstract mathematical theories to advance. The discussion highlights how mathematical concepts have enabled the development of physical theories like relativity and quantum theory, with modern physics often borrowing and benefiting from pre-existing mathematical frameworks. This reliance on mathematics has defined modern physics’ ability to predict and represent physical reality, emphasizing an asymmetry where mathematics has had a more profound influence on physics than the reverse. The chapter further discusses the historical and conceptual development of both fields, noting key contributions from figures like Dirac, who advocated for using pure mathematics to advance theoretical physics, and Heisenberg, whose formulation of quantum mechanics relied on abstract mathematical structures.

The chapter also considers the philosophical implications of this relationship, particularly in quantum theory and quantum mechanics, where the mathematical formalism does not necessarily correspond to physical reality but serves as a tool for probabilistic predictions. Plotnitsky challenges the view that theoretical physics can resolve pure mathematical problems, suggesting that such solutions would require theoretical physics to transform into a form of mathematics. The discussion also touches on the role of modern mathematics in addressing complex physical theories and its influence on conceptualizing the structure and behavior of physical phenomena.

The last chapter is "Pythagorean Book II of the *Elements* restored and Pythagorean Incommensurabilities reconstructed" by Stelios Negrepontis and Vassiliki Farmaki. The chapter delves into the historical and mathematical exploration of how the Pythagoreans might have originally proven the incommensurability of magnitudes, particularly focusing on the diameter and the side of a square. The authors argue against the widely accepted arithmetical proofs hinted at by Aristotle and later incorporated anonymously into Euclid's Elements as Proposition X.117. Instead, they propose that the original Pythagorean proof was anthyphairetic, meaning it used a method of reciprocal subtraction (anthyphairesis) to demonstrate the infinite nature of the ratio between the side and the diagonal, thereby proving their incommensurability. Negrepontis and Farmaki support their argument by reconstructing Book II of Euclid's Elements to reflect what they believe to be its original Pythagorean form. This reconstruction includes repositioning the Pythagorean theorem and related propositions to align with the philosophical principles of the Infinite and Finite. They also present the recursive definition of side and diameter numbers, providing evidence that these were used by the Pythagoreans to approximate incommensurability through an infinite anthyphairesis. The authors further bolster their argument by interpreting various ancient texts and stories, such as those involving Hippasus, to correlate incommensurability with the concept of the Infinite. Through this detailed historical and mathematical analysis, they conclude that the Pythagorean proof of incommensurability was fundamentally anthyphairetic and deeply intertwined with their philosophical views on infinity and finiteness.

Chapter 2
Interview with Athanase Papadopoulos

A. Muhammed Uludağ

Abstract Athanase Papadopoulos was interviewed for this chapter. Beginning with his multicultural upbringing and engineering experience, he went on to become a well-known figure in mathematics. In addition to his academic accomplishments, we talk about his collaborative work style and his keen interest in mathematical history, particularly as it relates to Greek passages found in Arabic manuscripts. He talks about his passion for music, how travel has influenced his outlook, and his opinions on contemporary problems affecting the mathematical community.

– Let me start from the very beginning. You were born to Syriac and Greek parents—who both spoke Turkish, I presume—who emigrated from Anatolia to Syria where you grew up, speaking Greek and Arabic.

This is correct. People around me used to speak Arabic, Greek, French, Italian, Armenian, Aramean and other languages. I grew up in a Levantine society where it was not unusual to speak six or seven languages. My grandparents spoke Turkish. They were among the people who were forced to emigrate from the Ottoman Empire. When I am in Turkey and listen to a conversation, I remember my maternal grandfather's voice. I recognize certain words and I understand the general meaning of the conversation. I hold a special affection for this part of the world that you call Anatolia. That is why I am so moved each time I come to Turkey.

– This sounds very much like my family history, who were expelled from Caucasia after the Russo-Ottoman war of 1877–1878 and ended up in Anatolia. Can you tell us how your family ended up in Italy?

When my father passed away, my mother's family (her mother and brothers) were all settled in Italy. My mother, with all of us (we were 5 children) moved there. It was in the 1970s.

– And what was your father doing in Syria?

A. M. Uludağ (✉)
Department of Mathematics, Galatasaray University, Beşiktaş, Istanbul, Türkiye
e-mail: muludag@gsu.edu.tr

A. M. Uludağ, A. Zeytin (eds.), *Essays on Geometry*,
https://doi.org/10.1007/978-3-031-76257-4_2

He had a shipping company, in Lattakia, the port city of Syria. Even though his business was going well and making life easy for us, none of us had the idea of taking over his business after his death, and in any case we were not prepared for the business world: we were all influenced by my mother, who was an artist—a painter.

– And what is the story of your decision to study engineering in Paris and how have you decided to do Ph.D. in mathematics afterwards?

Actually, I first hesitated between several subjects and eventually entered mathematics at Orsay. But as you know, in France there is this system of "grandes écoles", and it is tempting to try to get into one of them. I presented the entrance exam to École Centrale while I was in my 4th year of mathematics and I was accepted there. I finished the section of civil engineering two years later. It was also a kind of family tradition on my mother's side to study civil engineering.

It was during the last year internship, a 3 month training in a construction company (Grands Travaux de Marseille) that I decided not to work as an engineer, and came back to mathematics. During my two years at École Centrale, I had continued going to Orsay, and I had completed there the 5th year diploma, which permits you to start a PhD (it was called Diplôme d'Études Approfondies). During my last year at École Centrale, I followed a course that Gromov gave at Orsay; he had just arrived to France, after he spent a few years in the United States.

To come back to my internship: I was working on a site which was a project of extension of a metro line in Paris, line No. 7. I was officially in charge of calculating, every day, the volume of concrete needed for the next day. Of course, the site foreman knew very well how to do this without my help, and I was of no use on the construction site. I was bored the whole day long. You know, in that kind of work, in an office with others, you can read a newspaper, you can chat all the day with colleagues, but it is frowned upon to do mathematics, or something of this kind, or to read a novel. Beyond this fact, I realized that I wasn't cut out for working as an engineer.

By the end of this internship I contacted François Laudenbach, described my situation and told him I would like to enter a PhD program in mathematics. I already knew François from my undergraduate studies. He understood the situation, and with some phone call, he obtained for me a scholarship for a PhD. A few days later I went to some office and I signed the contract, for three years. Two years later I had completed my PhD with him, after I solved the problem he asked me.

– There should be a mathematical work (theorem, definition, lemma, conjecture) of yours that you must be particularly fond of. Which one is it?

– I think this is my first theorem, that every symplectic automorphism of the homology of a surface is realized by a pseudo-Anosov mapping class. It is the problem that Laudenbach gave me when I entered PhD with him. It is thanks to this question that I learned Thurston's theory of surfaces, and it was something I would carry with me for the rest of my career. My DEA diploma was in logic (model theory), nothing to do with surfaces. With Laudenbach's question in mind, I learned about mapping class groups, measured foliations, train tracks, dynamics, Teichmüller spaces, etc. I had already been introduced to hyperbolic geometry through an undergraduate course on complex analysis by Poénaru. It was a topic rarely discussed in those times.

– And your preferred paper?

– My recent paper about Pavel Florensky. It involves mathematics, history, poetry, philosophy, theology, literature, cinema and other things.

– Unbeknownst to outsiders, mathematics is very much about friendship, and I know you've had several long-lasting collaborations in your journey. Could you tell us about your style? I mean, how do you work with your colleagues?

– I always say that as mathematicians we have the chance of being able to choose our collaborators. There is a great quantity of good mathematicians around, but I choose my collaborators according to their human qualities. You mention friendship. It is something I value very much, in working with others. The response to your question (how I work with my colleagues): there is no special way, except maybe the fact that I usually work in cafés. Usually alone, but with several of my collaborators as well, we usually meet in cafés. When I think about a result, I always remember the café where the idea came out.

– Other values you like in your collaborators?

– Gentleness and modesty. The latter is, as you know, a quality which is quite rare among mathematicians.

– I can see that mathematics itself is not the main element in your choice of your collaborators.

– Indeed it is not.

– And how do you choose the problems you work on?

– I formulate them, guided by analogies.

– So you do not work on open problems?

– No. I do not like competition. I think that doing mathematics is not compatible with competition. In general I work on problems I formulate myself, or with my collaborators, during conversations. There is one exception: the problem that Laudenbach asked me when I entered PhD. Of course, later, you might learn that the problem you work on has been asked by others. I learned that Laudenbach's question was asked by Katok.

– And if you want to characterize in one word your activity as a mathematician?

– Understanding.

– You have met many prominent mathematicians in Europe and around the world. Is there a particular one who left his/her mark on you?

– Thurston.

– A great name indeed. I have never had the privilege of meeting him, though I consider him a friend through you... And how did you end up in Strasbourg?

After I finished my PhD, and after one year I spent in Princeton, I obtained a position at CNRS. In those times, when you were appointed at CNRS, you could choose the university you wanted to work in. Laudenbach suggested that I join the Strasbourg Mathematics Department. There were a few geometers from the older generation, but they were no more active in geometry. Godbillon was completely involved in administration, and Reeb was focused on non-standard analysis. The only geometer there was Bernard Morin (I am not talking about algebraic geometry), and the geometry group needed to be revived. Laudenbach, who cared about the state of French mathematics beyond Paris, told me it would be good if I join Morin in Strasbourg. His project was to set up a new geometry and topology team there.

– Did you work with Morin?

– No. He had a complicated personality, and he was difficult to work with. He used to monopolize everyone around him.

– Was this related to the fact that he was blind?

– In part. He was taking advantage of this situation.

– But he was a good geometer... He worked with Masaaki Yoshida, a friend of mine, on four dimensions.

– Morin was an excellent geometer.

– And you were not interested in working with him?

– Not after I had him as a colleague.

– Your first PhD student, Michel Coornaert, was also visually impaired, and you co-authored a book with him. Can you share any memorable moments or achievements from your collaboration with our joint friend Michel?

I coauthored two books with Michel. When I started working with him, his vision wasn't so bad; the disease attacked him gradually. When I arrived to Strasbourg, he was teaching in a technical high school near IRMA, as "professeur agrégé". He had studied at the École Normale Supérieure. Although teaching in a lycée, he was very much interested in mathematical research. He used to come to IRMA to attend seminars. We often chatted together, and at some point I suggested that he enroll for a doctorate. We were reading a paper by Sullivan on dynamics and measures at infinity, and we formulated together a problem in that direction. It worked out well. Michel is very talented. After his PhD we became collaborators, and for many years we were co-authors. We used to work together every day in one of the cafés, near the math institute, le Zouave or la Faluche.

Michel is at the same time serious and humorous. It was good to collaborate with him. It was a time when I was interested in geometric group theory. Then I developed other interests and realized that Michel had become much stronger than I in this field. I told him, "Michel, you should now work with others instead of me." I felt he was doing most of the work and that idea bothered me. I'm not sure he actually understood what I meant, but in any case, he continued on his own. He found new collaborators, especially Tullio Checcherini-Silberstein, with whom he did a great amount of work.

– Mathematics is about communication and travel, too. And you have an astonishing travel record. Why? Do you find it easier to work when you are away from home?

No. These are different matters. I like to travel for cultural and emotional reasons. I like Asia, Athens, Italian cities, some American campuses, Istanbul, a city straddling Europe and Asia, and many other places. I like the idea of merging with different cultures and working with people there. I feel at home at several places. Each time I leave Asia, I feel nostalgic. Istanbul is such a place. One of my great uncles was a teacher at the Phanar Greek College, before the Greeks were forced to leave, after what they call the Catastrophe. Of course I never met him, but sometimes I imagine him standing at the window of that building, looking at me as I walk past, when I am in Istanbul. Regarding mathematics, I can work the same at home, in a café, on a train, on a plane or in a hotel room. It turned out that all my collaborators live in different countries. So I travel also for mathematics. Traveling also gives a sense of freedom. The fact that I have a position at CNRS, with no obligation to teach, simplifies these matters. This also makes it easy for me to give courses in Varanasi, Tokyo, Saint Petersburg, Istanbul, Florence, Beijing, etc. I like this kind of temporary teaching abroad, with no preassigned program, no exam at the end, and where the students attend only if they are interested.

– Incidentally, I notice that when I asked you to send me pictures, to include in the printed version of this interview, you sent me pictures taken outside Strasbourg (Figs. 2.1–2.4). Papers about music start to appear in your cv from the year 2000 on; though I know that your interest in music precedes that since three of your children became professional musicians (I can't help noting that there are no mathematicians). How does music sits in the greater scheme of things in your life? And how does it relate to mathematics?

– In fact, the four of them are professional musicians. My eldest daughter first entered CNRS, in applied mathematics (signal analysis); she remained there for 5 or 6 years, then she realized she wanted to dedicate all her time to music, piano performance, so she resigned from CNRS. It is a kind of intellectual honesty. It is very rare that someone resigns from CNRS. It is like resigning from your pension.

Regarding my publications on music: In the 1990s, I started teaching a course called "mathematics and music", which was common to mathematics and music students at the University of Strasbourg. I was interested—first by curiosity—in reading works on music theory written by mathematicians, like Euler, Descartes, Huygens, and others. Then I started reading older authors: Aristoxenus, Euclid, Ptolemy, etc. Then I read modern authors: Boulez, Xenakis, and Olivier Messiaen's *Traité de rythme, de couleur, et d'ornithologie*. The treatise is all about mathematics, but you have to be a mathematician in order to see this. And then, when you are involved in such a course, you naturally want to write down what you know. This is how I started publishing on music theory.

– Were you involved yourself in music, I mean practical music ?

– Yes, for a dozen of years, I was choirmaster at the Russian orthodox church in Strasbourg. That was in the 1980s and 1990s. And I was heavily involved in my children's musical education.

– In mathematics, you have edited several books of surveys. What are the reasons for this work?

– Because I believe that the mathematical community is in need of good surveys, and also, because there are subjects I like and I want to promote. Furthermore, it is a collective effort, each book involving a number of colleagues and friends, and from that point of view it is a good experience. Personally, when I need specific information on a subject, I look through these books and see what a competent person has written on it.

Fig. 2.1 2016: Athanase Papadopoulos with his students Elena Frenkel, Guillaume Théret and Vincent Alberge in Montpellier (France)

Fig. 2.2 2019: Athanase Papadopoulos with Sumio Yamada, Krishnendu Gongopadhyay and Alena Zhukova—VV Nagar (Gujarat—India)

Fig. 2.3 2019: Athanase Papadopoulos with Gilles Courtois and Gérard Besson, CIMPA school, Varanasi (India)

Fig. 2.4 2013: Yantra Raj observatory in Jaipur (Rajasthan, India). Conference on history of mathematics. To the left of Athanase Papadopoulos, S.G. Dani

– **What are the topics you like?**

– Metric geometry, and the topics related to Thurston's theories (all of them).

– **Can you say a few words about your books related to music theory?**

– As I told you, I taught a course on mathematics and music, for several years. The course was shared with a colleague at the music department, Xavier Hascher, who later became a collaborator and a friend. I wrote notes, and it is natural that after you have done this, you want to write a book on the subject, especially that the literature on this field is very limited. In fact, I chose mathematics as a profession but music has always been in my background. With Xavier, Renzo Caddeo, Franck Jedrzejewski, Pierre Jehel, and my daughter Hélène, we published a commented edition of Euler's works on music theory. This was the beginning of a series of commented editions of old treatises on music by scientists, some of them not so well-known, like Félix Savart and Joseph Sauveur. We have other projects in progress, on the works of Chladni, the Galilei (father and son), the Pythagoreans, and Plato. Regarding the latter, I am learning a lot from Stelios Negrepontis. Plato is a difficult subject. Most people who read his dialogues do not realize that it is mainly mathematics. You have to interpret. His philosophy cannot be separated from his mathematics and music.

– **Is it true that music composers think mathematically?**

– Yes, absolutely. Bach, Rameau and others were very explicit about this. To understand this, you must first know what mathematical thinking is. Mathematicians can understand this.

– **Who are the mathematicians who influenced you?**

– Thurston, Busemann, and my co-authors. I learned a lot from my co-authors, the young and the older ones, and if I had to mention only one name, it would certainly be Norbert A'Campo. He explained to me, with great patience, many notions in mathematics. With him, I went through papers by Euler, Riemann, Teichmüller and others. On history of mathematics, I was also influenced by Roshdi Rashed. With him, I learned how to work on manuscripts. Then, among the recent collaborators, I would mention Stelios Negrepontis, who taught me Plato and philosophy of mathematics. Before I talked with him, I had no idea of what Plato's philosophy is about. From him, I learned that it is all about mathematics.

– **You mentioned Roshdi Rashed. By the year 2014 we start to see the results of your collaboration with him on history of mathematics. If there were one lesson to take from your study of Greek mathematicians' works in their Arabic treatises, what would that be?**

To start, let me say how I came to know Rashed. I first met him in Princeton. Coincidentally, we both spent the same years, 1984–1985 and 1993–1994 at the Institute For Advances Study. The first year, I was a post-doc there, it was my first position, and Rashed was working with André Weil, who, as you know, was very much involved in the history of number theory. Rashed was also working with Neugebauer, who used to come from Brown (he passed away before Rashed's second visit). At Princeton, with Rashed, I built up a cordial relationship. We used to go to concerts together and so on. Naturally, he talked to me about Egypt—his birth country—, and I was very much interested, but also about history of mathematics.

He told me what he had done during his career through several historical moments, in particular how he had discovered Arabic manuscripts of books of Diophantus' *Arithmetic* and Apollonius' *Conics* that were unknown before, and other things that awakened my curiosity. The last time we met at Princeton, he told me: "If some day you are tempted, we can discuss history of mathematics in France", and he gave me his phone number in Paris. Later, I became involved in history of mathematics. S. G. Dani was influential in this respect, he used to ask me articles for *Gaṇita Bhāratī* (and he still does). I published on Lobachevsky, Euler and Lambert, and I kept thinking about Rashed's proposal.

Finally, I decided to go and see him in Paris, and I told him that I would like to try to work with him on Arabic mathematics. He gave me a small work to do, based on a short manuscript on the astrolabe. I translated it with his help, and I started learning about this astronomical instrument, which gives quite a good idea of how astronomers in the medieval times understood the basic stellar motions. The instrument involves a serious amount of mathematics: You have to represent on a plane, at the same time, several motions on the sphere, taking into account that the rotation of axis of the Earth is not the one of the motion of the sun, etc., and from this you want to learn your location, the time of the day, etc. We were around 2008–2010. It was just an initiation to reading manuscripts and explaining the mathematics there. Then he started talking to me about the possibility of doing more substantial work. He suggested that I choose between three subjects: a project on the history of astronomy (a continuation of what we have started), a project on Euclid (Arabic manuscripts of the *Elements* that are still unpublished), and a project on spherical geometry. I was immediately tempted by the third subject. Having worked on Lobachevsky, I knew that the latter has been teaching spherical geometry, before developing what we call today Lobachevsky geometry. More importantly, I was involved in non-Euclidean geometry, because of Thurston, but also, more specifically, in spherical geometry, collaborating with Norbert A'Campo on a project on transitional geometry. I was actually reading with him some papers by Euler on spherical trigonometry.

Rashed gave me a first task on spherical geometry, to understand a proposition of Menelaus of Alexandria which did not survive in Greek but which existed in several Arabic manuscripts. Several Arab mathematicians, between the ninth and the thirteenth centuries, had commented on it but he told me that the theorem was still mysterious. Of course, all this is unknown to modern geometers. Some modern historians of mathematics knew about this proposition, but it was impossible for them to see the scope. The proposition is indeed difficult, expressing a subtle equality between two proportions, and the proof contained, among other things, the invariance of the spherical cross ratio by projectivity. We worked on it together and we published two 30-page articles on that proposition, explaining how several Arab mathematicians of the Middle Ages tackled it.

After these two papers were published, Rashed asked me if I wanted to work on an edition of the whole of Menelaus' *Spherics*. The work consists in 91 propositions, all of them interesting. I said yes, without hesitation. He told me that in that case, I would have to concentrate on this work for a few years, that it was a full-time job, that I could not work on that besides something else. I said that this was not possible, that I was a mathematician, that I had commitments with co-authors, working on several projects, and I cannot just stop them.

I came back to see him later, we discussed the project again, I explained that even if I was doing something else, I could spend several hours a day on this work. He asked me how many, I said 5 or 6. Although it is strange to talk about hours of work, I think this was important for him. He did not want to start with me a work and never complete it. He accepted, and in the end, we got the job done. This work took us 7 years. The final book is circa 900 pages, from an initial manuscript of fifty pages. Menelaus was one of the Thurstons of antiquity, with a geometrical vision beyond comprehension. In his treatise, he set out the 91 propositions, one after the other, giving each time a figure and a small hint of the proof, and for some, no indication at all; it is clear that for him they were very easy. Some of the propositions are stated without any hint for the proof, but all of them are correct. One day, about seven years after we started, Rashed told me, when I arrived at his place: "We should finish the book now." It was for some practical reasons: his secretary was about to retire. I told him that for the last 11 propositions, I still had no proof which I could write in Menelaus' style. He said that's fine, we shall mention this in the book.

Menelaus' work is amazing. He insisted on having synthetic proofs for the spherical propositions, without using the geometry of the ambient Euclidean space. He also wanted all his proofs to be constructive. All the Arab translators and commentators had stressed this fact, and this is why they found it difficult. I learned a lot of non-Euclidean geometry by working on Menelaus' treatise.

– I understand this, about your work with Rashed. My question was more general, about how Arabic manuscripts are used in our knowledge of Greek mathematics, and what is the role of mathematicians in this.

– There is still a titanic amount of work to be done, on Arabic manuscripts translated from the Greek, for which the Greek original did not survive. There are very few historians who can do this kind of work; you need to know Arabic, you need to know how to read manuscripts, and you need to understand the mathematics, to see what is new and what is already known, what is important and what is not, what is difficult and what is not; you need to know not just the basics, but the mathematics in depth, to understand the proofs and to be able to complete them, using the mathematics of the epoch. And that's something that an actual historian of mathematics cannot do if he is not a mathematician.

Now regarding Arabic manuscripts. I mentioned Christiaan Huygens above. I read his correspondence, while I was working on the edition of his works on music. We read there that he learned Arabic because he needed to read a manuscript of Apollonius he had in his possession, in Arabic translation (no Greek edition had survived). One of his uncles who was a great scholar, an arabist, had brought him some manuscripts on one of his trips to Aleppo. It is an intriguing story, involving a competition with the Medicis, for the publication of Apollonius' manuscript. Huygens learned by himself to decipher such manuscripts. He wrote in his correspondence that for him, such an exercise (translating from a remote foreign language) was analogous to solving a mathematical problem. In the book

I published with Caddeo, Hascher and Jedrzejewski on Huygens' musical theories, we talk about all this. To answer your question precisely: Arabic manuscripts are essential in completing our knowledge of Greek mathematics. I mentioned several prominent examples: Menelaus, Diophantus, Apollonius, Euclid, and there are many others. Unfortunately, there are very few mathematicians involved in such projects.

– I can see that being able to read Arabic manuscripts has given you unique insights or access to historical mathematical texts or concepts that might be overlooked by others. What about the rest of your work on history of mathematics?

Well, doing history of mathematics consists to a large extent in studying original texts. André Weil, in a letter to his sister Simone, declares that he learned the theory of algebraic functions by reading Riemann. Riemann's works on geometry are difficult, and they are being read by very few mathematicians. Many of Euler's works on geometry have not been read by mathematicians. There are many classical authors that are worth reading carefully. Geometry, in the sense we intend it today, exists since the Greeks. For various reasons, many of the Greek geometric texts survive only in Arabic, and many of them are still unread. In your country, there are several libraries of ancient Arabic manuscripts. You remember that several years ago I asked you to send me scans of a manuscript of al Harawī's edition of Menelaus' *Spherics* which belongs to the Topkapi Saray collection. You sent a student for this, and she came back with the scans of a collection consisting in a large number of manuscripts (it is not rare that in libraries, several old manuscripts are bound together), including the one I asked you for, but also other manuscripts on geometry, optics and astronomy. With Rashed, we worked on Menelaus *Spherics*, but some other texts in this collection are still not edited, and they were never read by modern mathematicians. This is just one example. There are many manuscripts spread all over the world, awaiting for someone to edit them.

– Her name is Merve Durmuş, and I remember you asked her to send you scans two manuscripts of Menelaus' *Spherics* that are kept in two different libraries.

– You are absolutely right. The second one is al Maghribī's edition of the *Spherics*; it contains the propositions, with no proofs. It belongs to the Nuruosmaniye Library. You sent me both manuscripts. With Rashed, I published a critical edition of al Harawī's version. Starting with that period, I used to visit you regularly and we became close friends.

– Your interest in history of mathematics is not restricted to Greek mathematicians and their Arabic successors. I know you developed very intimate bonds with names such as Euler. Can you tell us what wisdom you gathered from him?

At a personal level, I feel close to these authors, they became like friends. A large amount of Euler's results are contained in his correspondence. You know that there were no journals in those times, except the proceedings of some Academies. Euler's work is published in the Proceedings of the Russian and German Academies of Sciences to which he belonged, but he also used to write to his friends and colleagues, explaining what he was working on. His correspondence is enormous, and it is written mostly in French. When you read it, you learn about his mathematics, but also you know about his life, his everyday preoccupations. This is why I said you become friends with him. He had high moral values. I learned a lot of mathematics, reading Euler, Lobachevsky, Lagrange, Lambert, Huygens and other authors. Then you make connections. Recently, I published a book on geography. The idea was to explain works of Euler, Lagrange and Lambert on geography. But then, I realized the sources were again in Greek mathematics, and I discovered Arabic manuscripts, which I mention in the book. It was quite an experience. There are also evident connections with modern works on best Lipschitz maps between surfaces, in particular with works by Chebyshev, Darboux, Milnor, Thurston and others. This is a topic I have been working on since more than 35 years. Working on history of mathematics, you learn that mathematics is essentially the same as 2000 years ago, nothing has changed.

– Before moving to the next question, let me note that to my knowledge you haven't shown interest in Gauss yet... So, do you consider yourself as part of the community of historians of mathematics?

Gauss came to geometry because of his interest in geography. Regarding the other question: There is no community of historians of mathematics. There are groups, clans, gangs.

– So do you belong to one of these clans?

– Yes, to the clan of mathematicians working on history of mathematics.

– Can you give a few names of people in this clan?

– Yes, Stelios Negrepontis, Victor Pambuccian, S. G. Dani, Christian Houzel, Arkady Plotnitsky, Norbert A'Campo, ... Yuri Manin was a very good historian of mathematics. I cannot think of many others right now. I know a few mathematicians interested in the history of mathematics, but they work on it in an amateurish way. You know, there were times where history of mathematics was the domain of mathematicians, but then, gradually, mathematicians stopped being really interested in the history of their domain and history of mathematics became a specialty of "historians of mathematics" who, unfortunately, do not know much about mathematics. I think the real brake occurred in the 1970s, with a famous controversy, involving on the one hand, people like André Weil, Otto Neugebauer, Hans Freudenthal, and Bartel van der Waerden, and on the other hand, a group of historians of mathematics who were not mathematicians and who used to study mathematical texts as literary pieces. The fight was violent (in terms of the accusations and the words that were used). After that period, and for reasons that have nothing to do with the fact that arguments from one side or the other were convincing for the other side, history of mathematics became a field reserved to the second group. This is only due to the lack of interest by mathematicians for history of mathematics. Very few mathematicians work seriously on the history of their field, although, I think, there is a need for this.

– You've edited a monumental 7-volume Handbook on Teichmüller theory, which is now a very mature topic. What is its future?

– The seventh volume appeared in 2020. In these 7 volumes, there are more than 170 articles. Some authors wrote more than one article, but many articles were written by several authors, so let us say that the ones cancel the others, and we have about 170 persons who contributed to this collection. I have also included, in these volumes, translations of papers by Grötszch, Lavrentieff and Teichmüller. They contain invaluable results, sometimes unknown to the mathematicians working in this field. Some results of Teichmüller have been attributed to others. Some colleagues-friends helped in the translations. I am happy if I contributed in sorting out this matter. I enjoyed very much reading these classical works, and I am glad that people can read them now, if they want, in English translation. You ask about the future of the collection: the response is that I don't know. In the meanwhile, I have co-edited other collections, the *Handbook of group actions*, with Ji and Yau, of which 5 volumes already appeared, and the collection *In the tradition of Thurston*, which I am editing with Ohshika; we are presently working on Volume 4.

– I keep hearing some very pessimistic remarks about the future of mathematics in Europe. It is very hard for young talents to find jobs, and for faculty to find good students. What are your thoughts about this?

Before answering the question, let me say, at the expense of saying the obvious, that in general, the level of teaching in schools in France is low, not only of mathematics, since several years, and for many reasons. But mathematical research is still at a top level. How can we explain this? Maybe it is that the mathematics we learn at school has not much to do with the one we need for doing research. Moreover, our profession is such that it only takes a handful of remarkable people to make a laboratory of a hundred or more people into an excellent laboratory. Returning to your question: Yes, sure, the job opportunities for mathematicians in Western Europe and in the US are now very narrow, which gives a lot of room to things that have nothing to do with mathematics, things we call "politics". I understand that in countries like China, Russia and a few others, there are opportunities for good mathematicians. If I were young and couldn't find a job in Europe, I probably would have moved to one of these countries.

– There is now this not-quite-recent trend of measuring scientists and scientific institutions by their outputs especially by their original research papers published in 'high-quality' journals. In fact many of your later editorial work wouldn't even count as a proper output according to these so-called *metrics*, since they are not published in journals. Could you share your feelings about this?

– For me this question is irrelevant. I learned very recently that there are medium-, high- and top-quality journals. I used to send my articles almost randomly, only taking into account the fact that the journal has published papers on the same subject. I never took into account the prestige of the journal (and I did not know that this thing exists). For me the important thing is that the article is published. The same holds when I referee a paper, I do not take into account the name of the journal. It is also the same for books. I choose my editors for "sentimental" reasons. I publish my books in French with the editor Hermann, who was the editor of Weil, Cartan, Hadamard, but also Sartre, Cavaillès, Lautman, and who also published the philosophical writings of Poincaré, but I published also with Blanchard, whose store was located on rue de Médicis, right next to the café where I always go in Paris, the Rostand, in front of the Jardin du Luxembourg. Blanchard has recently closed, I think he was unable to manage the business and his store had become too antiquated. My last visit to this store was a few years ago, when I went in with Penner who wanted to buy a birthday present to Manin.

– I know you take great joy in expressing yourself by handwriting in your personal diaries. We talked about this one day we were together and you were buying a fountain pen. Should the mathematics community be afraid of them if they ever go public?

– I write only good things about mathematicians, because I write only about my friends and people I enjoy talking with. Writing a diary is like a conversation I keep, with myself and with people around me.

– Let us finish with a personal anecdote or mathematical story that you are known for.

If you want an anecdote, I will tell you a story I already recounted. A few years ago, with Norbert A'Campo and Sumio Yamada, we spent two weeks at a Research Institute, in Trento. We were working on a project (which is still ongoing) on Riemann surfaces. One day, one of the professors at the university there came to us and asked if we can give an informal talk to his students, explaining in simple words what we are working on. We fixed a date with him, and we decided that Norbert will give that talk. Norbert started by saying that we are working on a problem on Riemann surfaces, one-dimensional complex geometry. At this point, the professor stopped him—he thought something was wrong. He asked: "Dimension one?" Norbert said yes. The other one continued: "Are there still problems in dimension one?" He was very surprised. We learned later that he was an algebraic geometer. Norbert told me that for these people, a Riemann surface is just a one-dimensional scheme, and there is nothing to say about this object. This is how algebraic geometers look at the rest of mathematics. (I am exaggerating, but only a little bit.)

– As a former algebraic geometer, I concur. On the other hand I know you can do much better than hassling algebraic geometers. Can you try once more before we finish?

– Then let me tell you something on historians of mathematics. I once sent a paper to a journal of history of mathematics—one of the two well-known journals. The editor wrote me back, saying the paper is interesting and that they will publish it, but he advised me to send a new version in which I include some formulae and equations.

– A word of conclusion, about mathematics, or about history of mathematics?

– Then let me conclude with a word about the history of mathematics, why we study it, or rather, about history tout court: the past helps us understand the present. And this is precisely what we wish to do in mathematics, we would like to understand.

Chapter 3
A Glance at S. Novikov's Theory of Multivalued Morse Functions

François Laudenbach

To Athanase, with friendly admiration

Abstract This chapter provides a brief presentation of the so-called Morse-Novikov complex associated with a closed manifold, equipped with a closed non-exact one-form α and a descending gradient. We are interested in the *homoclinic* bifurcation of this gradient when crossing the stratum of gradients which have a *simple* homoclinic orbit, based at a non-degenerate zero of α. The effect of such a crossing on the Morse-Novikov complex is analyzed. Moreover, an original construction, described in this chapter, shows that the phenomenon of homoclinic bifurcation is very easy to make up. Finally, a surprising doubling phenomenon is associated to every such a bifurcation.

I first met Athanase Papadopoulos in September 1976. He was just arriving from Beirut, in war at that time. Athanase applied to enter the second year of the mathematical curriculum in the Faculty of Sciences at Orsay (University of Paris-Sud, France).[1] I was called to estimate Athanase's skill to this level. I do not remember how I accomplished my duty; I have still a vague memory of the scene where he was the unique candidate in a similar situation, which was not true as he recently recalled me. My memory was so, maybe, by the convincing strength that emanated from his look. A strength that is still alive after more than forty-five years.

A few years later, I became Athanase's advisor for his PhD thesis. This was in the early eighties, at a time where the topology team in Orsay was still influenced

[1] Today, this university is named Paris-Saclay.

F. Laudenbach (✉)
Laboratoire de Mathématiques Jean Leray, UMR 6629 du CNRS, Faculté des Sciences et Techniques, Université de Nantes, Nantes cedex 3, France
e-mail: francois.laudenbach@univ-nantes.fr

A. M. Uludağ, A. Zeytin (eds.), *Essays on Geometry*,
https://doi.org/10.1007/978-3-031-76257-4_3

by the recent seminar devoted to *Thurston's work on surfaces* [1, 2]. Very rapidly, Athanase appropriated Thurston's approach. He wrote a short but brilliant paper about post composing by a pseudo-Anosov diffeomorphism and related topics [16]. Thurston's influence marked Athanase for ever and ever. At the beginning, this was my fault but it turned to be a part of his good fortune. Today, in addition to his own mathematical work, Athanase develops an impressive editorial activity, often turned to the History of mathematics, including the relation to music. As an example, I extract the collective work on Christiaan Huygens, particularly sensitive, that collects the writings of Huygens on music and sound [17]. All of that makes me full of admiration.

Dear Athanase, I wish you the best to continue your route.

Sergei Novikov launched a new subject in his 1981 article, titled *Multivalued functions and functionals. An analogue of the Morse theory* [13]—see also the extended version [14]. A famous *Novikov ring*[2] was born at that time. Its first use was in infinite dimension in the setting of Floer theory of symplectic manifolds (M, ω) when the symplectic form ω does not vanish on the second homotopy group $\pi_2(M)$—for instance, [10]. This piece of work also inspired topologists for various problems: For example, existence of non-singular closed one-form [6], circle valued functions [18], Generalized Reidemeister torsion [4]. Surprisingly enough, this topic has remained unaddressed from the point of view of bifurcations in dynamics.

I am very pleased to present here a summary of a piece of work, which I share with my former student Carlos Moraga Ferrándiz, about a dynamical aspect of Novikov's work in the setting of Morse theory [8]. This completely answers one question, among many others, that Carlos raised in a seminar paper [12]. The present paper is a mine for new topological subjects in the line of Novikov, whose crowning could be a *pseudo-isotopy theorem* for multi-valued functions, or likely, its obstructed form.

3.1 From the Morse Complex to the Morse-Novikov Complex

3.1.1 Preamble

Note, to begin with, that these two algebraic complexes deal with two different settings. To define a *Morse complex*, we should have a smooth manifold M, that we assume to be *closed* in the present paper, endowed with a *Morse function*[3] $f : M \to \mathbb{R}$ and a *gradient* (or pseudo-gradient) vector field X satisfying some generic condition—personally, I prefer the descending (or negative) gradient which, in particular, means $df(X) \leq 0$.

[2] A final form of this ring was given by J.-C. Sikorav [20] in 1987.

[3] By definition, such a function has a non-degenerate Hessian at every critical point.

To have a *Morse-Novikov complex*, the function f should be replaced with a non-exact closed differential form of degree one on M, which we denote by α. Locally, $\alpha = df_{loc}$ and there is an indeterminate additive constant; Novikov speaks of a *multivalued* function. A magnetic field—once a volume form is prescribed—gives a famous example of such a one-form through the Hodge duality.

When α is locally the differential of a Morse function, α is said to be a *Morse one-form*. In this case, as for a Morse function, the set $Z(\alpha)$ of zeroes of α is finite. And each zero of α, as its corresponding critical point of a local primitive, has a Morse index $i(p)$, namely the maximal dimension of a linear subspace in T_pM on which the second derivative $D^2_p f$ is negative definite. Note that, for a given Riemannian metric, the descending gradient X is uniquely defined though α has no global primitive.

Moreover, though the dynamics of X could be very complicated, there is an injectively immersed *unstable* (resp. *stable*) manifold, denoted by $W^u(p, X)$ (resp. $W^s(p, X)$) which is the set of points $x \in M$ whose trajectory $\big(t \mapsto X^t(x)\big)$ under the flow of X goes to p as t goes to $-\infty$ (resp. $+\infty$). Note that the local picture near p of $W^{u/s}(p, X)$ determines their global property to be injectively immersed, just by the flow invariance.

A Morse function f or a Morse one-form α being given, endowed with a descending gradient X, the matter is to organize an intelligent counting of orbits of X connecting two critical points of f (or two zeroes of α) of consecutive Morse index. This requires some generic conditions which are explained in the next subsection. These conditions will allow us to introduce the twisted Morse complex $C_*(f, X)$ (resp. the Morse-Novikov complex $C_*(\alpha, X)$).

We will continue our presentation assuming that M is not 1-connected. If not, by the Stokes formula, α would have a global primitive.

3.1.2 Generic Conditions

The gradient of a Morse function has a very simple dynamics. Namely, every orbit goes from one critical point to another, this is very common. What Smale proved in [21] is less banal. This states as follows:

Let M be a closed manifold and let $f : M \to \mathbb{R}$ be a Morse function of class C^{r+1}, $r \geq 1$. In the set of gradient vector fields of class C^r for f, the subset of those that are today named Morse-Smale *is open and dense.*

A gradient vector field X of f is said to be Morse-Smale if, for every pair (p, q) of critical points, the unstable manifold $W^u(p, X)$ is transverse to the stable manifold $W^s(q, X)$, which reads $W^u(p, X) \pitchfork W^s(q, X)$. This transversality reduces to finitely many applications of the basic transversality theorem of two submanifolds in some level set of f. With the same proof, the result remains the same in the space of gradients that are imposed near crit$f$.

The next theorem, proved independently by Kupka [5] and Smale [22], is more subtle (see, for instance, [15]):

Theorem (See [15]) *For a given closed one-form α on the closed manifold M, there is a residual subset of C^r vector fields in the set of C^r gradients of α, $r \geq 1$, which are* Kupka-Smale *in the sense that, for every pair of zeroes (p, q) of α, one has $W^u(p, X) \pitchfork W^s(q, X)$.*

For short, one says that the property of being Kupka-Smale is *generic*. We recall that a subspace P of a complete metric space E is said to be *residual* if it is a countable intersection of open and dense subsets in E, and hence, P is dense. As above, the Kupka-Smale theorem still holds if the considered gradients have an imposed form near the zeroes of α. Incidently, other types of constraints could be imposed, as it is shown in [19].

3.1.3 The Fundamental Groupoid and the Novikov Ring

I am indebted to Carlos Moraga who showed me the advantage of the groupoid point of view. This avoids the choice of a base point in M. Recall that a groupoid is a *small category* where every arrow is invertible. The *fundamental groupoid* Π of M is defined as follows: its objects are the points of M and, if (p, q) is a pair of points, the set of arrows whose *source* is p and *target* is q, denoted by $\mathrm{Hom}(p, q)$, is the set of homotopy classes of paths from p to q. The *identity* arrow at p, denoted by $1_p \in \mathrm{Hom}(p, p)$, embeds M into the set of arrows.

The set of finite linear combinations of arrows with integral coefficients, denoted by $\mathbb{Z}[\Pi]$, is a ring (not unital). The product of two arrows $g \in \mathrm{Hom}(p, q)$ and $g' \in \mathrm{Hom}(p', q')$, denoted by gg', is zero if $q \neq p'$ and is equal to the composed arrow if not. Here, this is the homotopy class of the composition of the two paths representing g and g' respectively.

If α is a Morse one-form on M, one denotes by Π_α the full subcategory of Π whose set of objects is $Z(\alpha)$. In that case, $\mathbb{Z}[\Pi_\alpha]$ is a unital ring whose unit is $1 := \sum_{p \in Z(\alpha)} 1_p$. There is also a groupoid morphism that we name the *α-length*,

$$u_\alpha : \Pi_\alpha \to \mathbb{R}, \qquad g \mapsto -\int_\gamma \alpha,$$

where γ is a path in the homotopy class g and $\mathbb{R}$ is viewed as a groupoid with 0 as the unique object.

A formal series $\lambda \in \mathbb{Z}[[\Pi_\alpha]]$ fulfills the *Novikov Condition* if,

$$\text{for every } L \in \mathbb{R}, \text{ the set } \mathrm{supp}(\lambda) \cap u_\alpha^{-1}(-\infty, L) \text{ is finite.} \tag{3.1}$$

Here, $\mathrm{supp}(\lambda)$ denotes the *support* of λ, that is, the subset of Π_α whose elements appear in λ with a non-zero coefficient. The set of formal series fulfilling the Novikov Condition is denoted by Λ_α. One easily checks that Λ_α is a unital ring with the same identity element as $\mathbb{Z}[\Pi_\alpha]$. For $a, b \in \Lambda_\alpha$, the product is first performed by truncating the factors modulo $L > 0$, that is, by erasing from a and b all arrows whose α-length is larger than L. Then, one leaves L go to $+\infty$. The limit exists in Λ_α since there is some positive L_0, depending only on a and b, such that, for every L, the truncation $mod(L - L_0)$ of the product remains constant if the factors are truncated at any α-length larger than L. The ring Λ_α is named the *Novikov ring* associated with α.

Example 3.1 Let $p \in Z(\alpha)$ and $g \in \pi_1(M, p)$ with $u_\alpha(g) < 0$. Then the sum of $\sum_{j=0}^{+\infty} g^j$ belongs to the Novikov ring. Actually, it is the inverse of $1 - g$.

When $\alpha = df$, the same language is still appropriate but, no series, and hence no Novikov condition, is appearing since the df-length is bounded. Here, Π_{df} is the full subcategory of Π whose set of objects is $\mathrm{crit} f$, the critical set of f. In this case, $\mathbb{Z}[\Pi_{df}]$ is a unital ring.

3.1.4 The Twisted Morse Complex

We have a Morse function f and a descending gradient X. This gradient fulfills the generic condition to be Morse-Smale. The set of critical points of Morse index k is denoted by $\mathrm{crit}_k f$. Their unstable manifolds are arbitrarily oriented, and hence, the stable manifolds are co-oriented. The $\mathbb{Z}[\Pi_{df}]$-module generated by $\mathrm{crit}_k f$ is a submodule of degree k in the complex $C_*(f, X)$, which actually does not depend on X.

All unstable manifolds are arbitrarily oriented and hence, the stable manifolds are co-oriented. The differential ∂^X, which deeply depends on X, is given by the following formula on each generator of degree k:

$$\partial^X(p) = \sum_{q \in \mathrm{crit}_{k-1}(f)} n(p, q)^X q. \tag{3.2}$$

The coefficient $n(p, q)^X$, named the *incidence* of p to q, is a linear combination of arrows in $\mathrm{Hom}(p, q)$. More precisely, this is a sum of signed arrows in $\mathrm{Hom}(p, q)$

$$n(p, q)^X = \sum_\ell \varepsilon_\ell g_\ell$$

where the sum ranges over the X-orbits ℓ from p to q; the arrow g_ℓ is the Π-value of ℓ and the sign ε_ℓ is given by the next equation along ℓ:

$$\varepsilon_\ell X_{|\ell} \wedge \text{co-or}(W^s(q, X)) = \text{or}(W^u(p, X)). \tag{3.3}$$

The sum is finite since X is Morse-Smale.

The twisted Morse-complex informs us about the topology of M and the topology of its covers associated with the representations of Π; for instance, about the Whitehead torsion when some cover is acyclic [11].

3.1.5 The Morse-Novikov Complex

Here, we have a closed one-form α whose zeroes are non-degenerate, that is, the local primitives f_{loc} are Morse functions. For any Riemannian metric, let X be the descending gradient of f_{loc}. This gradient is globally defined. Generically, X fulfills the generic Kupka-Smale transversality condition.

Under this condition, for every pair (p, q) of zeroes whose Morse indices satisfy $\mathrm{ind}(p) = \mathrm{ind}(q) + 1$, the set of connecting orbits from p to q fulfills the Novikov Condition (3.1). This is a *key fact* which follows from standard estimates (see appendix to [8]).

The Morse-Novikov complex $C_*(\alpha, X)$ is the graded differential module over the ring Λ_α generated by the zeroes of α. The grading is the Morse index and the differential ∂^X is formally similar to the differential given in (3.2). The reason for using the Novikov ring is that infinitely many connecting orbits could exist if the α-length is not bounded[4]. The formula reads

$$\begin{cases} \partial^X(p) &= \sum\limits_q n(p,q)^X q\,, \qquad\qquad q \in Z(\alpha),\ \mathrm{ind}(q) = \mathrm{ind}(p) - 1,\\ n(p,q)^X &= \lim\limits_{L \to +\infty} \sum\limits_{\{\ell \mid u_\alpha(g_\ell) < L\}} \varepsilon_\ell g_\ell\,, \end{cases}$$

where the sign ε_ℓ is defined as in (3.3)—this requires that the unstable manifolds are oriented, which co-orients every stable manifolds. The homology of $C_*(\alpha, X)$ depends only on the cohomology class of α [6]. Hence, it gives a first obstruction to find a non-singular one-form cohomologous to α. This is not the unique obstruction as it is made explicit in [6].

In the remainder of this paper, the one-form α will be kept as above. We are interested in some bifurcations of X, namely homoclinic bifurcations, which is the main topic of [8]. One of the goals of this note is to describe the effect on the Morse-Novikov complex by moving X along a path of α-gradients which crosses a *simple homoclinic bifurcation*.

A general principle, that Thom has defended in his book *Structural Stability and Morphogenesis* [23], states that a non-generic situation contains more information than any generic one in its *universal unfolding*. We have applied the same principle by requiring the germ of gradients near the zeroes of α to be *adapted to* α (one will say *α-adapted*). The precise definition will be given in Sect. 3.2. An important point, which we immediately emphasize, is that this new requirement has no effect on the genericness arguments. For instance, by Peixoto's article [19] or by their own proofs, the Kupka-Smale property is generic among the gradients which are α-adapted.

4 Such an example is given in [9].

3.2 Morse Model, α-Adapted Gradients and Homoclinic Bifurcation

Definition 3.1 Given a closed Morse one-form α, a gradient X of the local primitives of α is said to be α-adapted if, for every zero p of α, X is linear in some Morse coordinates about p,[5] and radial in each invariant manifold—stable or unstable. The space of α-adapted gradients will be denoted by $\mathcal{F}_\alpha$.

As announced in the end of the previous section, it is not at all generic that X be α-adapted. We are going to describe the Morse model for such a hyperbolic linear vector field and point out its main property.

3.2.1 *The Morse Model in the Setting of an Adapted Gradient*

If p is the involved zero of α and j is the Morse index of p, this model will be denoted by $\mathcal{M}(p)$. In ambient dimension n, its lower boundary $\partial^-\mathcal{M}(p)$ is diffeomorphic to $S^{j-1} \times D^{n-j}$ and its upper boundary is diffeomorphic to $D^j \times S^{n-j-1}$. The *local* unstable (resp. stable) manifold is the disc $D^j \times \{0\}$ (resp. $\{0\} \times D^{n-j}$,) both discs having the same radius (as small as we need, say 1 for simplicity). The boundary of the unstable (resp. stable) disc is denoted by Σ^- (resp. Σ^+) and named the *attaching* (resp. the *co-sphere*)—see Fig. 3.1.

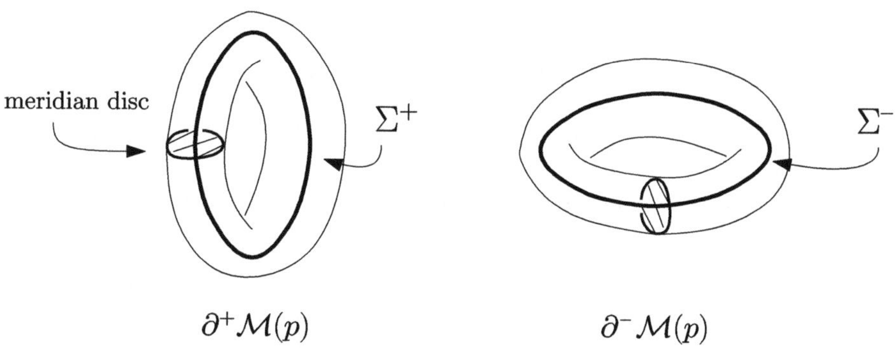

Fig. 3.1 Here, $n = 4$, $j = 2$. Only the lower and upper boundaries are represented. A generic point has multi-spherical coordinates (ϕ, r, ψ)

[5] Morse coordinates about p are coordinates where the Hessian of a local primitive of α has a constant diagonal matrix.

If the unstable manifolds are oriented, as it is required for having a Morse-Novikov complex, then Σ^- is oriented by the following convention:

$$X \wedge \mathrm{or}(\Sigma^-) = \mathrm{or}(W^u(p, X)). \tag{3.4}$$

Note that, unlike the attaching sphere, the co-sphere is co-oriented in $\partial^+\mathcal{M}(p)$ without using any convention.

As we are interested in homoclinic bifurcation, we may assume that the Morse index of p is different from 0 and n. So, $\partial^\pm\mathcal{M}(p)$ is provided with multi-spherical coordinates $(\phi, r, \psi) \in S^{j-1} \times [0, 1] \times S^{n-j-1}$. Here, neither ϕ nor ψ is a number, but it is a point on its respective sphere! The α-adapted gradient X allows us to derive the following proposition the first item of which is the main property of every α-adapted gradient; the second item is for further use.

Proposition 3.1

(1) *The* descent map $Desc : \partial^+\mathcal{M}(p) \setminus \Sigma^+ \to \partial^- M(p) \setminus \Sigma^-$ *that consists of following the flow of X reads Id in multi-spherical coordinates.*
(2) *A C^1 diffeomorphism K of $\partial^+\mathcal{M}(p)$ of the form $(\phi, r, \psi) \mapsto \big(\phi, r, k(\phi, r, \psi)\big)$ with $k(\phi, 0, \psi) = \psi$ extends to a C^1 diffeomorphism of $\mathcal{M}(p)$ preserving the leaves of α.*[6]

We are ready for speaking of homoclinic orbit and the bifurcation in the space of α-gradients associated with this phenomenon.

Definition 3.2 An orbit ℓ of an α-adapted gradient X is said to be *homoclinic at p* if it goes from p to p. In addition, it is said to be *simple* if the following conditions are fulfilled.

(i) ℓ is not a *broken orbit*, that is a union of orbits issued from some others zeroes and making a loop at p.
(ii) For every point $m \in \ell$, the span $T_m W^u(p, X) + T_m W^s(p, X)$ is of codimension one in $T_m M$.
(iii) There is no other connecting orbit from p to p in the same homotopy class in $\pi_1(M, p)$.

Having such an ℓ, let a^- (resp. a^+) denote its *last* intersection with $\partial^-\mathcal{M}(p)$ (resp. $\partial^+\mathcal{M}(p)$) before reaching p. The arc of ℓ from a^- to a^+ (in the sense of X) is named the *restricted homoclinic* orbit and is denoted by $\underline{\ell}$. Let T denote a small tubular neighborhood of $\underline{\ell}$ in $M \setminus int(\mathcal{M}(p))$ whose boundary over $int(\underline{\ell})$ is tangent

[6] See a proof in [7, Chap 6]. There is a similar statement about $\partial^-\mathcal{M}(p)$ by changing ψ to ϕ. This item 2 does not hold in class C^2.

to X. Such a tube has coordinates $(v, x, y, z) \in [-1, +1] \times D^{j-1} \times D^{n-j-1} \times [0, 1]$ such that the following holds:

- X is positively colinear to ∂_z.
- $T \cap \partial^- M(p) = \{z = 0\}$ and $T \cap \partial^+ M(p) = \{z = 1\}$.
- $T \cap \Sigma^- = \{v = 0, y = 0, z = 0\}$ and $T \cap \Sigma^+ = \{v = 0, x = 0, z = 1\}$.
- $\underline{\ell} = \{v = 0, x = 0, y = 0\}$.
- The frame $(\partial_v, \partial_x, \partial_y)$ is tangent to the leaves of α at every point of T.

At every point $(0, 0, 0, z_0) \in \underline{\ell}$, one can orient ∂_x with the convention used to orient Σ^- in (3.4); and hence, ∂_y is co-oriented in $\{z = z_0\}$—a fiber of T. The vector ∂_v, not yet chosen, co-orients the hyperplan $\mathrm{span}(\partial_x, \partial_y)$ in this fiber. Its *preferred* co-orientation is given by the next equation which specified a half space for ∂_v in $\{z = z_0\}$.

$$\partial_v \wedge \mathrm{or}(\partial_x) = \text{co-or}(\partial_y). \tag{3.5}$$

On a unit Euclidean sphere $\mathbb{S}^j$, equipped with a northern pole ν, the function $\mathbb{S}^j \to [-1, +1]$ defined by the scalar product with ν will be named the ν-*latitude*. When this naturally happens, one speaks of a *preferred latitude*. This will be the case in the following setting:

Proposition 3.2 *If ℓ is a simple homoclinic orbit at $p \in Z(\alpha)$ of an α-adapted gradient, then both the co-sphere Σ^+ and the attaching sphere Σ^- have preferred latitudes, respectively named the* ψ-latitude *and the* ϕ-latitude.

Proof Let $H : \{z = 0\} \to \{z = 1\}$ denote the holonomy along $\underline{\ell}$ defined by the foliation of X-orbits. It reads Id in the (v, x, y)-coordinates and $H(a^-) = a^+$. Let $(-, 0, \psi_0)$ be the multi-spherical coordinate of a^+; the first entry is not defined since the radius is null. Let D denote the image of $\Sigma^- \cap \{z = 0\}$ through H. The projection π^{ψ_0} of $\partial^+ \mathcal{M}(p)$ onto the meridian disc at a^+, that is $(\phi, r, \psi) \mapsto (\phi, r, \psi_0)$, embeds D if T is fine enough. The tangent space at a^+ to $D^{\psi_0} := \pi^{\psi_0}(D)$ is a hyperplane of the meridian disc. Equation (3.5) prescribes one side of $T_{a^+} D^{\psi_0}$ in $T_{a^+}\{\psi = \psi_0\}$. Let ν^ϕ denote the northern pole of the hemisphere as above prescribed in $\partial(\{\psi = \psi_0\})$.

By $Desc$, this point of $\partial^+ \mathcal{M}(p)$ descends to a point in a longitude of $\partial^- \mathcal{M}(p)$ that has the same multi-spherical coordinates and that we still denote by ν^ϕ. The projection of the longitude $\{(\phi, 1, \psi_0)\}$ onto $\Sigma^- = \{(\phi, 0, -)\}$ defines a northern pole, again denoted by ν^ϕ, and hence, a northern hemisphere in Σ^-. The scalar product with this ν^ϕ is the desired latitude function on Σ^-. Of a particular interest will be the ϕ-latitude $\langle \nu^\phi, a^- \rangle$.

For the preferred latitude on Σ^+, one performs a similar cycle starting from $\Sigma^+ \cap \{z = 1\}$. Then, this $(n - j - 1)$-disc is mapped by H^{-1}, the inverse of the holonomy; then, it is projected to the meridian disc of a^-. The resulting tangent hyperplane at a^- is co-coriented—rule (3.5)—and finally mapped by $(Desc)^{-1}$. Whence a series of three northern poles ν^ψ, the first by the above-mentioned rule, the second by $(Desc)^{-1}$ and the third by projecting to Σ^+.

3.2.2 Final Choice of ∂_v and Holonomic Factor

By (3.5), we are allowed to choose

$$\partial_v = \nu^\phi \in T_{a^+}(\{\psi = \psi_0\}).$$

This determines ∂_v in $T_{a^-}\big(\partial^- \mathcal{M}(p)\big)$, without any prescribed orthogonality. Only its independence from $\mathrm{span}(\partial_x, \partial_y)$ is fulfilled. Then, we have only a decomposition

$$\partial_v = v_x + v_y + \bar{\eta}\, \nu^\psi \tag{3.6}$$

where v_x (resp. v_y) is a vector of the x-axis (resp. y-axis) and where $\bar{\eta} > 0$ by the preferred side of the hyperplane $\mathrm{span}(\partial_x, \partial_y)$. The *holonomic factor* η is defined as the inverse of the coefficient $\bar{\eta}$ from Eq. (3.6):

$$\eta = \bar{\eta}^{-1}.$$

For $g \in \pi_1(M, p)$, let $\mathcal{S}_g$ denote the set of smooth α-adapted gradients having a simple homoclinic orbit in the homotopy class g. This set will soon be a smooth hypersurface in the set of α-adapted gradients.

For $X \in \S_g$, the orbit ℓ is its unique homoclinic orbit in the homotopy class $g \in \pi_1(M, p)$; $\underline{\ell}$ is its corresponding restricted orbit whose end points are a^- and a^+; ν^ϕ (resp. ν^ψ) is the preferred northern pole of Σ^- (resp. Σ^+); η is the holonomic factor. All of these are depending on X.

Definition 3.3 The *character function* is the real function $\chi : \mathcal{S}_g \to \mathbb{R}$ defined by

$$\chi(X) = \eta\, \langle \nu^\psi, a^+ \rangle + \langle \nu^\phi, a^- \rangle.$$

Theorem 3.1

(1) The set $\mathcal{S}_g$ is a codimension-one, co-oriented submanifold in the space $\mathcal{F}_\alpha$ of C^∞ α-adapted gradients.
(2) Assume $n > 2$ and $\mathcal{S}_g \neq \emptyset$. Then, the vanishing locus $\mathcal{S}_g^0$ of χ is a non-empty codimension-one smooth submanifold in $\mathcal{S}_g$ meeting each of its connected components.

Proof

(1) In order to prove that a subset of a manifold is a submanifold there is a local aspect with respect to the set and then, a global aspect to exclude self-recurrence. For instance, a leaf of a linear foliation on the torus with an irrational-slope is not a submanifold though any of its compact arcs is.

Let $X_0 \in \mathcal{S}_g$, let ℓ be its unique homoclinic orbit and T be a tube around ℓ, with its coordinates as previously explained in Sect. 3.2. For $X \in \mathcal{F}_\alpha$ close to X_0, its holonomy H i s close to that of X_0 in the C^1 topology. Therefore, in

$\{z = 1\}$, the image D of the x-axis through H projects, parallelly to ∂_v, onto an $(i-1)$-disc in span(∂_x, ∂_y) that intersects the y-axis exactly in one point only and transversely. This point has a unique lifting $b(X) \in D$. The equation

$$v(b(X)) = 0$$

is an equation translating the existence of a homoclinic orbit of X in T. One easily checks that this equation is regular. This answers the "local" aspect. The co-orientation is given by $\partial_v b(X) > 0$ which does not depend of the chosen orientation for $W^u(p, X)$ as it is visible on Eq. (3.5).

The "global" aspect is solved by the classical Arzela-Ascoli Theorem. Consider a sequence (X_k) in $\mathcal{S}_g$ converging to X_0 in the C^0 topology and assume $v(b(X_k)) \neq 0$ for every k. Then, $\|X_k\|$ is uniformly bounded. Hence, the family (ℓ_k) of their respective homoclinic orbits in the homotopy class g is equicontinuous. Therefore, up to taking a subsequence, ℓ_k goes to some line ℓ_∞ as k goes to ∞. It is easy to prove that ℓ_∞ is a homoclinic orbit—possibly broken—of X_0 whose homotopy class is g. By assumption, ℓ_k is located outside of T and the same holds about the limit ℓ_∞. So, we get two distinct homoclinic orbits for X_0, namely ℓ and ℓ_∞ in the class g, which is a contradiction.

(2) If $\mathcal{S}_g$ is not empty, the Morse index of p is not extremal, neither 0 nor n. If $n > 2$, then either the attaching sphere or the co-sphere is not 0-dimensional, and hence, at least one of the points a^- and a^+ is free to move in its sphere. The holonomic factor is also free to vary in the positive real numbers. This is enough to move every point in $\S_g$ and to reach the vanishing locus, denoted by $\S_g^0$, of the character function χ. One checks that $\chi = 0$ is a global regular equation. The side $\chi > 0$ gives the co-orientation of $\mathcal{S}_g^0$ in $\mathcal{S}_g$.

3.3 Effect of Crossing $\mathcal{S}_g$ on the Morse-Novikov Complex

3.3.1 A Natural Question

Given a generic α-gradient X_0 in the stratum $\mathcal{S}_g$ and a germ at X_0 of a path $(X_s)_{s\in Op(0)}$ crossing $\mathcal{S}_g$ transversely, compute the Morse-Novikov $C(\alpha, X_s)$, $s \in O_+$, from the complex when $s \in O_-$. Here, $Op(0)$ stands for an open interval centered at 0 in $\mathbb{R}$ whose length is not specified; O_- (resp. O_+) denotes its negative (resp. positive) part.

The question makes sense only if X_0 is isolated among the X_s that are not Kupka-Smale. This will be the case in the example presented in Sect. 3.4. Moreover, this seems reasonable in a very general situation. Indeed, the set of codimension-one strata where the Kupka-Smale property fails is countable and an argument *à la* Ascoli should generically prevent from an accumulation of such strata to X_0.

In order to circumvent this difficulty, we work *mod* L, in the sense of Sect. 3.1.3. It is worthy to note that reducing *mod* L the Morse-Novikov complex $C_*(\alpha, X)$ yields a complex, that is, ∂_L denoting the reduced differential, we have $\partial_L \circ \partial_L = 0$.[7]

We are going to explain the main result in [8]. It deals with the change of the Morse-Novikov complexes, reduced *mod* L, by crossing $\mathcal{S}_g$. The statement needs some definitions and notations.

Definition 3.4

(1) An α-gradient X is said to be *Kupka-Smale up to* $L > 0$ if, for every connecting orbit ℓ of α-length not larger that L, the respective unstable and stable manifolds of its extremities are transverse along ℓ. The subspace of $\mathcal{F}_\alpha$ formed by these α-gradients is denoted by $\mathcal{KS}_L$; it is an open and dense subspace of $\mathcal{F}_\alpha$.

(2) An α-gradient $X \in \mathcal{S}_g$ is said to be *almost up to* L if it is Kupka-Smale up to L except for its unique homoclinic orbit in the homotopy class g. The subspace of $\mathcal{S}_g$ formed with such α-gradients is denoted by $\mathcal{S}_{g,L}$ and one sets $\mathcal{S}_{g,\infty} := \bigcap_L \mathcal{S}_{g,L}$.

Proposition 3.3 *The subspace $\mathcal{S}_{g,L}$ is open and dense in $\mathcal{S}_g$, and $\mathcal{S}_{g,\infty}$ is residual in $\mathcal{S}_g$.*

Given $X_0 \in \mathcal{S}_{g,\infty}$ and $L > 0$, it is easily seen that there exists a path $(X_s)_{s\in Op(0)}$, which crosses $\mathcal{S}_g$ positively and such that, for every $s \neq 0$, the α-gradient X_s is in $\mathcal{KS}_L$. Therefore, it is not useful to specify the interval $Op(0)$, the incidence $n(p,q)$ *mod* L is constant for $s \in O_-$, denoted $n(p,q)_L^-$; and it is constant for $s \in O_+$, denoted by $n(p,q)_L^+$. Be careful, increasing L should require to shrink the interval $Op(0)$, except when X_0 is isolated among the non-Kupka-Smale X_s. It is time to define $\mathcal{S}_g^+ := \{\chi > 0\}$ and $\mathcal{S}_g^- := \{\chi < 0\}$. In this setting, we proved the following theorem [8].

Theorem 3.2 *Let $(X_s)_{s\in Op(0)}$ be a path of α-gradients that positively crosses $\mathcal{S}_g$ at time $s = 0$. If $X_0 \in \mathcal{S}_{g,\infty}$, the following holds for every $L > 0$.*

(1) If $X_0 \in \S_g^+$, then $n(p,q)_L^+ = (1 + g + g^2 + \cdots) \cdot n(p,q)_L^- \quad (mod\ L)$.
(2) If $X_0 \in \S_g^-$, then $n(p,q)_L^+ = (1 + g) \cdot n(p,q)_L^- \quad (mod\ L)$.

Remark 3.1

(1) In the first item, the left multiplier is a series belonging to the Novikov ring Λ_α—truncated at L—where each summand is a homotopy class of loops based at $p \in Z(\alpha)$. The product with the L-incidence to the right makes sense since the summands in $n(p,q)_L^-$ is a linear combination of arrows in Π_α from p to q. So, the product of arrows is non-zero. And similarly in 2).

[7] As every summand in the incidence has a positive α-length, every summand of length not larger than L in the reduction *mod* L of $\partial \circ \partial$ has its two factors with such α-lengths. This is not true for general Novikov complexes.

(2) It is not difficult to prove that 1) is equivalent to 2) by changing ∂_v into its opposite, which introduces a sign in Formula (3.5).
(3) By the action of C^1-diffeomorphisms whose restriction to $\mathcal{M}(p)$ are given by Proposition 3.1, it is possible to *normalize* the crossing path, in the sense of the next definition.

Definition 3.5 A path $(X_s)_{s\in Op(0)}$ crossing $\mathcal{S}_g$ at $s=0$ is said to be *normalized* if it fulfills the following properties.

(1) If H_0 denotes the holonomy along the restricted unique homoclinic orbit $\underline{\ell}$ of X_0, the disc $D_1(0) := H_0(\Sigma^-)\cap\{z=1\}$ is contained in the meridian disc $\{\psi = \psi_0\}$ and, more precisely, in its hyperplane through x a^+ which is orthogonal to the vector ν^ϕ.
(2) If H_s denotes the *perturbed holonomy* along $\underline{\ell}$, $H_s : Op\{z=0\}\to Op\{z=1\}$, the disc $D_1(s) := H_s(\Sigma^-)$ moves in the meridian disc $\{\psi=\psi_0\}$.
(3) The point $a^+(s) := D_1(s)\cap\mathbb{R}\nu^\phi$ moves on this ray with velocity x $V_1 = 1$.

An α-adapted gradient X_s in a crossing path and its normalized representative have the same dynamics up to C^1 conjugation, and hence, the same Morse-Novikov complex x x$mod\ L$ if $s\neq 0$.

Sketch of Proof of Theorem 3.2 Without loss of generality, it is sufficient to prove Item (1) in the case of a normalized crossing. We are going to give a few ideas, skipping some calculations. This will be subdivided into three lemmas and the actual proof.

For simplicity, we add a generic assumption, namely that $\langle\nu^\phi, a^-(0)\rangle\neq 0$.[8] For $s\neq 0$, we set $C_1(s) := Desc(D_1(s))\subset\partial^-\mathcal{M}(p)$. Since $D_1(s)$ avoids Σ^+ when $s\neq 0$, the submanifold $C_1(s)$ is still a $(j-1)$-disc. In the remainder of this section $a^\pm$ will stand for $a^\pm(0)$.

Lemma 3.1 *Let $\pi^- : \partial^-\mathcal{M}(p)\to\Sigma^-$ be the natural projection. Let K be any compact disc in the open southern hemisphere of Σ^-. Then, for $s\in O_-$, the intersection $C_1(s)\cap\pi^{-1}(K)$ is a section over K of π^- that C^1 converges to the zero-section over K with its reversed orientation.*

A similar statement holds for $s\in O_+$ and K in the open northern hemisphere, except that the convergence is orientation preserving in that case.

This easy lemma is implied in Morse theory when one speaks of handle slide (see for instance the figure in Hatcher-Wagoner's book [3, p. 90]). From this lemma, we keep the information that $C_1(s)\cap\{z=0\}$ accumulates to the x-axis in the C^1-topology if and only if

New notations are needed for the next lemma. Let $p_v : \{z=0\}\to\{v=0, z=0\}$ be the projection parallel to the v-axis onto the (x,y)-space. One sets $D_1'(s) := H_s^{-1}(\Sigma^+)\cap\{z=0\}$; this $(n-j-1)$-disc accumulates to the y-axis when s goes to 0. Then, for every $s\in Op(0)$ such that $s\cdot\langle\nu^\phi, a^-\rangle > 0$, the projected discs $p_v(C_1(s))$

[8] One gets the missing case by intertwining ϕ and ψ and changing ∂_v into its opposite.

and $p_v(D'_1(s))$ intersects transversely in one point only, denoted by $b_1(s)$. So, this point lifts to $c_1(s) \in C_1(s)$ and to $d'_1(s) \in D'_1(s)$. Consider The following statement gives a kinematic meaning of the character function.

Lemma 3.2 *Assuming that the crossing is normalized and* $\langle \nu^\phi, a^- \rangle \neq 0$, *then the derivative of* $v_1(s)$ *at* $s = 0$ *satisfies*

$$\dot v_1(0)\, \langle \nu^\phi, a^- \rangle = \chi(X_0). \tag{3.7}$$

Hint The proof consists of elementary differential calculus on the objects we have previously introduced. One separately computes the derivatives at $s = 0$ of $v(c_1(s))$ and $v(d'_1(s))$. One uses the following claim.

If $\bar c_1(s)$ *is another point in* $C_1(s)$ *depending* C^1 *of* s *and which goes to* a^- *when* s *goes to 0, then*

$$\frac{d}{ds} v(c_1(s)|_{s=0} = \frac{d}{ds} v(\bar c_1(s))|_{s=0}$$

and similarly in analogous settings.

Here, one chooses $\bar c_1(s) := C_1(s) \cap \{\phi = \phi_0\}$. One derives

$$\frac{d}{ds} v(c_1(s)|_{s=0} = \frac{\langle \nu^\psi, a^+ \rangle}{\langle \nu^\phi, a^- \rangle}\eta.$$

From Item (3) in Definition 3.5, it follows that

$$\frac{d}{ds} v(d'_1(s)) = -1.$$

Combining the previous two equations yields the desired formula. □

Lemma 3.3

(1) *In the previous setting, suppose the* ϕ*-latitude* a^- *is positive—and hence* $\dot v_1(0) > 0$. *Then, for* $s \in O_+$, *there are sequences of non-empty* $(j-1)$*-discs* $(D_k(s))_{k>1}$ *and* $(C_k(s))_{k>1}$ *inductively defined from the previous* $D_1(s)$ *and* $C_1(s)$ *by in this statement*

$$\begin{cases} D_k(s) := H_s\,(C_{k-1}(s)) \cap \{z = 1\} \\ C_k(s) := Desc\,(D_k(s)) \subset \partial^- M_p \end{cases} \tag{3.8}$$

Moreover, as s *goes to* 0, *the disc* $C_k(s)$ *tends to the northern hemisphere* $\mathcal{H}^+(\Sigma^-)$ *in the* C^1*-topology, uniformly over every compact set of* $\mathcal{H}^+(\Sigma^-)$. *When* $s \in O_-$, *both previous sequences are empty when* $k > 1$.

(2) *If* $\dot v_1(0)$ *and* $\cos_\phi(a^-)$ *are negative, then for* $s \in O_-$, *the disc* $C_2(s)$ *is well defined as in* (3.8) *and the subsequent discs,* $D_3(s)$, ..., *are empty. Moreover,*

$C_2(s)$ tends to $\mathcal{H}^+(\Sigma^-)$ in the C^1-topology with the reversed orientation. When $s \in O_+$, all discs in (3.8) *are empty when $k > 1$.*

Hint In both items, the sequence of discs $D_k(s)$ stops to be defined at rank k just for the brutal reason that the annulus $C_k(s)$ does not meet $\{z = 0\}$, the origin of the tube T.

The disc $D_k(s)$ intersects the v-axis through a^+ at a point whose velocity V_k at $s = 0$ is given by a recursive formula, analogous to (3.7):

$$V_k = \frac{\langle \nu^\psi, a^+\rangle}{\langle \nu^\phi, a^-\rangle} \eta\, V_{k-1} + 1.$$

By Lemma 3.1 the sign of V_k determines the hemisphere of Σ^- to which the disc $C_k(s)$ is adhering when s goes to 0_+. If this sign guaranties that $C_k(s)$ meets $\{z = 0\}$, the recursion goes on. □

Sketch of Proof of Theorem 3.2 Continued Let $X_\pm$ denote any X_s with $s \in O_\pm$; this is an α-adapted gradient which is Kupka-Smale up to L. The incidence $n(p, q)^\pm_L$ is a finite sum of signed connecting orbits from p to q; the sign ε_ℓ of ℓ is determined by the rule (3.3) and all orbits in the same homotopy class are algebraically added.

There is an equivalence relation $\sim$ on the homotopy classes of connecting orbits defined by the following rule: $[\gamma] \sim [\gamma']$ if and only if $[\gamma] = g^k \cdot [\gamma']$ for some $k \in \mathbb{Z}$. This equivalence relation defines a partition $P(p, q)$ of the set of connecting orbits from p to q. Each class $[\gamma]_\sim \in P(p, q)$ is represented by its—unique up to homotopy and multiplicity—connecting orbit $[\gamma_0]$ of minimal α-length, actually a positive number—except if $n(p, q)^\pm_L = 0$.

Let $\lambda_\pm$ be the contribution to $n(p, q)^\pm_L$ of the whole set of connecting $X_\pm$-orbits in $[\gamma_0]_\sim$. To prove Theorem 3.2, it is sufficient to prove

$$\lambda_+ = (1 + g + g^2 + \cdots) \cdot \lambda_- . \tag{3.9}$$

Recall that the crossing is through $\S^+_g$. Let b denote the point of Σ^- where γ_0 gets out the Morse model $\mathcal{M}(p)$. Let $\Delta_s \subset \big(W^s(q, X_s) \cap \partial^-\mathcal{M}(p)\big)$ be a small $(n-j)$-disc centered at b and transverse to X_s.

We limit ourselves to the case $\langle \nu^\phi, a^-(X_0)\rangle \neq 0$. There are still four cases to consider. We are going to deal with two of them which are very informative. In these two cases, namely,

(i) $\langle \nu^\phi, a^-(X_0)\rangle > 0$
(ii) $\langle \nu^\phi, a^-(X_0)\rangle < 0$,

the point b is assumed to lie in the northern hemisphere $\mathcal{H}^+$ of Σ^-.

The union of all X_0-orbits passing through the positive (resp. negative) hemisphere of Σ^- will be called the *positive part* (resp. *negative part*) of $W^u(p, X_0)$. It will be denoted by $W^u(p, X_0)^\pm$ depending on the hemisphere sign.

For a given $k > 0$, one says that the manifolds $W^u(p, X_s)$ accumulate to $g^k \cdot W^u(p, X_0)^{\pm}$ when s goes to 0_- (resp. 0_+) if it is topologically true when lifting to the universal cover. As the convergence is C^1, one can compare the orientations. The accumulations we are interested in are all predicted by Lemma 3.3.

Case (i) When s goes to 0_-, the manifolds $W^u(p, X_s)$ accumulate to $-g \cdot W^u(p, X_0)^-$ and nothing else. As $b \in \mathcal{H}^+$, we have $\lambda_- = [\gamma_0]$. When s goes to 0_+, by Lemma 3.3 (1), then the manifolds $W^u(p, X_s)$ accumulate to $+g^k \cdot W^u(p, X_0)^+$ for every $k > 0$ and hence intersect $g^k \cdot \Delta_0$. equa The change from λ_- to λ_+ is then given by Formula (3.9).

Case (ii) When s goes to 0_-, by Lemma 3.3 (2), the manifolds $W^u(p, X_s)$ accumulate to $-g \cdot W^u(p, X_0)^-$ and also to $-g^2 \cdot W^u(p, X_0)^+$, but to nothing else. When s goes to 0_+, the manifolds $W^u(p, X_s)$ accumulate to $+g \cdot W^u(p, X_0)^+$ and nothing else. As $b \in \mathcal{H}^+$, we have $\lambda_- = (1-g^2)\cdot[\gamma_0])$ and $\lambda_+ = (1+g)\cdot[\gamma_0]$. Formula (3.9) is again fulfilled.

□

3.4 Isolated Homoclinic Bifurcations

We are going to explain that an isolated homoclinic bifurcation in Morse-Novikov theory is far from being exceptional on a closed manifold M of dimension $n > 2$.

3.4.1 Some Basics on Closed One-Forms

Here, we give a large family of examples of isolated homoclinic bifurcations, in the spirit of [9]. We first recall the following result well known to experts [6, Lemme 4.1].

Every non-exact closed one-form is cohomologous to a Morse one-form without zeroes of extremal Morse index.

In the remainder of this section, α will be such a differential form on the closed manifold M. Therefore, if X is an α-adapted gradient, for almost every point $x \in M$—in the sense of a volume measure on M—the X-orbit of x does not go to a zero of α when the time goes to $+\infty$ or to $-\infty$.

Such an orbit crosses at least twice some flow box $\tilde{B}$ which is a piece of a finite cover of M by Morse models about the zeroes of α, and by cylinders which are bifoliated by X and the leaves of α. Some classical modification of X in $int(\tilde{B})$ creates a periodic orbit R. Let g denote its homotopy class—non-zero by Stokes' formula.

Then, an isotopy preserving α and supported in $int(\tilde{B})$ reduces the first holonomy diffeomorphism along R to be the identity of a small transversal to R. So, a tube T of periodic orbits—with a same period, say 1—is created around R. A third modification makes R a repeller and ∂T an attractor, strong enough so that the forthcoming construction works. We still name X such an α-adapted gradient. Generically, X is Kupka-Smale.

3.4.2 *Birth of Two Zeroes*

Let $B = D^{n-1} \times [-\varepsilon, +\varepsilon]$ be a box in $int(T)$ bifoliated by X and α. The vector field X points inwards along the upper boundary $\partial^+ B = D^{n-1} \times \{+\varepsilon\}$ and outwards along the lower boundary $\partial^- B = D^{n-1} \times \{-\varepsilon\}$. The local primitive h of α in B, valued in $[-\varepsilon, +\varepsilon]$, is modified to a smooth function h' with a pair (p, q) of critical points of respective indexes $k+1$ and k in cancelling position. Hence, (α, X) is changed into (α', X') where $\alpha' := \alpha + d(h' - h)$ and where X' is equal to X outside of B and is an α'-adapted gradient in B (Fig. 3.2).

The leaf of $\alpha|_T$ containing $\partial^\pm B$ is denoted by $L^\pm$. Let $\rho : L^- \to L^+$ be the first holonomy diffeomorphism of X (or X'). Set $D := \rho(\Delta^-)$. The repulsion of R is so strong that D is disjoint from B.

One chooses the orientation of $W^u(q, X')$ arbitrarily and the orientation of $W^u(p, X')$ so that, by (3.3), *the incidence* $n(p, q)^{X'}$ *is equal to* $+1$.

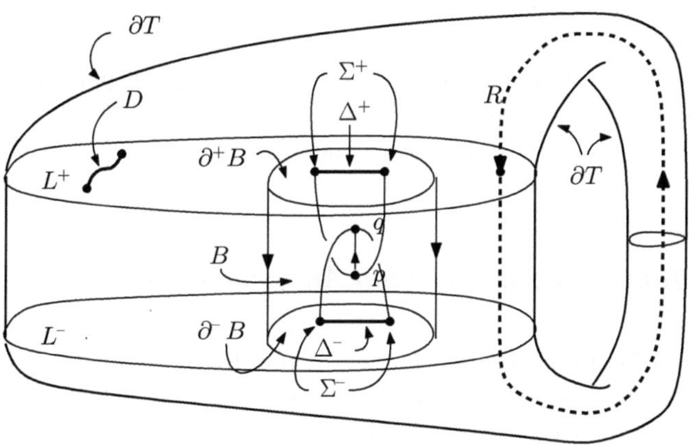

Fig. 3.2 A symbolic drawing of the tube T containing the birth box B of a pair of critical points p of index $k+1$ and q of index k. Some gradient lines are drawn

3.4.3 Descent Map

The unstable manifold $W^u(q, X')$ meets $\partial^- B$ along a $(k-1)$-sphere Σ^- that bounds a disc $\Delta^- \subset \partial^- B$ contained in the *local* unstable manifold $W^u_{loc}(p, X')$. The stable manifold $W^s(p, X')$ meets $\partial^+ B$ along an $(n-k-2)$-sphere Σ^+. This sphere bounds a disc Δ^+ that is contained in the local stable manifold $W^s_{loc}(q, X')$.

Let $Desc : L^+ \dashrightarrow L^-$ be the descent map by the flow of X', well defined in the complement of $\Sigma^+ \cup \Delta^+$. For every $a^+ \in \Sigma^+$, we denote by m_{a^+} a fiber of the normal disc bundle to Σ^+ in L^+. The intersection $m_{a^+} \cap \Delta^+$ is a ray of m_{a^+}, denoted by $[a^+, v_{a^+}]$. The hyperplane P orthogonal to $[a^+, v_{a^+}]$ in the planar meridian disc m_{a^+} determines a *co-oriented equator* E in the sphere ∂m_{a^+}. Understanding the descent of m_{a^+} is a key point of the expected construction.

One checks that $\Gamma^- := Desc(m_{a^+})$ is a closed-interval bundle over Δ^- *pinched* over Σ^- and whose fibers are the images of the rays of $m_{a^+} \setminus \{a^+\}$. The pinching comes from the ray $[a^+, v_{a^+}]$ contained in Δ^+ whose interior descends to q. One sets $\Gamma^+ := \rho(\Gamma^-)$, the fibration being carried by ρ simultaneously.

The hyperplane P descends to the restriction of Γ^- over some $(k-1)$-sphere $E^- \subset int(\Delta^-)$. This E^- yields a decomposition

$$\Delta^- = \delta^- \cup_{E^-} \delta^+,$$

where the sign in $\delta^\pm$ reflects the co-orientation of E, and hence of E^-, determined by v_{a^+}. Therefore, δ^- is a disc and δ^+ is an annulus bounded by $\Sigma^- \cup E^-$. The half disc bounded by P and containing v_{a^+} (also named the *positive* half disc of m_{a^+}) descends to the restriction of Γ^- over δ^+.

For what follows, it will be convenient to consider also another planar meridian disc m'_{a^+} over a^+ with a slightly larger radius than m_{a^+}. Equipped with its rays, it descends to an interval bundle Γ'^- over Δ^- (pinched over Σ^-) containing Γ^- fiberwise. The image of Γ'^- through ρ is denoted by Γ'^+.

3.4.4 A Homoclinic Bifurcation Attached to p

The operation of cutting M along L_+ and gluing the so-created two lips by some diffeomorphism φ supported in $int(L^+)$ preserves α' but changes X' into a new α'-adapted gradient that we denote by $\varphi * X'$.[9]

To have a path of such α'-adapted gradients, one uses some flow $(Y^s)_{s\in[0,1]}$ on L^+ and we set $X'_s := Y^s * X'$. Let D denote the image $\rho(\Delta^-)$. If $Y^{1/2}(D)$ meets

[9] Without more specifications, this vector field is not uniquely defined but its first return dynamics to L^+ is.

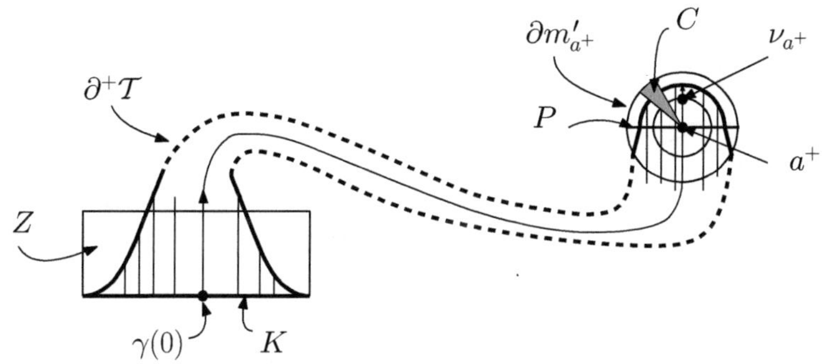

Fig. 3.3 A scheme for a tongue. A few arcs of Y-trajectories are drawn

Σ^+ in one point a^+, the α'-adapted gradient $X'_{1/2}$ has a homoclinic orbit making a loop at p in the homotopy class g.

One chooses any k-dimensional compact disc K in the interior of $\rho(\delta^+)$. Let Z be the cylinder made by restricting Γ'^+ over K. Its inverse image by ρ is denoted by Γ'^-_K. The inverse image of this latter by $Desc$ is a compact spherical sector C in the positive half meridian m'_{a^+}, with no ray in Δ^+.

Definition 3.6 A *tongue* $\mathcal{T}$ issued from K and snapping up the meridian disc m_{a^+} is a $(k+1)$-submanifold of L^+ which fulfills the following (Fig. 3.3).

(1) $\mathcal{T}$ is a cylinder whose lower boundary $\partial^-\mathcal{T}$ is K and upper boundary $\partial^+\mathcal{T}$ is infinitely tangent to K along ∂K.
(2) $\mathcal{T}$ contains m_{a^+} in its interior and is glued to a domain of m'_{a^+} bounded by a k-disc in $\partial^+\mathcal{T}$ and another k-disc in $\partial m'_{a^+}$.
(3) A *trunk* of the tongue is glued to the cylinder Z from K up to a sub-disc in the upper boundary of Z—as shown on Fig. 3.3.[10]
(4) There is a vector field Y in L^+, non-zero along $int(K)$—where it is pointing inwards $\mathcal{T}$—and whose trajectories in $\mathcal{T}$ go from K to $\partial^+\mathcal{T}$ in infinite time.
(5) The support S of Y is a small enlargement of the tongue and its frontier contains $\partial^+\mathcal{T}$.

3.4.5 The Construction Continued

Here are some conditions that guarantee the one-parameter family $(Y^s * X')_{s\in[0,1]}$ to be a normalized positive crossing of $\mathcal{S}_g$ through its positive part $\mathcal{S}_g^+ = \{\chi > 0\}$.

[10] The coordinate in each fiber of Z is nothing but the radial coordinate in the spherical sector C.

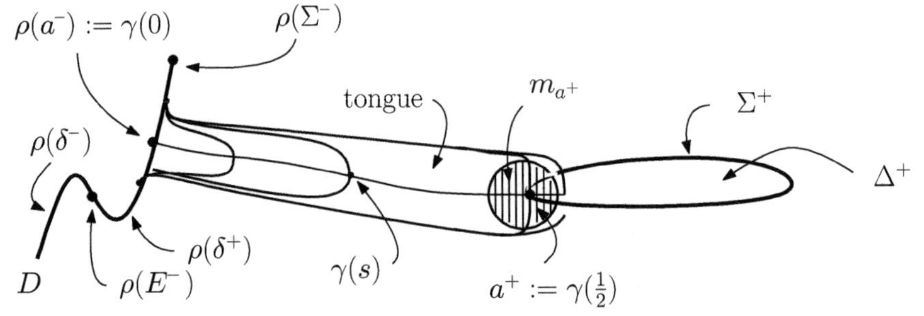

Fig. 3.4 Here, the ambient space is L^+. The thick line on the left is $D := \rho(\Delta^-)$. This is the *first holonomy intersection* of $W^u(p, X')$ with L^+. Its boundary is $\rho(\Sigma^-)$, the first holonomy intersection of $W^u(q, X')$ with L^+

Moreover, they will assure that this crossing is *isolated*. We insist that a^+ may be chosen arbitrarily in Σ^+.

(i) Let $\gamma(s)$ be the Y-trajectory passing through a^+ at time $s = \frac{1}{2}$. One needs $\gamma(0) \in D$.
(ii) The point $a^- := \rho^{-1}(\gamma(0))$ is located in the annulus δ^+.
(iii) The vector field Y is parallel to $[a^+, v_{a^+}]$ in the positive half disc m'_{a^+}.
(iv) The support S of Y is a neighborhood of $\gamma([0, 1])$ that is disjoint from $\rho(\delta^-)$ and from the image $(\rho \circ Desc)^i(\partial^+ B)$ for every $i > 0$.
(v) From $s = 0$ to $+\infty$, $D \cap S$ swepts a *tongue* $\mathcal{T}$ that covers the disc m_{a^+} (see Figs. 3.3 and 3.4); when $Y^s(D)$ meets m_{a^+}, their intersection is planar and parallel to P.
(vi) The holonomy factor,[11] denoted by η, is small enough so that $\chi(X'_{1/2}) > 0$.
(vii) The velocity of $Y^s(D)$ is so small for $s \in (\frac{1}{2}, 1]$ that $Y^s(D) \cap m'_{a^+}$ separates a^+ from the base $C \cap \partial m_{a^+}$ of the spherical sector C in m'_{a^+}.

Every item is easily fulfilled, except items (iv) and (vi). The second requirement in item (iv) is fulfilled as soon as ∂T is a strong attractor.

Item (vi) is satisfied once the spatial derivative of $Y^{\frac{1}{2}}$ at $\gamma(0)$ has a small norm. Indeed, its one-jet controls the holonomy one-jet $\mathfrak{h}$ from a^- to a^+ along the homoclinic orbit. The holonomy factor is small if $\mathfrak{h}$ has a small norm. Recall from Definition 3.3 that the character function reads $\chi = \eta\omega_+ + \omega_-$, where $\omega_\pm$ is a latitude valued in $[-1, +1]$. Each term of this formula depends on the one-jet $\mathfrak{h}$ only. By the choice of $a^- \in \delta^+$, we have $\omega_- > 0$. So, for every ω_+, if η is small enough, χ is positive.

Theorem 3.3 *On a closed manifold M of dimension $n > 2$, the above-mentioned construction provides an isolated bifurcation leading to an incidence coefficient*

[11] See Definition 3.3.

$n(p,q)^{X_1'}$, in the Morse-Novikov complex $C(\alpha, X_1')$, that reads

$$n(p,q)^{X_1'} = 1 + g + g^2 + g^3 + \cdots \tag{3.10}$$

Here, α is a Morse closed one-form, X_1' is some α-adapted gradient, p and q are two zeroes of α in cancelling position and g is the homotopy class of some loop based in p such that $\int_g \alpha < 0$.

Proof Once we know that the bifurcation is isolated, Theorem 3.2 is available. At the beginning of the path which crosses the bifurcation stratum $\S_g$, the incidence $n(p,q)^{X_0'}$ is equal to $+1$ as the two zeroes are in cancelling position. Since the crossing of $\S_g$ is positive and happens at a point of $\S_g^+$, the first item of Theorem 3.2 applies and yields the desired incidence (3.10).

After Sect. 3.4.5, we have a complete construction fulfilling almost all demands except that the bifurcation is isolated. It remains to check this property.

Nothing happens outside of T where X' is independent of s. Over there, X' is a Kupka-Smale α-adapted gradient, tangent to ∂T and foliating ∂T with periodic orbits. So, the affair is localized in T where there are only two zeroes p and q in cancelling position. The zero q of lower index has an unstable manifold that is sucked up by ∂T. So, it never could cross the stable manifold of q (see Fig. 3.4).

About p, the matter is to prove that, for $s \in [0, 1]$, $s \neq \frac{1}{2}$, there is no homoclinic connecting orbit at p in the homotopy class g^k, $k > 0$. We first prove it for $k = 1$. We are going to prove that,[12]

$$\text{for every } s \in (0, 1/2) \cup (1/2, 1], \text{ one has } \quad Y^s(D) \cap \Sigma^+ = \emptyset. \tag{3.11}$$

The part of D in the complement of the support S of Y follows the dynamics of $Desc \circ \rho$ and never meets Σ^+. The remaining part $D \cap S$ moves in the tongue $\mathcal{T}$ whose interior is foliated by discs, namely $Y^s(D) \cap \mathrm{int}(S)$. The only leaf which meets Σ^+ is $Y^{\frac{1}{2}}(D) \cap \mathrm{int}(S)$ and (3.11) is fulfilled.

The same question about the class g^k, $k > 1$, will be proved recursively. For $k \geq 1$, we introduce the following subset of M:

$$H_k(s) := \left[(Y^s \rho(Desc))^{k-1} Y^s(D) \right] \cap S.$$

The recursive assumption reads:

$$(*)_k \quad \begin{cases} \text{a) When } s \text{ goes to } \frac{1}{2} \text{ from below, then } Desc(H_k(s)) \text{ is empty if } k \geq 2. \\ \text{b) For every } s \in (\frac{1}{2}, 1],\ H_k(s) \text{ is located in the tongue beyond } a^+. \\ \text{Moreover, when } s \text{ goes to } \frac{1}{2} \text{ from above, } H_k(s) \text{ descends into the bundle} \\ \Gamma^- \text{ as a section over } \rho^{-1}(K) \text{ which goes to the 0-section.} \end{cases}$$

---

[12] We recall from Sect. 3.4.4 that D stands for $\rho(\Delta^-)$.

We refer to Sect. 3.4.4 for the definition of K and the bundles associated with. We have to prove $(*)_{k+1}$. When s goes to $\frac{1}{2}$ from below, Lemma 3.1 tells us that $\rho(Desc)Y^s(D)$ accumulates to every compact subset of $\rho(\delta^-)$, and hence does not approach the support of Y. In this case, $H_{k+1}(s)$ becomes empty.

By $(*)_k$, one sees, when s goes to $1/2$ from above, that $Desc(H_k(s))$ accumulates to $\rho^{-1}(K)$ as a section of Γ'^- over it. Hence, $\rho(Desc)(H_k(s)) \cap S$ is located in the tongue $\mathcal{T}$ for s close to $(1/2)_+$ and accumulates to $\rho(\delta^+) \cap S$. Since the tongue is invariant by the flow, $Y^s\rho(Desc)(H_k(s))$ is located in $\mathcal{T}$ for every $s > \frac{1}{2}$ and beyond of $Y^s(D)$ with respect to the flow of Y. This proves the first part of $(*)_{k+1}$.

For the second part, one uses Items (v) and (vii) from Sect. 3.4.5. Then, when $s \in (\frac{1}{2}, 1]$, $H_{k+1}(s)$ intersects every ray of C in one point. By applying $Desc$, we get a complete proof of $(*)_{k+1}$.[13]

3.5 The Doubling Phenomenon

We go back to the article [8] to comment on the doubling phenomenon that first appeared to us as a way to avoid a contradiction. Actually, the stratum $\mathcal{S}_{g^2}$ is the sole stratum near a point X_0 in $\mathcal{S}_g^0 = \{\chi = 0\}$, under the mild condition that X_0 is not in the ϕ-axis (see Fig. 3.5).

3.5.1 An Apparent Contradiction

We recall the stratum $\S_g$ of α-adapted gradients that have a *simple* homoclinic orbit at p. Here, g belongs to Hom(p, p), in the groupoidic language, where p is a zero of α that is both the source and the target of g. By Theorem 3.1, we know that, if the ambient dimension n is greater than 2, the locus $\S_g^0 \subset \S_g$ is a smooth codimension-two submanifold of $\mathcal{F}_\alpha$, meeting every connected component of $\S_g$.

Consider a disc $\mathcal{D}$ in $\mathcal{F}_\alpha$, centered at $X_0 \in \S_g^0$, transverse to $\S_g$ and to $\S_g^0$. Parametrize its boundary by $\gamma : \mathbb{S}^1 \to \partial\mathcal{D}$ so that γ crosses $\{\chi < 0\}$ positively at $\gamma(\pi/2)$ and crosses $\{\chi > 0\}$ negatively at $\gamma(-\pi/2) = \gamma(3\pi/2)$.

Assume this disc could be small enough so that it avoids all bifurcation strata different from $\mathcal{S}_g$. Thus Theorem 3.2 is available without working *mod L*. Starting from $\theta = 0$ and turning counterclockwise, the differential of the Morse-Novikov complex $\partial^{\gamma(0)}$, evaluated at p, is first modified at $\gamma(\pi/2)$ by left multiplication by $(1+g)$ and then, at $\gamma(3\pi/2)$, by left multiplication by $(1+g+g^2+\cdots)^{-1} = (1-g)$. Finally, we get the absurd result that the product of modifications after one turn around γ is not Id but the product by $(1 - g^2)$. This made us think that the crossing

[13] This arguing should clarify [9, Lemma 1].

of $\mathcal{S}_{g^2}$ is unavoidable; this is the doubling phenomenon in question that we are going to state precisely.

Definition 3.7

(1) The open set $\mathcal{S}_g^{0,\pm} \subset \mathcal{S}_g^0$ is defined by the sign of the ϕ-latitude, that is: $X \in \mathcal{S}_g^{0,\pm} \Leftrightarrow \langle \nu^\phi(X), a^-(X)\rangle \in \mathbb{R}_\pm$. The complementary closed subset in $\mathcal{S}_g^0$ where the ϕ-latitude vanishes—and hence, the ψ-latitude does the same—is denoted by $\mathcal{S}_g^{0,0}$.

(2) Let $X_{0,0} \in \mathcal{S}_g^0 \setminus \mathcal{S}_g^{0,0}$. Let $\big(\mathcal{D}(s,t) := X_{s,t}\big)_{s,t}$ be a *germ* at $X_{0,0}$ of a two-parameter family in $\mathcal{F}_\alpha$, the space of α-adapted gradients. This germ is said to be *adapted* to the pair $(\mathcal{S}_g, \mathcal{S}_g^0)$ if the following conditions are fulfilled:

   (i) The one-parameter family $\big(\mathcal{D}(0,t)\big)_{t\in\mathcal{O}p(0)}$ is contained in $\mathcal{S}_g$, transverse to $\mathcal{S}_g^0$ and $\partial_t\mathcal{D}(0,0)$ is non-zero and points towards $\mathcal{S}_g^+$.

   (ii) The partial derivative $\partial_s\mathcal{D}(0,0)$ is transverse to $\mathcal{S}_g$ and points towards its positive side.

Theorem 3.4 *Let $\mathcal{D}$ be a germ of 2-discs transverse to $\mathcal{S}_g^0 \setminus \mathcal{S}_g^{0,0}$ and adapted to the pair $(\mathcal{S}_g, \mathcal{S}_g^0)$. Then $\mathcal{D}$ intersects $\mathcal{S}_{g^2}$ transversely along an arc of $\mathcal{S}_{g^2}^+$. The trace on $\mathcal{D}$ of the strata $\{\mathcal{F}_\alpha\,,\ \mathcal{S}_g \cup \mathcal{S}_{g^2}^+\,,\ \mathcal{S}_g^{0,\pm}\}$ is C^1-diffeomorphic to*

$$\{\mathbb{R}^2,\ \mathbb{R}\times\{0\}\cup\{0\}\times\mathbb{R}_\pm\,,\ (0,0)\}\,.$$

Moreover, the natural co-orientation of $\mathcal{S}_{g^2}$ restricts to the natural co-orientation of $\mathcal{S}_g^0$ in $\mathcal{S}_g$ or to its opposite depending upon whether $\mathcal{S}_{g^2}$ approaches $\mathcal{S}_g^{0,+}$ or $\mathcal{S}_g^{0,-}$ respectively (Fig. 3.5).

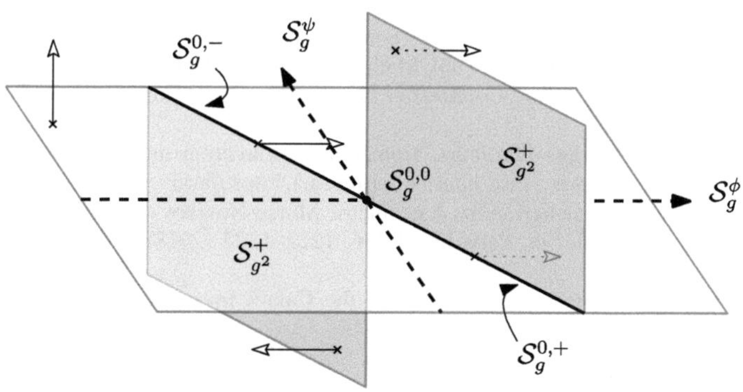

Fig. 3.5 The ϕ-axis, that is the locus where the ψ-latitude vanishes, is denoted by $\mathcal{S}_g^{\phi}$

Finally, if $X_{0,0}$ also fulfills some generic property $\mathcal{S}_{g,\infty}$,[14] *then the germ $\mathcal{D}$ does not meet $\mathcal{S}_{g^k}$ for $k > 2$.*

This theorem states, in particular, that $\mathcal{S}_g^0 \setminus \mathcal{S}_g^{0,0}$ looks like a boundary to $\mathcal{S}_{g^2}$. We refer to [8, Section 4] for the proof. We do not even give an idea of the proof of the main part. This needs some work, analogous to our Sect. 3.3, except that it deals with a 2-dimensional disc instead of a path. However, the reader could easily find the reason why it is only the positive part of $\mathcal{S}_{g^2}$ that is adhering to $\mathcal{S}_g^0$ (hint: see Sect. 3.5.1).

As a conclusion, Theorem 3.4 emphasizes the evidence that the stratification of the space of α-adapted gradients of a Morse one-form α could not be horrible, at least in codimension one and two.

Acknowledgments I am grateful to Carlos Moraga for his comments on a preliminary version of this text. I also thank the projects ANR-11-LABX-0020-01 and COSY ANR-21-CE40-0002 for their support.

References

1. A. Fathi, F. Laudenbach, V. Poénaru, *Travaux de Thurston sur les surfaces,* Séminaire Orsay 1976 Astérisque, 66–67 (Société Mathématique de France, Paris, 1979), 284 pp.
2. A. Fathi, F. Laudenbach, V. Poénaru, *Thurston's work on surfaces*. Translated from the 1979 French original by Djun M. Kim and Dan Margalit. Mathematical Notes, 48 (Princeton University Press, Princeton, 2012), xvi+254 pp. ISBN: 978-0-691-14735-2
3. A. Hatcher, J. Wagoner, *Pseudo-isotopies of compact manifolds*. Astérisque 6 (Soc. Math. de France, Paris, 1973)
4. M. Hutchings, Reidemeister torsion in generalized Morse theory. Forum Math. **14**, 209–244 (2002)
5. I. Kupka, Contribution à la théorie des champs génériques. Contrib. Differential Equations **2**, 457–484 (1963)
6. F. Latour, Existence de 1-formes fermées non singulières dans une classe de cohomologie de de Rham. Inst. Hautes Études Sci. Publ. Math. **80**, 135–194 (1994)
7. F. Laudenbach, *Transversalité, Courants et Théorie de Morse* (Éditions de l'École polytechnique, Palaiseau, 2011)
8. F. Laudenbach, C. Moraga Ferrándiz, Homoclinic bifurcation in Morse-Novikov theory, a doubling phenomenon. Ann. Inst. Fourier (to appear). https://arxiv.org/abs/1604.01809
9. F. Laudenbach, C. Moraga Ferrándiz, A geometric Morse-Novikov complex with infinite series coefficients. C. R. Acad. Sci. Paris Ser. I **356**, 1222–1227 (2018). https://doi.org/10.1016/j.crma.2018.09.008
10. H.V. Lê, K. Ono, Symplectic fixed points, the Calabi invariant and Novikov homology. Topology **34**(1), 155–176 (1995)
11. J. Milnor, Whitehead torsion. Bull. Am. Math. Soc. **72**(3), 358–426 (1966)

[14] See Definition 3.4.

12. C. Moraga Ferrándiz, An extended Steinberg group. A tool to detect non-singular closed 1-forms which are non-isotopic, in *61st Annual Japanese Topology Symposium*. Available at https://www.mathsoc.jp/~topology/topsymp/2014/ts2014Moraga0810.pdf
13. S.P. Novikov, Multivalued functions and functionals. An analogue of the Morse theory. Soviet Math. Dokl. **24**(2), 222–226 (1981)
14. S.P. Novikov, The Hamiltonian formalism and a multi-valued analogue of Morse theory. Russ. Math. Surv. **87**, 1–56 (1982)
15. J. Palis, W. de Melo, *Geometric Theory of Dynamical Systems: An Introduction* (Springer, Berlin, 1982)
16. A. Papadopoulos, Difféomorphismes pseudo-Anosov et automorphismes symplectiques de l'homologie. Ann. Sci. École Norm. Sup. (4) **15**(3), 543–546 (1982)
17. A. Papadopoulos et al., Christiaan Huygens, Écrits sur la musique et le son, (Ed. Hermann, Paris, 2021). ISBN: 9791037002952
18. A. Pazhitnov, On the Novikov complex for rational Morse forms. An. Fac. Sci. Toulouse Math. (6) **4**, 297–338 (1995)
19. M. Peixoto, On an approximation theorem of Kupka and Smale. J. Differ. Equ. **3**, 214–227 (1966)
20. J.-C. Sikorav, Points fixes de difféomorphismes symplectiques, intersections de sous-variétés lagrangiennes, et singularités de 1-formes fermées, thèse d'État, Paris 11, Orsay (1987)
21. S. Smale, On gradient dynamical systems. Ann. Math. (2) **74**, 199–206 (1961)
22. S. Smale, Stable manifolds for differential equations and diffeomorphisms. Ann. Scuola Norm. Sup. Pisa Cl. Sci. (3) **17**(3), 97–116 (1963)
23. R. Thom, *Structural Stability and Morphogenesis* (Addison-Wesley Publishing Company, Redwood City, 1989)

Chapter 4
A Twisted Invariant of a Compact Riemann Surface

Nariya Kawazumi

Abstract We introduce a twisted version of the Kawazumi–Zhang invariant $a_g(C) = \varphi(C)$ of a compact Riemann surface C of genus $g \geq 1$, and discuss how it is related to the first Mumford–Morita–Milller class $e_1 = \kappa_1$ on the moduli space of compact Riemann surfaces and the original Kawazumi–Zhang invariant.

4.1 Introduction

A compact Riemann surface is still a fascinating research subject even in the twenty-first century. In fact, various mathematicians and even physicists are now studying the Teichmüller space $\mathcal{T}_g$ of genus g, a universal covering space of the moduli space $\mathbb{M}_g$ of compact Riemann surfaces of genus g, from their own various viewpoints. In this chapter we study functions on the moduli space $\mathbb{M}_g$, i.e., invariant functions on the Teichmüller space $\mathcal{T}_g$ under the mapping class group action. They are regarded also as analytic invariants of a compact Riemann surface. There are no non-constant holomorphic functions on $\mathbb{M}_g$ if $g \geq 3$, while the $3g-3$-dimensional orbifold $\mathbb{M}_g$ has an enormous amount of C^∞ functions. Hence we would like to find functions which have some geometric and/or arithmetic meanings. Hyperbolic geometry is a highly effective tool for studying Riemann surfaces and Teichmüller spaces. But, in the context of hyperbolic geometry, the author knows no interesting functions on the moduli space $\mathbb{M}_g$ of compact Riemann surfaces except the injective radius of a compact hyperbolic surface.

Other effective tools for studying compact Riemann surfaces are the classical theory of Abelian integrals and its powerful descendant, arithmetic geometry including the Arakelov geometry. Theta functions provide meaningful functions on the moduli space $\mathbb{M}_g$. Moreover, based on his own Arakelov geometry, Faltings [10] introduced an analytic invariant of a compact Riemann surface. Hain and Reed

N. Kawazumi (✉)
Department of Mathematical Sciences, University of Tokyo, Tokyo, Japan
e-mail: kawazumi@ms.u-tokyo.ac.jp

A. M. Uludağ, A. Zeytin (eds.), *Essays on Geometry*,
https://doi.org/10.1007/978-3-031-76257-4_4

[11] introduced another invariant by the Hodge line bundle and the intermediate Jacobian on the moduli space $\mathbb{M}_g$. Zhang [26] and the author [15, 16] introduced the same invariant $a_g = \varphi$ independently. Zhang's approach comes from the Arakelov geometry, while the author's from the Johnson homomorphism similar to that of Hain and Reed, but based on the classical theory of Abelian integrals. Later de Jong [4] proved that the Faltings invariant, the Hain-Reed invariant and the Kawazumi–Zhang invariant are linearly dependent. Moreover he has clarified various aspects of the invariant a_g in papers including [5–7]. In particular, de Jong and Shokrieh [8, Theorem 7.1] gave a complete description of the asymptotic behavior of the invariant a_g along 1-parameter families of compact Riemann surfaces degenerating into a stable curve. D'Hoker and Green [1] found some relation of the genus 2 invariant a_2 to high energy physics. D'Hoker et al. [3] proved the function a_2 is an eigenfunction with respect to the Laplacian on the moduli space of principally polarized Abelian varieties $\mathbb{A}_2$. In Sect. 4.3.2 we will show this does not hold in the case $g = 3$ by using a twisted version of the invariant a_g. The obstruction for a_3 to be an eigenfunction is the (extended) first Johnson homomorphism [13, 19].

In this paper, we introduce a twisted version of the invariant a_g, a map A from the 4-fold tensor product of the first cohomology group of the surface to the complex numbers. One can recover the invariant a_g from the map A by using the intersection pairing on the cohomology group, and can construct new invariants by a similar way to a_g. Moreover, from the invariant A, we obtain an explicit $(1, 1)$-form on the moduli space $\mathbb{M}_g$ representing the first Mumford–Morita–Miller class $e_1 = \kappa_1$ in Sect. 4.3.1.

The invariant A is given as an integral over the 2-fold product of the surface. D'Hoker and Schlotterer [2] generalized our construction to integrals over the n-fold product of the surface for $n \geq 3$, and proved highly nontrivial identities among them. It is amazing that their proof of the identities is based only on formal natures of the Arakelov Green function $G(-, -)$.

This paper is organized as follows. The twisted version A of the invariant a_g of a compact Riemann surface C is introduced in Sect. 4.2. In Sect. 4.3 we give two kinds of interpretation of the map A as tensor fields on the moduli space $\mathbb{M}_g$. First it defines a $(1, 1)$-form on $\mathbb{M}_g$. As is proved in Sect. 4.4, it is closed and represents the first Mumford–Morita–Miller class $e_1 = \kappa_1$. In Sect. 4.3.2, using the second interpretation, we discuss how the invariant a_g behaves in the cases $g = 2, 3$. The proof of Proposition 4.1, which we need for our consideration in the case $g = 3$, is given in Sect. 4.5.

4.2 A Twisted Invariant of a Compact Riemann Surface

Throughout this paper, we assume the genus g of a compact Riemann surface C is not smaller than 1, $g \geq 1$. Then the vector space $H' := H^0(C; \Omega^1_C)$ of holomorphic 1-forms on C does not vanish. Its complex conjugate $H'' := \overline{H^0(C; \Omega^1_C)}$ is the vector space of anti-holomorphic 1-forms on C. The first cohomology group $H := H^1(C; \mathbb{C})$ is naturally identified with the direct sum $H' \oplus H''$, the vector space of harmonic 1-forms on the surface C. The intersection pairing

$$(\phi_1, \phi_2) \in H \times H \mapsto \phi_1 \cdot \phi_2 := \int_C \phi_1 \wedge \phi_2 \in \mathbb{C}, \tag{4.1}$$

is non-degenerate. Hence we identify H with its dual through the pairing. Under this identification, the subspaces H' and H'' are dual to each other.

For $\phi_1', \phi_2' \in H' = H^0(C; \Omega_C^1)$ we can define a positive definite Hermitian pairing

$$(\phi_1', \phi_2') \in H' \otimes H' \mapsto \frac{\sqrt{-1}}{2} \int_C \phi_1' \wedge \overline{\phi_2'} \in \mathbb{C}. \tag{4.2}$$

In fact, any holomorphic 1-form ϕ' locally given by $\phi' = f(z)dz$ in terms of a local complex coordinate $z = x + \sqrt{-1}y$ satisfies $\frac{\sqrt{-1}}{2}\phi' \wedge \overline{\phi'} = |f(z)|^2 dx \wedge dy$. We take an orthonormal basis $\{\psi_i\}_{i=1}^g \subset H'$ with respect to the Hermitian pairing

$$\frac{\sqrt{-1}}{2} \int_C \psi_i \wedge \overline{\psi_j} = \delta_{ij}, \quad (1 \le i, j \le g).$$

Then the element

$$\widehat{\omega} := \sum_{i=1}^{g} \psi_i \otimes \overline{\psi_i} \in H' \otimes H''(\subset H \otimes H) \tag{4.3}$$

is independent of the choice of the basis $\{\psi_i\}_{i=1}^g$. In fact, the set $\{\psi_i, \frac{\sqrt{-1}}{2}\overline{\psi_i}\}_{1 \le i \le g}$ is a symplectic basis of $H = H^1(C; \mathbb{C})$ with respect to the pairing (4.1). Hence the symplectic form on H, which depends only on the pairing (4.1), equals

$$\frac{\sqrt{-1}}{2} \sum_{i=1}^{g} \psi_i \otimes \overline{\psi_i} - \frac{\sqrt{-1}}{2} \sum_{i=1}^{g} \overline{\psi_i} \otimes \psi_i \in H^1(C; \mathbb{C})^{\otimes 2},$$

whose component in $H' \otimes H''$ equals $\frac{\sqrt{-1}}{2}\widehat{\omega}$. In particular, a (1, 1)-form B on the surface C defined by

$$B := \frac{\sqrt{-1}}{2g} \sum_{i=1}^{g} \psi_i \wedge \overline{\psi_i},$$

is independent of the choice of the basis $\{\psi_i\}_{i=1}^g$. We have $\int_C B = 1$. Moreover the form B is a real volume form on the surface C, since the 1-forms ψ_i's have no common zeroes because of the Riemann-Roch formula.

We denote by $\Omega^p(C)$ the vector space of $\mathbb{C}$-valued p-forms on C for $p = 0, 1, 2$. Then the Hodge $*$-operator $* : \Omega^1(C) \to \Omega^1(C)$ depends only on the complex structure of the surface C. In fact, in terms of a local complex coordinate z, we have $*dz = -\sqrt{-1}dz$ and $*d\bar{z} = \sqrt{-1}d\bar{z}$. Then, since the operator $d * d$ is equivalent to the Laplacian by means of the volume form B, we have an exact sequence

$$0 \to \mathbb{C} \to \Omega^0(C) \xrightarrow{d*d} \Omega^2(C) \xrightarrow{\int_C} \mathbb{C} \to 0,$$

where the left $\mathbb{C}$ means the constant functions. Hence we have a unique linear map $\widehat{\Phi} : \Omega^2(C) \to \Omega^0(C)$ such that

$$d * d\widehat{\Phi}(\Omega) = \Omega - \left(\int_C \Omega\right) B \quad \text{and} \quad \int_C \widehat{\Phi}(\Omega)B = 0$$

for any $\Omega \in \Omega^2(C)$. It is nothing but the Green operator associated with the volume form B. The operator $\widehat{\Phi}$ is real, $\overline{\widehat{\Phi}} = \widehat{\Phi}$, and the induced map

$$\widehat{\Phi}|_{\mathrm{Ker}(\int_C)} : \mathrm{Ker}\left(\int_C : \Omega^2(C) \to \mathbb{C}\right) \to \Omega^0(C)/\mathbb{C}$$

depends only on the complex structure on the surface C. Moreover we have

$$\int_C \widehat{\Phi}(\Omega)\Omega' = \int_C \Omega\widehat{\Phi}(\Omega') = \frac{1}{4\pi} \iint_{(P_1,P_2)\in C\times C} (\Omega)_{P_1}(\log G(P_1, P_2))(\Omega')_{P_2} \tag{4.4}$$

for any $\Omega, \Omega' \in \Omega^2(C)$, where we denote by $G(-, -)$ the Arakelov Green function. Here we remark $\widehat{\Phi}(B) = 0$. In fact, the function $\widehat{\Phi}(B)$ is constant since $d*d\widehat{\Phi}(B) = 0$, and so $\widehat{\Phi}(B) = \int_C \widehat{\Phi}(B)B = 0$.

Now we define a map A by

$$A : H' \otimes H'' \otimes H' \otimes H'' \to \mathbb{C}, \quad \phi_1 \otimes \overline{\phi_2} \otimes \phi_3 \otimes \overline{\phi_4} \mapsto \int_C \phi_1 \wedge \overline{\phi_2}\,\widehat{\Phi}(\phi_3 \wedge \overline{\phi_4}), \tag{4.5}$$

which is independent of the choice of the basis $\{\psi_i\}_{i=1}^g$, and so is an invariant of the compact Riemann surface C. As will be shown for the rest of this paper, the invariant A is a twisted version of the invariant a_g. One can extend it to a map $H\otimes H\otimes H\otimes H \to \mathbb{C}$ by the same formula (4.5), but with no additional information.

In order to compute the map A explicitly, it is convenient to introduce a complex number $A_{i\bar{j}k\bar{l}}$ defined by

$$\begin{aligned} A_{i\bar{j}k\bar{l}} &:= A(\psi_i \otimes \overline{\psi_j} \otimes \psi_k \otimes \overline{\psi_l}) = \int_C \psi_i \wedge \overline{\psi_j}\, \widehat{\Phi}(\psi_k \wedge \overline{\psi_l}) \\ &= \frac{1}{4\pi} \iint_{(P_1, P_2) \in C \times C} (\psi_i \wedge \overline{\psi_j})_{P_1} (\log G(P_1, P_2)) (\psi_k \wedge \overline{\psi_l})_{P_2} \end{aligned} \tag{4.6}$$

for $1 \le i, j, k, l \le g$ by using the orthonormal basis $\{\psi_i\}_{i=1}^g$. Then we have

$$\begin{aligned} \overline{A_{i\bar{j}k\bar{l}}} &= \int_C \psi_j \wedge \overline{\psi_i}\, \widehat{\Phi}(\psi_l \wedge \overline{\psi_k}) = A_{j\bar{i}l\bar{k}}, \\ A_{i\bar{j}k\bar{l}} &= \int_C \widehat{\Phi}(\psi_i \wedge \overline{\psi_j})\, \psi_k \wedge \overline{\psi_l} = A_{k\bar{l}i\bar{j}}. \end{aligned} \tag{4.7}$$

The latter follows from (4.4). We can contract the indices $i, k, \ldots$ and the ones $\bar{j}, \bar{l}, \ldots$ to get a new complex number. For example, the Kawazumi–Zhang invariant[1] $a_g = a_g(C)$ is obtained by

$$a_g = \sum_{i,j=1}^{g} A_{i\bar{j}j\bar{i}} = A((24)(\widehat{\omega} \otimes \widehat{\omega})),$$

which is a real number by the formula (4.7). Here $(24) : H' \otimes H'' \otimes H' \otimes H'' \to H' \otimes H'' \otimes H' \otimes H''$ is the switch map of the 2nd and the 4th components of the tensor product $H' \otimes H'' \otimes H' \otimes H''$.

We can describe the numbers $A_{i\bar{j}k\bar{l}}$ and a_g in the diagrams

$A_{i,\bar{j},k,\bar{l}} =$

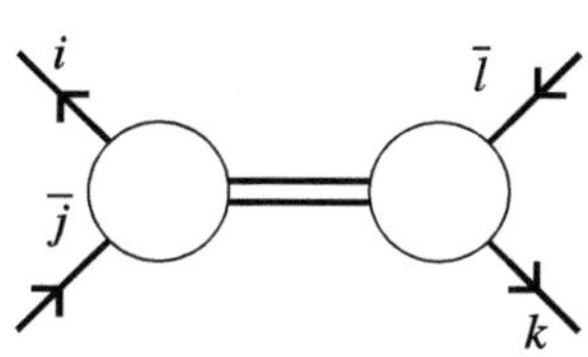

$a_g =$

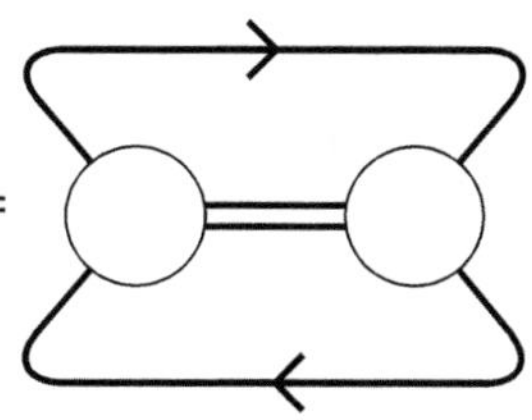

.

Here the circle with indices i and $\bar{j}$ means $\psi_i \wedge \overline{\psi_j}$, the double line without arrows $\log G(-,-)$, and a line with an arrow the contraction. For example, since

[1] The author uses the symbol $a_g = a_g(C)$, while Zhang and others use the symbol $\varphi = \varphi(C)$.

$\frac{\sqrt{-1}}{2g}\widehat{\Phi}(\sum_{i=1}^{g} \psi_i \wedge \overline{\psi_i}) = \widehat{\Phi}(B) = 0$, we have $\sum_{i,j,l=1}^{g} A_{i\bar{j}l\bar{l}} = \sum_{j,k,l=1}^{g} A_{j\bar{j}k\bar{l}} = 0$, that is,

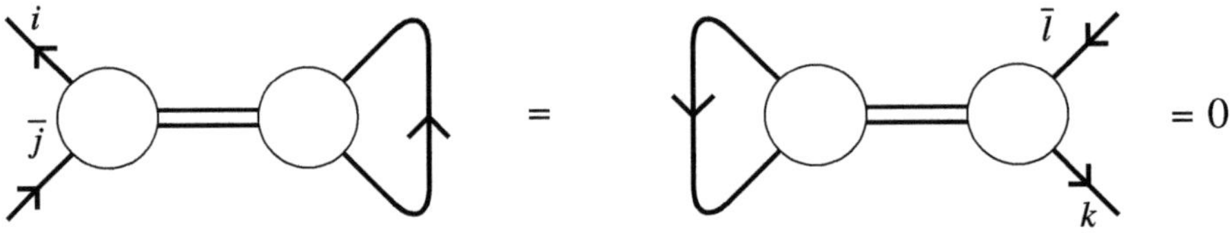

Similarly, we can introduce complex numbers associated with the Riemann surface C, i.e., functions on the moduli space $\mathbb{M}_g$ of compact Riemann surfaces. For example, we can consider a function on $\mathbb{M}_g$ defined by the diagram

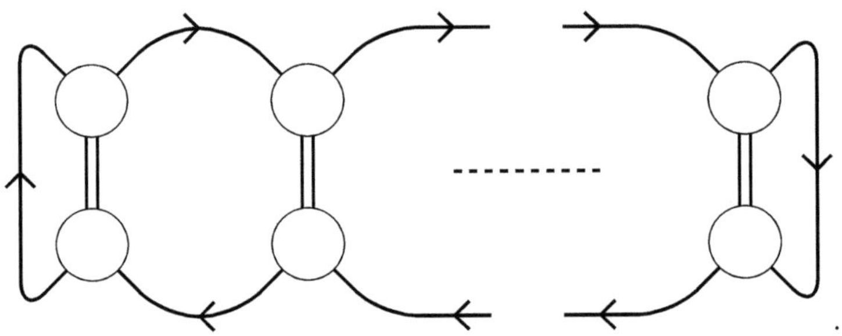

The number $A_{i\bar{j}k\bar{l}}$ is given by an integral over the 2-fold direct product $C^{\times 2} = C \times C$. We can generalize this construction to the 3-fold direct product and more. For example, we can consider

$$\int_C \widehat{\Phi}(\psi_i \wedge \overline{\psi_j})\psi_m \wedge \overline{\psi_n}\,\widehat{\Phi}(\psi_k \wedge \overline{\psi_l}) = \int_C \psi_i \wedge \overline{\psi_j}\,\widehat{\Phi}((\psi_m \wedge \overline{\psi_n})\,\widehat{\Phi}(\psi_k \wedge \overline{\psi_l}))$$

$$= \frac{1}{(4\pi)^2} \iiint_{(P_1,P_2,P_3)\in C\times C\times C} (\psi_i \wedge \overline{\psi_j})_{P_1} (\log G(P_1, P_2))$$

$$(\psi_m \wedge \overline{\psi_n})_{P_2} (\log G(P_2, P_3))(\psi_k \wedge \overline{\psi_l})_{P_3}$$

for $1 \leq i, j, k, l, m, n \leq g$. This can be described in the diagram

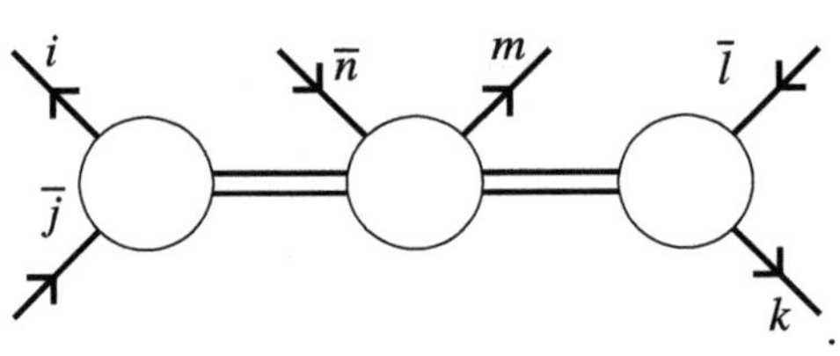

D'Hoker and Schlotterer [2] gave a systematic generalization of $A_{i\bar{j}k\bar{l}}$ where they constructed modular graph tensors using integrals on the direct product $C^{\times n}$ for any $n \geq 1$, and proved highly nontrivial identities among them. It is amazing that their proof of the identities is based only on formal natures of the Arakelov Green function $G(-,-)$.

4.3 Tensor Fields on the Moduli Space $\mathbb{M}_g$

When the Riemann surface C runs over the moduli space $\mathbb{M}_g$ of compact Riemann surfaces of genus g, the transpose ${}^tA$ of the linear map A in (4.5) can be regarded as some tensor fields on the moduli space $\mathbb{M}_g$. Now recall the vector spaces H' and H'' are dual to each other through the pairing (4.1). In particular, the basis $\{\psi_i\}_{i=1}^g$ and $\{\frac{\sqrt{-1}}{2}\overline{\psi_i}\}_{i=1}^g$ are dual to each other. Then the transpose ${}^tA : \mathbb{C} \to H''\otimes H'\otimes H''\otimes H'$ is determined by its own value at $1 \in \mathbb{C}$

$$ {}^tA(1) = \frac{1}{16}\sum_{i,j,k,l=1}^{g} A_{i\bar{j}k\bar{l}}\overline{\psi_i} \otimes \psi_j \otimes \overline{\psi_k} \otimes \psi_l \in H'' \otimes H' \otimes H'' \otimes H'. \tag{4.8} $$

When the surface C runs over the moduli space $\mathbb{M}_g$, the collections of H' and H'' are holomorphic and anti-holomorphic vector bundles, respectively. Then ${}^tA(1)$ is a C^∞ section of a tensor product of these vector bundles.

4.3.1 The First Mumford–Morita–Miller Class $e_1 = \kappa_1$

Here we recall the holomorphic cotangent space $T^*_{[C]}\mathbb{M}_g$ of the moduli space $\mathbb{M}_g$ at the equivalence class $[C]$ of the Riemann surface C is canonically isomorphic to the space of holomorphic quadratic differentials

$$ T^*_{[C]}\mathbb{M}_g = H^0(C; (\Omega^1_C)^{\otimes 2}). $$

For example, we may regard $\psi_l\psi_j = \psi_j\psi_l \in T^*_{[C]}\mathbb{M}_g$ and $\overline{\psi_i\psi_k} = \overline{\psi_k\psi_i} \in \overline{T^*_{[C]}\mathbb{M}_g}$. Hence we can define a $(1,1)$-form $e_1^{\widehat{\Phi}}$ on the moduli space $\mathbb{M}_g$ by[2]

$$ e_1^{\widehat{\Phi}}|_{[C]} := -3\sqrt{-1}\sum_{i,j,k,l=1}^{g} (\psi_j\psi_l A_{i\bar{j}k\bar{l}}\overline{\psi_k\psi_i} - \psi_j\psi_l A_{i\bar{j}k\bar{i}}\overline{\psi_k\psi_l}). $$

---

[2] In [17, 18] we denote $e_1^{\widehat{\Phi}}$ by e_1^D, whose coefficients was $48\sqrt{-1}$. It is not true. This mistake comes from computation of the intersection number $Y_i \cdot \overline{Y_i}$.

In fact, from (4.8), the RHS is independent of the choice of the basis $\{\psi_i\}_{i=1}^g$. Then we have

Theorem 4.1 *The* $(1,1)$*-form* $e_1^{\widehat{\Phi}}$ *is closed on the moduli space* $\mathbb{M}_g$*, and represents the first Mumford–Morita–Miller class* $e_1 \in H^2(\mathbb{M}_g; \mathbb{C})$.

The theorem was stated implicitly in the equation (5.5) of the preprint [15] without proof. We will give its proof in the next Sect. 4.4.

Very recently, de Jong and van der Lugt [9, 24] introduced a subalgebra $\mathcal{R}(\mathbb{M}_g)$ of the de Rham complex $\Omega^*(\mathbb{M}_g)$, called *the ring of tautological forms* on $\mathbb{M}_g$. It consists entirely of closed forms, and its image in the cohomology algebra $H^*(\mathbb{M}_g; \mathbb{C})$ is generated by all the Mumford–Morita–Miller classes, i.e., equals the image of the tautological ring of $\mathbb{M}_g$. It is remarkable that the ring $\mathcal{R}(\mathbb{M}_g)$ is finite-dimensional, which suggests us that study of differential forms on $\mathbb{M}_g$ could help us to understand the tautological ring of $\mathbb{M}_g$. As was shown in [9, 24], the degree 2 part of the ring $\mathcal{R}(\mathbb{M}_g)$ is closely related to the invariant a_g. So it would be very interesting if we could find further relations of the ring $\mathcal{R}(\mathbb{M}_g)$ with the complex numbers $A_{i\bar{j}k\bar{l}}$'s or generalizations by D'Hoker and Schlotterer [2].

4.3.2 The Jacobian Map Jac : $\mathbb{M}_g \to \mathbb{A}_g$

The transpose ${}^tA(1)$ in (4.8) has another interpretation: it can be regarded as a C^∞ section of the pullback $\mathrm{Jac}^*(T^*\mathbb{A}_g \otimes \overline{T^*\mathbb{A}_g})$ over the moduli space $\mathbb{M}_g$. Here $\mathbb{A}_g$ is the moduli space of principally polarized Abelian varieties of genus g, and Jac : $\mathbb{M}_g \to \mathbb{A}_g, [C] \mapsto [\mathrm{Jac}(C)]$, the Jacobian map. The moduli space $\mathbb{A}_g$ is the quotient space of the Siegel upper half space $\mathfrak{H}_g$ by the Siegel modular group $Sp_{2g}(\mathbb{Z})$. The space $\mathfrak{H}_g = Sp_{2g}(\mathbb{R})/U_g$ is the space of all complex structures J on the standard symplectic vector space $\mathbb{R}^{2g}$ with positivity condition. The tautological vector bundle $E_g := \coprod_{J\in\mathfrak{H}_g}(\mathbb{R}^{2g}, J)$ over $\mathfrak{H}_g$ is an $Sp_{2g}(\mathbb{R})$-equivariant holomorphic vector bundle with canonical Hermitian metric, and so descends to the moduli space $\mathbb{A}_g$. Through a canonical isomorphism $T\mathbb{A}_g = \mathrm{Sym}^2(E_g) \subset {E_g}^{\otimes 2}$, one can define a Kähler metric on the moduli space $\mathbb{A}_g$. See, for example, [21, Chapter II]. In particular, we may consider the Laplacian Δ on the moduli space $\mathbb{A}_g$. The pullback $\mathrm{Jac}^*E_g^*$ of the Hermitian vector bundle E_g under the Jacobian map Jac at the equivalence class $[C] \in \mathbb{M}_g$ is isometric to the holomorphic 1-forms $H' = H^0(C; \Omega_C^1)$ with metric given in (4.2). Hence the transpose ${}^tA(1)$ in (4.8) can be regarded as a C^∞ section of the pullback $\mathrm{Jac}^*(T^*\mathbb{A}_g \otimes \overline{T^*\mathbb{A}_g}))$.

By Rauch's formula [23], the differential $d\mathrm{Jac}$ of the Jacobian map is given by the commutative diagram

$$\begin{array}{ccc} T^*_{[C]}\mathbb{M}_g & \xleftarrow{(d\mathrm{Jac})^*} & T^*_{[\mathrm{Jac}(C)]}\mathbb{A}_g \\ \| & & \| \\ H^0(C;(\Omega^1_C)^{\otimes 2}) & \xleftarrow{\text{multiplication}} & \mathrm{Sym}^2 H^0(C;\Omega^1_C). \end{array} \tag{4.9}$$

In particular, from M. Noether's theorem, the differential $d\mathrm{Jac}_{[C]}$ is injective if and only if $g = 2$ or ($g \geq 3$ and C is non-hyperelliptic). On the other hand, $\dim \mathbb{M}_g = \dim \mathbb{A}_g$ if and only if $g = 2$ or 3. Hence the Laplacian Δ on $\mathbb{A}_g$ acts on functions on $\mathbb{M}_2 \overset{\text{open}}{\subset} \mathbb{A}_2$ and $\mathbb{M}_3 \setminus \mathbb{H}_3 \overset{\text{open}}{\subset} \mathbb{A}_3$. Here we denote by $\mathbb{H}_g \subset \mathbb{M}_g$ the locus of hyperelliptic Riemann surfaces of genus $g \geq 2$.

D'Hoker et al. [3] proved the following remarkable theorem.

Theorem 4.2 ([3] (4.23)) $\Delta a_2 = 5a_2$. *In particular, the invariant a_2 is an eigenfunction of the Laplacian Δ.*

Based on the theorem, Pioline [22] gave an explicit formula for a_2 in terms of theta functions. Later, for any $g \geq 1$, Wilms [25] gave an explicit formula for a_g in terms of a function introduced by de Jong as well as theta functions. But the methods of Wilms [25] seem to be different from those discussed in this paper.

There remains the case $g = 3$. In this paper we prove

Theorem 4.3 *In the case $g = 3$, the invariant a_3 is not an eigenfunction of the Laplacian Δ.*

The obstruction for a_3 to be an eigenfunction is the first Johnson homomorphism [13, 19], which is represented by a twisted 1-form on $\mathbb{M}_g$ as follows. We denote by $U = U(C)$ the kernel of the contraction map $\Lambda^3 H \to H$, $\phi_1 \wedge \phi_2 \wedge \phi_3 \mapsto (\phi_2 \cdot \phi_3)\phi_1 + (\phi_3 \cdot \phi_1)\phi_2 + (\phi_1 \cdot \phi_2)\phi_3$. It should be remarked that U vanishes if and only if $g \leq 2$. For $\phi_i = \phi_i' + \phi_i'' \in H = H' \oplus H''$, $1 \leq i \leq 3$, we define $Q(\phi_1, \phi_2, \phi_3) \in C^\infty(C; (T^*C)^{\otimes 2})$ by

$$\begin{aligned} Q(\phi_1, \phi_2, \phi_3) := & -\sqrt{-1}\phi_1'\partial\widehat{\Phi}(\phi_2 \wedge \phi_3) - \sqrt{-1}\phi_2'\partial\widehat{\Phi}(\phi_3 \wedge \phi_1) \\ & - \sqrt{-1}\phi_3'\partial\widehat{\Phi}(\phi_1 \wedge \phi_2). \end{aligned} \tag{4.10}$$

The tensor $Q(\phi_1, \phi_2, \phi_3)$ is alternating in ϕ_1, ϕ_2, ϕ_3, but is not holomorphic in general. Since $d * d = -2\sqrt{-1}\partial\overline{\partial}$, we have

$$\begin{aligned} & 2\overline{\partial}Q \\ = & \phi_1'(\phi_2 \wedge \phi_3 - (\int_C \phi_2 \wedge \phi_3)B) + \phi_2'(\phi_3 \wedge \phi_1 - (\int_C \phi_3 \wedge \phi_1)B) \end{aligned}$$

$$
\begin{aligned}
&\qquad\qquad + \phi_3'(\phi_1 \wedge \phi_2 - (\int_C \phi_1 \wedge \phi_2)B) \\
=&\phi_1'\phi_2'\phi_3'' - \phi_1'\phi_3'\phi_2'' + \phi_2'\phi_3'\phi_1'' - \phi_2'\phi_1'\phi_3'' + \phi_3'\phi_1'\phi_2'' - \phi_3'\phi_2'\phi_1'' \\
&\qquad - (\phi_1'(\phi_2 \cdot \phi_3) + \phi_2'(\phi_3 \cdot \phi_1) + \phi_3'(\phi_1 \cdot \phi_2))B \\
=& - (\phi_1'(\phi_2 \cdot \phi_3) + \phi_2'(\phi_3 \cdot \phi_1) + \phi_3'(\phi_1 \cdot \phi_2))B.
\end{aligned}
$$

This implies the restriction of Q to the subspace U has its value in the space $H^0(C; (\Omega_C^1)^{\otimes 2})$ of holomorphic quadratic differentials

$$
Q : U = U(C) \to H^0(C; (\Omega_C^1)^{\otimes 2}) = T^*_{[C]}\mathbb{M}_g.
$$

We remark the map Q is equivariant under the action of the holomorphic automorphism group $\mathrm{Aut}(C)$. In particular, if the Riemann surface C is hyperelliptic, we have

$$
Q(U(C)) \subset (\text{the } (-1)\text{-eigenspace of the hyperelliptic involution in } T^*_{[C]}\mathbb{M}_g), \tag{4.11}
$$

since the involution acts on the space $U(C)$ by $-1_{U(C)}$.

We may identify the map Q with its transpose[3] $Q : T_{[C]}\mathbb{M}_g \to (U(C))^*$, which can be regarded as a twisted $(1,0)$ form on the moduli space $\mathbb{M}_g$. Here the pairing

$$
\begin{aligned}
&\langle -, - \rangle : \Lambda^3 H \times \Lambda^3 H \to \mathbb{C}, \\
&(\phi_1^{(1)} \wedge \phi_2^{(1)} \wedge \phi_3^{(1)}, \phi_1^{(2)} \wedge \phi_2^{(2)} \wedge \phi_3^{(2)}) \\
&\quad \mapsto \sum_{\sigma,\tau \in \mathfrak{S}_3} (\phi^{(1)}_{\sigma(1)} \cdot \phi^{(2)}_{\tau(1)})(\phi^{(1)}_{\sigma(2)} \cdot \phi^{(2)}_{\tau(2)})(\phi^{(1)}_{\sigma(3)} \cdot \phi^{(2)}_{\tau(3)})
\end{aligned} \tag{4.12}
$$

is non-degenerate on $U = U(C)$, and so we identify $U(C)$ with its dual $(U(C))^*$ through the pairing (4.12). We denote by the same symbol U the flat vector bundle $\coprod_{[C]\in\mathbb{M}_g} U(C)$ over the moduli space $\mathbb{M}_g$. The tensor Q equals the $(1,0)$-part of the first variation of the harmonic volumes introduced by Harris [12, Theorem 5.8]. In particular, the sum $Q + \overline{Q}$ is a twisted closed 1-form with values in the flat vector bundle U. As was proved in [14, §8], the cohomology class of $-(Q + \overline{Q})$ equals the extended first Johnson homomorphism $\tilde{k}$ [19]

$$
\tilde{k} = -[Q + \overline{Q}] \in H^1(\mathbb{M}_g; U).
$$

The cohomology group $H^1(\mathbb{M}_g; U)$ is canonically isomorphic to the first group cohomology of the mapping class group of a genus g Riemann surface with values

[3] For simplicity, we use the same symbol Q, not ${}^tQ$.

in the vector space $U = U(C)$. The restriction of $\tilde{k}$ to the Torelli group equals the first Johnson homomorphism [13]. As was proved by Morita [20], the first Mumford–Morita–Miller class e_1 equals the contraction $\langle \tilde{k}, \tilde{k} \rangle \in H^2(\mathbb{M}_g; \mathbb{C})$ of the cup product $\tilde{k} \cup \tilde{k} \in H^2(\mathbb{M}_g; U^{\otimes 2})$ up to non-zero constant factor. By the same recipe using the pairing in (4.12), we obtain a (1, 1)-form e_1^J on $\mathbb{M}_g$ representing the class e_1, which equals $\langle Q + \overline{Q}, Q + \overline{Q} \rangle$ up to non-zero constant factor. From [15, Theorem 6.1], we have

$$\frac{1}{12} e_1^{\widehat{\Phi}} = \frac{-2\sqrt{-1}g}{2(2g+1)} (\partial \overline{\partial} a_g) + \frac{1}{12} e_1^J. \tag{4.13}$$

An alternative proof of the formula (4.13) using the Deligne pairing is given by de Jong [7]. If $g = 2$, then $U = 0$, so that $e_1^J = 0$. But $e_1^J \neq 0$ for $g \geq 3$.

In [3, Appendices B and C], D'Hoker, Green, Pioline and Russo gave suggestive computations for any genus $g \geq 2$. In the equations (B.4) and (C.1), they divide the exact (1, 1) form $\overline{\partial}\partial a_g$ into two parts $\psi_A^1 + \psi_B$ and $\psi_A^2 + \psi_C$. Their formulae (C.3) and (C.5) say that $\psi_A^1 + \psi_B$ and $\psi_A^2 + \psi_C$ equal $e_1^{\widehat{\Phi}}$ and e_1^J up to non-zero constant factor, respectively. These formulae are equivalent to our equation (4.13). The formula (C.9) says that $\psi_A^2 + \psi_C$ vanishes if $g = 2$. In order to understand the formula (C.4)

$$\Delta a_g \Big|_{\psi_A^1 + \psi_B} = (2g+1) a_g,$$

we need to introduce a lift of the differential form $e_1^{\widehat{\Phi}}$ to an element $\widehat{E_1}$ of $C^\infty(\mathbb{M}_g; \mathrm{Jac}^*(T^*\mathbb{A}_g \otimes \overline{T^*\mathbb{A}_g}))$ defined by

$$\widehat{E_1}|_{[C]} := -3\sqrt{-1} \sum_{i,j,k,l=1}^{g} ((\psi_j \bullet \psi_l) A_{i\bar{j}k\bar{l}} (\overline{\psi_k} \bullet \overline{\psi_i}) - (\psi_j \bullet \psi_l) A_{i\bar{j}k\bar{i}} (\overline{\psi_k} \bullet \overline{\psi_l})).$$

Here $\psi_j \bullet \psi_l := \frac{1}{2}(\psi_j \otimes \psi_l + \psi_l \otimes \psi_j) \in \mathrm{Sym}^2 H'$ and so on. Now we recall the Laplacian Δ on any Kähler manifold X is given by

$$\Delta f = 2\sqrt{-1} \Lambda \partial \overline{\partial} f$$

for any $f \in C^\infty(X; \mathbb{C})$. Here $\Lambda : T^*X \otimes \overline{T^*X} \to X \times \mathbb{C}$ is the Kähler contraction. In our case $X = \mathbb{A}_g$, we have

$$\Lambda \widehat{E_1}|_{[C]} = 6 \left(\sum_{j,l=1}^{g} A_{l\bar{j}j\bar{l}} + \sum_{j,l=1}^{g} A_{j\bar{j}l\bar{l}} - g \sum_{i,j=1}^{g} A_{i\bar{j}j\bar{i}} - \sum_{i,l=1}^{g} A_{i\bar{l}l\bar{i}} \right) = -6g a_g,$$

since $\int_C \psi_i \wedge \overline{\psi_j} = -2\sqrt{-1}\delta_{ij}$ and $\sum_{j,k,l=1}^{g} A_{j\bar{j}k\bar{l}} = 0$. Applying the Kähler contraction to the equation (4.13), we obtain $-\frac{g}{2}a_g = -\frac{g}{2(2g+1)}\Delta a_g + \frac{1}{12}\Lambda e_1^J$, and so

$$\Delta a_g = (2g+1)a_g + \frac{g}{6}\Lambda e_1^J, \tag{4.14}$$

if $g = 2, 3$. In the case $g = 3$, the Kähler contraction $|\Lambda e_1^J|$ with respect to the metric of $\mathbb{A}_3$ satisfies the following.

Proposition 4.1 *There is a hyperelliptic Riemann surface C_0 of genus 3 such that*

$$|\Lambda e_1^J| \longrightarrow +\infty$$

as a non-hyperelliptic Riemann surface C of genus 3 goes to the surface C_0.

The proof will be given in Sect. 4.5, and is based on a theorem of Harris [12, Theorem 6.5].

Now the invariant a_3 is C^∞ near $[C_0] \in \mathbb{M}_3$, so is locally bounded near $[C_0]$. Proposition 4.1 together with (4.14) implies that Δa_3 is not bounded near $[C_0]$. Hence the function a_3 is not an eigenfunction of the Laplacian Δ. This completes the proof of Theorem 4.3 modulo that of Proposition 4.1. □

Since the Jacobian map $\mathrm{Jac}|_{\mathbb{H}_g} : \mathbb{H}_g \to \mathbb{A}_g$ restricted to the hyperelliptic locus $\mathbb{H}_g$ is an embedding, the Kähler metric on $\mathbb{A}_g$ induces a Kähler metric on $\mathbb{H}_g$ and its Laplacian $\Delta_{\mathbb{H}_g}$. Since the (1, 1)-form e_1^J vanishes along $\mathbb{H}_g$ from (4.11), one can clarify $\Delta_{\mathbb{H}_g}(a_g|_{\mathbb{H}_g})$ if the Kähler contraction Λ on $\mathbb{H}_g$ can be computed.

4.4 A Proof of Theorem 4.1

In this section we prove Theorem 4.1 based on computations given in [15], where a closed (1, 1)-form E_1^D was introduced on the moduli space $\mathbb{M}_g$. As was proved in the equation (5.3), it represents the class $\frac{1}{12}e_1$. We will prove the equality $E_1^D = \frac{1}{12}e_1^{\widehat{\Phi}}$ to obtain the theorem.

Our starting point is Lemma 5.5 in [15] which says that

$$E_1^D(\lambda, \overline{\mu}) = 4M \int_C \mathcal{H}(\omega'_{(1)}\lambda) \wedge \omega'_{(1)} \widehat{\Phi} d * (\omega''_{(1)}\overline{\ell^\mu} - \overline{\ell^\mu}\omega''_{(1)}) \tag{4.15}$$

for any Beltrami differentials $\lambda, \mu \in C^\infty(C; TC \otimes \overline{T^*C})$.

First of all we need to explain the notation in the equation (4.15): M, $\omega'_{(1)}$, $\omega''_{(1)}$, $\mathcal{H}$ and ℓ^μ. The first complex homology group $H_1(C; \mathbb{C})$ admits the intersection pairing which we denote by $\cdot : H_1(C; \mathbb{C}) \times H_1(C; \mathbb{C}) \to \mathbb{C}$. The integrand of the RHS is an $H_1(C; \mathbb{C})^{\otimes 4}$-valued (1, 1)-form. The symbol $M : H_1(C; \mathbb{C})^{\otimes 4} \to \mathbb{C}$ is given by $M(Z_1Z_2Z_3Z_4) := (Z_2 \cdot Z_3)(Z_4 \cdot Z_1)$ for any $Z_i \in H_1(C;\mathbb{C})$. Here and throughout this section we omit the symbol $\otimes$. The first complex cohomology group $H = H^1(C;\mathbb{C})$, is embedded into the space of 1-forms $\Omega^1(C)$ as the subspace of harmonic 1-forms. The embedding is an element of $\mathrm{Hom}(H^1(C;\mathbb{C}), \Omega^1(C)) = \Omega^1(C)\otimes H^1(C;\mathbb{C})^* = \Omega^1(C)\otimes H_1(C;\mathbb{C})$, which can be regarded as an $H_1(C;\mathbb{C})$-valued real harmonic 1-form. We denote it by $\omega_{(1)} \in \Omega^1(C)\otimes H_1(C;\mathbb{C})$, and its $(1,0)$- and $(0,1)$-parts by $\omega'_{(1)}$ and $\omega''_{(1)}$, respectively. If homology classes $Y_i \in H_1(C;\mathbb{C})$, $1 \le i \le g$, satisfy

$$\int_{Y_i} \psi_j = \delta_{i,j}, \quad \text{and} \quad \int_{Y_i} \overline{\psi_j} = 0$$

for $1 \le i, j \le g$, in other words, $\{Y_i, -2\sqrt{-1}\overline{Y_i}\}_{1\le i\le g}$ is the dual basis of the symplectic basis $\{\psi_i, \frac{\sqrt{-1}}{2}\overline{\psi_i}\}_{1\le i\le g}$, then we have

$$\omega_{(1)} = \omega'_{(1)} + \omega''_{(1)}, \quad \omega'_{(1)} = \sum_{i=1}^{g} \psi_i Y_i, \quad \text{and} \quad \omega''_{(1)} = \sum_{i=1}^{g} \overline{\psi_i Y_i}. \tag{4.16}$$

Here we remark $Y_i \cdot \overline{Y_j} = \frac{\sqrt{-1}}{2}\delta_{ij}$. The Hodge decomposition on the 1-forms on C is given by

$$\varphi = \mathcal{H}\varphi + *d\widehat{\Phi}d\varphi + d\widehat{\Phi}d*\varphi$$

for any $\varphi \in \Omega^1(C)$. Here $\mathcal{H}$ is the harmonic projection, which is written by

$$\mathcal{H}(\varphi) = \frac{\sqrt{-1}}{2}\sum_{j=1}^{g}\left(\left(\int_C \varphi\wedge\overline{\psi_j}\right)\psi_j - \left(\int_C \varphi\wedge\psi_j\right)\overline{\psi_j}\right).$$

For example, $\omega'_{(1)}\lambda$ can be regarded as an $H_1(C;\mathbb{C})$-valued $(0,1)$-form by contracting T^*C in $\omega'_{(1)}$ and TC in λ, so that we have

$$\mathcal{H}(\omega'_{(1)}\lambda) = \sum_{l=1}^{g}\mathcal{H}(\psi_l\lambda)Y_l = \frac{\sqrt{-1}}{2}\sum_{l,j=1}^{g}\left(\int_C \psi_j\psi_l\lambda\right)\overline{\psi_j}Y_l, \tag{4.17}$$

since $\psi_l\lambda\wedge\psi_j = -\psi_j\wedge\psi_l\lambda = -\psi_j\psi_l\lambda$. On the other hand, we have

$$\varphi'' - \mathcal{H}(\varphi'') = *d\widehat{\Phi}d\varphi'' + d\widehat{\Phi}d*\varphi'' = 2\overline{\partial}\widehat{\Phi}d*\varphi'' \tag{4.18}$$

for any $(0,1)$-current φ''.

Finally $\ell^\mu \in \Omega^0(C) \otimes H_1(C; \mathbb{C})$ is defined by

$$\ell^\mu := 2\widehat{\Phi} d * (\omega'_{(1)}\mu).$$

From (4.18) we have

$$\begin{aligned}
& d * (\omega'_{(1)}\ell^\mu - \ell^\mu \omega'_{(1)}) \\
&= -\sqrt{-1}\partial(\omega'_{(1)}\ell^\mu - \ell^\mu \omega'_{(1)}) \\
&= \sqrt{-1}(\omega'_{(1)} \wedge \overline{\partial}\ell^\mu - (\overline{\partial}\ell^\mu) \wedge \omega'_{(1)}) \\
&= 2\sqrt{-1}\omega'_{(1)} \wedge \overline{\partial}\widehat{\Phi} d * (\omega'_{(1)}\mu) + 2\sqrt{-1}\overline{\partial}\widehat{\Phi} d * (\omega'_{(1)}\mu) \wedge \omega'_{(1)} \\
&= \sqrt{-1}\omega'_{(1)} \wedge (\omega'_{(1)}\mu - \mathcal{H}(\omega'_{(1)}\mu)) + \sqrt{-1}(\omega'_{(1)}\mu - \mathcal{H}(\omega'_{(1)}\mu)) \wedge \omega'_{(1)} \\
&= -\sqrt{-1}\omega'_{(1)} \wedge \mathcal{H}(\omega'_{(1)}\mu) - \sqrt{-1}\mathcal{H}(\omega'_{(1)}\mu) \wedge \omega'_{(1)}.
\end{aligned}$$

The last equality follows from

$$\begin{aligned}
& \omega'_{(1)} \wedge \omega'_{(1)}\mu + \omega'_{(1)}\mu \wedge \omega'_{(1)} \\
&= \sum_{i,j=1}^{g} (\psi_i \wedge \psi_j\mu + \psi_i\mu \wedge \psi_j) Y_i Y_j \\
&= \sum_{i,j=1}^{g} (\psi_i\psi_j - \psi_j\psi_i)\mu Y_i Y_j \\
&= 0.
\end{aligned}$$

Hence, from (4.15), we have

$$\begin{aligned}
E_1^D(\lambda, \overline{\mu}) = 4\sqrt{-1}M \int_C & \mathcal{H}(\omega'_{(1)}\lambda) \wedge \omega'_{(1)} \\
& \widehat{\Phi}\left(\mathcal{H}(\omega''_{(1)}\overline{\mu}) \wedge \omega''_{(1)} + \omega''_{(1)} \wedge \mathcal{H}(\omega''_{(1)}\overline{\mu})\right),
\end{aligned}$$

which is exactly the equation (5.5) in [15].

From (4.17) we have

$$\begin{aligned}
& 4\sqrt{-1}M \int_C \mathcal{H}(\omega'_{(1)}\lambda) \wedge \omega'_{(1)} \widehat{\Phi}(\mathcal{H}(\omega''_{(1)}\overline{\mu}) \wedge \omega''_{(1)}) \\
&= \sqrt{-1} \sum_{j,l,i,k,m,n=1}^{g} \left(\int_C \psi_j\psi_l\lambda\right)\left(\int_C \overline{\psi_k\psi_i\overline{\mu}}\right)
\end{aligned}$$

$$\int_C \overline{\psi_j} \wedge \psi_m \widehat{\Phi}(\psi_k \wedge \overline{\psi_n}) Z(Y_l Y_m \overline{Y_i Y_n})$$

$$= -\frac{\sqrt{-1}}{4} \sum_{j,l,i,k,m,n=1}^{g} \left(\int_C \psi_j \psi_l \lambda\right) \left(\int_C \overline{\psi_k \psi_i \mu}\right) A_{m\bar{j}k\bar{n}} \delta_{mi} \delta_{nl}$$

$$= -\frac{\sqrt{-1}}{4} \sum_{j,l,i,k=1}^{g} \left(\int_C \psi_j \psi_l \lambda\right) \left(\int_C \overline{\psi_k \psi_i \mu}\right) A_{i\bar{j}k\bar{l}}.$$

On the other hand, we have

$$4\sqrt{-1}M \int_C \mathcal{H}(\omega'_{(1)}\lambda) \wedge \omega''_{(1)} \widehat{\Phi}(\omega''_{(1)} \wedge \mathcal{H}(\omega''_{(1)}\overline{\mu}))$$

$$= \sqrt{-1} \sum_{j,l,i,k,m,n=1}^{g} \left(\int_C \psi_j \psi_l \lambda\right) \left(\int_C \overline{\psi_k \psi_i \mu}\right)$$

$$\int_C \overline{\psi_j} \wedge \psi_m \widehat{\Phi}(\overline{\psi_n} \wedge \psi_k) Z(Y_l Y_m \overline{Y_n Y_i})$$

$$= \frac{\sqrt{-1}}{4} \sum_{j,l,i,k,m,n=1}^{g} \left(\int_C \psi_j \psi_l \lambda\right) \left(\int_C \overline{\psi_k \psi_i \mu}\right) A_{m\bar{j}k\bar{n}} \delta_{mn} \delta_{il}$$

$$= \frac{\sqrt{-1}}{4} \sum_{j,l,k,m=1}^{g} \left(\int_C \psi_j \psi_l \lambda\right) \left(\int_C \overline{\psi_k \psi_l \mu}\right) A_{m\bar{j}k\bar{m}}$$

$$= \frac{\sqrt{-1}}{4} \sum_{i,j,l,k=1}^{g} \left(\int_C \psi_j \psi_l \lambda\right) \left(\int_C \overline{\psi_k \psi_l \mu}\right) A_{i\bar{j}k\bar{i}}$$

Consequently we obtain

$$E_1^D = -\frac{\sqrt{-1}}{4} \sum_{i,j,k,l=1}^{g} \left(\psi_i \psi_l A_{i\bar{j}k\bar{l}} \overline{\psi_k \psi_i} - \psi_j \psi_l A_{i\bar{j}k\bar{i}} \overline{\psi_k \psi_l}\right) = \frac{1}{12} e_1^{\widehat{\Phi}},$$

as was to be shown. □

The present proof is based on direct computations given in [15]. It is desirable to have a conceptual proof of Theorem 4.1.

4.5 Proof of Proposition 4.1

In this section we prove Proposition 4.1 based on the following theorem of Harris.

Theorem 4.4 ([12], Theorem 6.5) *For any $g \geq 3$ there exists a hyperelliptic Riemann surface C_0 such that the inclusion* (4.11) *is the equality*

$$Q(U(C_0)) = (\textit{the } (-1)\textit{-eigenspace of the hyperelliptic involution in } T^*_{[C_0]}\mathbb{M}_g).$$

In other words, the extended Johnson homomorphism $\tilde{k} = -[Q + \overline{Q}]$ at $[C_0]$ is non-degenerate in the normal direction to $\mathbb{H}_g$ in $\mathbb{M}_g$.

The contraction map $\Lambda^3 H \to H$ vanishes on $\Lambda^3 H'$ and $\Lambda^3 H''$, while it maps $\Lambda^2 H' \otimes H''$ and $H' \otimes \Lambda^2 H''$ to H' and H'', respectively. Hence the kernel $U(C) \subset \Lambda^3(H' \oplus H'') = \bigoplus_{p+q=3} \Lambda^p H' \otimes \Lambda^q H''$ of the contraction map is decomposed into

$$U(C) = \bigoplus_{p+q=3} U(C)^{p,q}, \quad U(C)^{p,q} := U(C) \cap (\Lambda^p H' \otimes \Lambda^q H'').$$

From the definition of the map Q (4.10) we have

$$Q|_{U(C)^{p,q}} = 0 \tag{4.19}$$

if $(p, q) \neq (2, 1)$. Through the pairing $\langle -, - \rangle$ in (4.12), $U(C)^{p,q}$ and $U(C)^{3-p,3-q}$ are dual to each other. Hence the image of the transpose $Q : T_{[C]}\mathbb{M}_g \to U(C)$ is included in $(U(C)^{2,1})^* = U(C)^{1,2}$, i.e., we have $Q(T_{[C]}\mathbb{M}_g) \subset U(C)^{1,2}$. From the positivity of the Hermitian pairing (4.2), we have

$$\forall v \in U(C)^{1,2} \setminus \{0\}, \quad -\sqrt{-1}\langle v, \overline{v} \rangle > 0. \tag{4.20}$$

In fact, $\{\psi_i \wedge \overline{\psi_j} \wedge \overline{\psi_k};\ 1 \leq i, j, k \leq g, j < k\}$ is a orthogonal basis of $H' \otimes \Lambda^2 H'' (\supset U(C)^{1,2})$.

Now we go back to the hyperelliptic Riemann surface $[C_0] \in \mathbb{M}_3$ given by Theorem 4.4 of Harris. We choose a marking m_0 of C_0 to get a point $[C_0, m_0]$ in the Teichmüller space $\mathcal{T}_3$ of genus 3. Its Jacobian $\mathrm{Jac}(C_0)$ with the marking m_0 defines a point $[\mathrm{Jac}(C_0), m_0] \in \mathfrak{H}_3$. Then there are complex coordinates (z_1, z_2, z_3) of $\mathcal{T}_3$ centered at $[C_0, m_0]$, and (w_1, w_2, w_3) of the Siegel upper half space $\mathfrak{H}_3$ centered at $[\mathrm{Jac}(C_0), m_0]$ such that the locus $\{z_1 = 0\}$ coincides with the inverse image of $\mathbb{H}_3$ locally, and that

$$\mathrm{Jac}(z_1, z_2, z_3) = ({z_1}^2, z_2, z_3) = (w_1, w_2, w_3),$$

since the Jacobian map Jac is a 2 to 1 map in the normal direction of the submanifold $\mathbb{H}_3$. There are $U(C)^{1,2}$-valued functions v_1, v_2, v_3 defined around $[C_0, m_0] \in \mathcal{T}_3$ such that $\tilde{k} = v_1 dz_1 + v_2 dz_2 + v_3 dz_3$ on $T\mathcal{T}_3$. Then we have $v_1(0, *, *) \neq 0$ from

Theorem 4.4, $v_2(0, *, *) = v_3(0, *, *) = 0$ from (4.11), and the differential form e_1^J equals

$$\sum_{i,j=1}^{3} \langle v_i, \overline{v_j} \rangle dz_i \wedge d\overline{z_j}$$

up to non-zero constant factor. Since $\frac{\partial}{\partial w_1} = \frac{1}{2z_1}\frac{\partial}{\partial z_1}$, the amount $z_1 \cdot \Lambda e_1^J$ is a combination of the amounts

$$\frac{1}{z_1}\langle v_1, \overline{v_1} \rangle, \quad \langle v_1, \overline{v_j} \rangle, \quad \langle v_i, \overline{v_1} \rangle \text{ and } z_1 \langle v_i, \overline{v_j} \rangle \quad (i, j = 2, 3)$$

with coefficients in non-zero locally bounded functions near $[C_0, m_0]$. The amounts $\langle v_1, \overline{v_j} \rangle$, $\langle v_i, \overline{v_1} \rangle$ and $z_1 \langle v_i, \overline{v_j} \rangle$ go to 0 as $z_1 \to 0$. Together with (4.20), this implies

$$|z_1| \cdot |\Lambda e_1^J| \geq \frac{\text{(positive constant)}}{|z_1|} |\langle v_1, \overline{v_1} \rangle| \longrightarrow +\infty$$

as $z_1 \to 0$. This completes the proof of Proposition 4.1 and that of Theorem 4.3. □

Acknowledgments First of all the author expresses his sincere thanks and respect to Athanase Papadopoulos for his friendship and outstanding leadership in world-wide research networks on Teichmüller theory. As well as his own research, many books he has edited illustrate his great insight. The author also thanks M. Uludag for giving him a chance to take part in a tribute to Athanase Papadopoulos. This paper is based on the author's talks at OIST, University of Geneva and University of Strasbourg in 2016 and 2017. The slides for OIST and Strasbourg are uploaded in [17, 18]. The author thanks S. Hikami, A. Alekseev and C. Vespa for giving him these opportunities. In particular, discussions at OIST with Pioline aroused the author's interest in the genus 3 invariant a_3. He deeply thanks B. Pioline. Furthermore the author thanks R. de Jong for his helpful comments on the first version of this paper. The present research is supported in part by the grants JSPS KAKENHI 18KK0071, 18K03283, 19H01784, 20H00115 and 22H01120.

References

1. E. D'Hoker, M.B. Green, Zhang-Kawazumi invariants and superstring amplitudes. J. Number Theory **144**, 111–150 (2014)
2. E. D'Hoker, O. Schlotterer, Identities among higher genus modular graph tensors. Commun. Number Theory Phys. **16**, 35–74 (2022)
3. E. D'Hoker, M.B. Green, B. Pioline, R. Russo, Matching the $D^6\mathcal{R}^4$ interaction at two-loops. J. High Energy Phys. **2015** Article Number 31 (2015)
4. R. de Jong, Second variation of Zhang's λ-invariant on the moduli space of curves. Am. J. Math. **135**, 275–290 (2013)
5. R. de Jong, Asymptotic behavior of the Kawazumi-Zhang invariant for degenerating Riemann surfaces. Asian J. Math. **18**, 507–524 (2014)
6. R. de Jong, Point-like limit of the hyperelliptic Zhang-Kawazumi invariant. Pure Appl. Math. Q. **11**, 633–653 (2015)

7. R. de Jong, Torus bundles and 2-forms on the universal family of Riemann surfaces, in *Handbook of Teichmuller Theory*, ed. by A. Papadopoulos, vol. VI (EMS Publishing House, Zurich, 2016), pp. 195–227
8. R. de Jong, F. Shokrieh, Jumps in the height of the Ceresa cycle. Preprint, arXiv:2104.10060 (2021)
9. R. de Jong, S. van der Lugt, Rings of tautological forms on moduli spaces of curves. Preprint, arXiv:2207.05428 (2022)
10. G. Faltings, Calculus on arithmetic surfaces. Ann. Math. **119**, 387–424 (1984)
11. R. Hain, D. Reed, On the Arakelov geometry of the moduli space of curves. J. Differential Geom. **67**, 195–228 (2004)
12. B. Harris, Harmonic volumes. Acta Math. **150**, 91–123 (1983)
13. D. Johnson, An abelian quotient of the mapping class group $\mathcal{I}_g$. Math. Ann. **249**, 225–242 (1980)
14. N. Kawazumi, Harmonic Magnus expansion on the universal family of Riemann surfaces. Preprint, math.GT/0603158 (2006)
15. N. Kawazumi, Johnson's Homomorphisms and the Arakelov-Green function. Preprint, math.0801.4218 (2008)
16. N. Kawazumi, Canonical 2-forms on the moduli space of Riemann surfaces, in *Handbook of Teichmuller Theory*, ed. by A. Papadopoulos, vol. II (EMS Publishing House, Zurich, 2009), pp. 217–237
17. N. Kawazumi, Some tensor field on the Teichmüller space. Lecture at MCM2016, OIST (2016). https://www.ms.u-tokyo.ac.jp/~kawazumi/OIST1610_v1.pdf
18. N. Kawazumi, Differential forms and functions on the moduli space of Riemann surfaces, in *Séminaire Algèbre et topologie*. Université de Strasbourg (2017). https://www.ms.u-tokyo.ac.jp/~kawazumi/1701Strasbourg_v1.pdf
19. S. Morita, The extension of Johnson's homomorphism from the Torelli group to the mapping class group. Invent. Math. **111**, 197–224 (1993)
20. S. Morita, A linear representation of the mapping class group of orientable surfaces and characteristic classes of surface bundles, in *Topology and Teichmüller Spaces (Katinkulta, 1995)* (World Sci. Publ., River Edge, 1996), pp. 159–186
21. D. Mumford, *Tata Lectures on Theta I*. Progr. Math., vol. 28 (Birkhäuser, Boston, 1983)
22. B. Pioline, A Theta lift representation for the Kawazumi-Zhang and Faltings invariants of genus-two Riemann surfaces. J. Number Theory **163**, 520–541 (2016)
23. H.E. Rauch, On the transcendental moduli of algebraic Riemann surfaces. Proc. Natl. Acad. Sci. U. S. A. **41**, 42–49 (1955)
24. S. van der Lugt, Tautological differential forms on moduli spaces of curves. Thesis (Ph.D.), Leiden University (2022)
25. R. Wilms, New explicit formulas for Faltings' delta-invariant. Invent. Math. **209**, 481–539 (2017)
26. S.-W. Zhang, Gross-Schoen cycles and dualising sheaves. Invent. Math. **179**, 1–73 (2010)

Chapter 5
Directional Moduli and Pseudoconvexity

Sachiko Hamano

Abstract This is a survey chapter on our works related to the rigidity of pseudoconvex domains fibered by open Riemann surfaces according to directional moduli. For a marked open Riemann surface R of finite genus g and a real g-vector $\mathbf{a} = (a_1, \ldots, a_g) \neq \mathbf{0}$, we introduce the $\mathbf{a}$-span $\rho_{\mathbf{a}}$, and establish a new relation between $\rho_{\mathbf{a}}$ and the set of period matrices of all closings of R. From the viewpoint of several complex variables, a variational formula of $\rho_{\mathbf{a}}(t)$ is obtained for a smooth family $\mathcal{R}$ of open Riemann surfaces $R(t)$ with a complex parameter t in a disk Δ. As an application, we state the subharmonicity of the diameter of the $\mathbf{a}$-directional moduli disk for higher genera when $\mathcal{R}$ is a two-dimensional pseudoconvex domain fibered by open Riemann surfaces of the same topological type.

5.1 Introduction

Any open Riemann surface R of finite genus g can be conformally embedded into a closed Riemann surface S of the same genus g. The existence of such S for a given R is a classical result of Bochner [4]. Indeed, in the case of $g = 0$, i.e., R is a planar Riemann surface, from the general uniformization theorem of Koebe, R can be conformally mapped onto the so-called *extremal horizontal* (resp.*vertical*) *slit domain* on the Riemann sphere $\hat{\mathbb{C}}$. While a closed Riemann surface of genus zero is only $\hat{\mathbb{C}}$, there are many different closed Riemann surfaces of positive genus. Then it is natural to study the totality of such closed Riemann surfaces for a fixed R. This has been studied by many authors in the framework of moduli and Teichmüller spaces (see, e.g., Heins [10], Oikawa [17], Shiba [25], Masumoto [15], Kahn–Pilgrim–Thurston [12], and [9]; cf. A'Campo–Ji–Papadopoulos [1] and [11] for the early history).

S. Hamano (✉)
Department of Mathematics, Faculty of Science, Kyoto Sangyo University, Kyoto, Japan
e-mail: sachiko@cc.kyoto-su.ac.jp

A. M. Uludağ, A. Zeytin (eds.), *Essays on Geometry*,
https://doi.org/10.1007/978-3-031-76257-4_5

The span is first defined by Schiffer [22] as an important conformal invariant for multiply connected plane domains. The definition is in terms of conformal mappings onto domains bounded by horizontal slits and vertical slits, respectively, and it has been generalized in many different directions. (For an analytic and geometric characterization, see Ahlfors–Beurling [2]. See Mori [16] and Oikawa [18] associated with null-sets. For principal functions with L_1- and L_0-behavior in the sense of Sario, see [3, 19–21].)

From the viewpoint of several complex variables, we are interested in the problem of how conformal invariants for $R(t)$ change when an open Riemann surface $R(t)$ varies with a complex parameter t. More precisely, let $\pi : \widetilde{\mathcal{R}} \to \Delta$ be a holomorphic submersion of a two-dimensional complex manifold $\widetilde{\mathcal{R}}$ into a disk $\Delta = \{t \in \mathbb{C} \mid |t| < r\}$ such that $\widetilde{R}(t) := \pi^{-1}(t)$, $t \in \Delta$, is open.We consider a family $\mathcal{R}$ of open Riemann surfaces $R(t) := (\pi|_{\mathcal{R}})^{-1}(t)$, $t \in \Delta$, of genus g $(0 \leq g < \infty)$ such that $\partial R(t)$ consists of C^{ω} smooth closed curves in $\widetilde{R}(t)$. We identify $\mathcal{R}$ with the *smooth variation* of $R(t)$ in $\widetilde{\mathcal{R}}$. Then, one can ask: how does the conformal invariants for $R(t)$ move with t? The aim of this paper is to state a short survey on the parameter dependency of conformal invariants for $R(t)$ when $\mathcal{R}$ is pseudoconvex in $\widetilde{\mathcal{R}}$.

In the case of $g = 0$, this problem can be formulated in terms of the Schiffer spans $s(t)$ for planar Riemann surfaces $R(t)$, $t \in \Delta$, and we proved in [5] that the parameter dependency of $s(t)$ is *logarithmically* subharmonic on Δ if $\mathcal{R}$ is a pseudoconvex domain in $\widetilde{\mathcal{R}}$ (see Theorem 5.1). The proof depends on the variational formula of $s(t)$, and from the form (see (5.4)) the theorem is valid only for planar Riemann surfaces. As an application of the parameter dependency of $s(t)$, we showed the simultaneous uniformization of moving planar Riemann surfaces $R(t)$ of class O_{AD} (cf. [5, Main Theorem 1.1]).

In the case of $g = 1$, the problem can be formulated in terms of the *hyperbolic span* with which we are concerned in [8], related to *closings* of an open Riemann surface (see Sect. 5.4.1). It is defined only for Riemann surfaces of genus one, but shares interesting function-theoretic properties with the Schiffer span (see Theorem 5.4). However, the method is valid only for Riemann surfaces R of genus one because the L_1- and L_0-closings of R are not unique except for $g = 1$ (see *Remark* 5.2 in Sect. 5.4.1).

To extend the above results to the cases of higher genus, in [7], we defined a new span: for a marked open Riemann surface R of genus g $(1 \leq g < \infty)$ and a real g-vector $\mathbf{a} = (a_1, \ldots, a_g) \neq \mathbf{0}$, the **a**-*span* $\rho_{\mathbf{a}}$ of R is defined in terms of L_1- and L_0-canonical semi-exact differentials normalized by **a** (see Definition 5.1). From the viewpoint of several complex variables, a variational formula of $\rho_{\mathbf{a}}(t)$ is obtained for a smooth family $\mathcal{R}$ of open Riemann surfaces $R(t)$ with a complex parameter t in a disk Δ (see (5.7)), which is generalization of the formulations [8, Lemma 2.2] in the genus one case. As an application, we proved that if $\mathcal{R}$ is a two-dimensional pseudoconvex domain fibered by open Riemann surfaces of the same topological type, then $\rho_{\mathbf{a}}(t)$ is subharmonic on Δ (see Theorem 5.3). This means the subharmonicity of the diameter of the **a**-*directional moduli disk* for higher genera when $\mathcal{R}$ is pseudoconvex (see Theorem 5.5).

This survey is organized as follows. In Sect. 5.2 we will recall the Schiffer span and define the **a**-span as preparations from one complex variable. In Sect. 5.3 we will consider a smooth variation of open Riemann surfaces and state the parameter dipendency of them. In Sect. 5.4 we will clarify the geometric meaning of the **a**-span $\rho_{\mathbf{a}}$ of R which is related to the period matrices of closings of R. Lemma 5.3 is the key to the relation between them. We will then give a restatement of Theorem 5.3 in terms of the **a**-directional moduli disk $\mathfrak{M}_{\mathbf{a}}(R(t), \chi(t))$. We will see that the uniqueness of the variational formula of $\rho_{\mathbf{a}}(t)$ is very effective in showing Theorem 5.3 in higher genus cases (see *Remark* 5.5 in Sect. 5.4.3).

5.2 Preparations from One Complex Variable

We first recall the L_1-and L_0-principal functions on an open Riemann surface which are related to vertical and horizontal slit mappings, respectively, and define the Schiffer span on the potential theory of one complex variable (cf. Ahlfors–Sario [3], Sario–Nakai [20] for general references). Secondly, we introduce the **a**-span which is defined in terms of L_1- and L_0-canonical semi-exact differentials normalized by **a**. Each differential is associated with functions, respectively.

5.2.1 The L_1- and L_0-principal Functions

Let R be the interior of a compact bordered Riemann surface conformally embedded in a larger Riemann surface $\widetilde{R}$ such that the boundary ∂R of R consists of a finite number of C^ω-smooth contours C_j $(j = 1, \dots, \nu)$ in $\widetilde{R}$. Let $a \in R$ and let $U_a = \{z \in \mathbb{C} \mid |z-\zeta| < r_a\}$ be the local coordinate of a neighborhood of a in R, where z is a variable and ζ is a fixed point. We consider all harmonic functions $p(z)$ on $R \setminus \{a\}$ with singularity $\mathrm{Re}\,\frac{1}{z-\zeta}$ at a normalized so that $\lim_{z\to\zeta}(p(z) - \mathrm{Re}\,\frac{1}{z-\zeta}) = 0$. Then there are two uniquely determined functions $p_i(z)$ $(i = 1, 0)$ with the following boundary conditions (L_i): for $j = 1, \dots, \nu$,

$$(L_1) \quad p_1(z) = c_j \text{ (constant) on } C_j \quad \text{and} \quad \int_{C_j} \frac{\partial p_1(z)}{\partial n_z} ds_z = 0;$$

$$(L_0) \quad \frac{\partial p_0(z)}{\partial n_z} = 0 \quad \text{on } C_j.$$

We have $p_i(z) = \mathrm{Re}\left\{\frac{1}{z-\zeta} + \sum_{n=1}^{\infty} A_n^i (z-\zeta)^n\right\}$ at ζ $(i = 1, 0)$. $p_i(z)$ and $\alpha_i := \mathrm{Re}\{A_1^i\}$ are called the *L_i-principal function* and the *L_i-constant* for (R, a) with respect to the local coordinate U_a, respectively (simply, for (R, ζ)), and $s := \alpha_0 - \alpha_1$ is called the *Schiffer span* for (R, ζ).

5.2.2 The Schiffer Span for (R, ζ)

In the special case when R is a *planar* Riemann surface, the Schiffer span s for (R, ζ) has the geometric meaning as follows. Let $\mathcal{P}(R)$ be the set of all univalent functions $P(z)$ on R with the given singularity $\frac{1}{z-\zeta}$ at ζ which are normalized so that $\lim_{z\to\zeta}(P(z) - \frac{1}{z-\zeta}) = 0$. Namely, $P(z)$ has the following expression

$$P(z) = \frac{1}{z-\zeta} + 0 + \sum_{n=1}^{\infty} A_n(z-\zeta)^n \qquad \text{at } \zeta.$$

For $w = P(z) \in \mathcal{P}(R)$, let $\mathcal{E}_P$ denote the Euclidean area of $\mathbb{P}_w \setminus P(R)$, and set $\mathcal{E}(R) = \sup\{\mathcal{E}_P \mid P(z) \in \mathcal{P}(R)\}$. Koebe constructed two special functions $P_0(z)$ and $P_1(z)$ in $\mathcal{P}(R)$ which are the horizontal slit mapping and the vertical slit mapping for (R, a) with respect to U_a (simply, for (R, ζ)), respectively. Grunsky introduced the functions $P_0(z) \pm P_1(z)$, and studied their extremal properties. Schiffer discovered that the quantity $s := A_1^0 - A_1^1$ is always real and nonnegative, where $P_i(z) = \frac{1}{z-\zeta} + \sum_{n=1}^{\infty} A_n^i(z-\zeta)^n$ at ζ $(i = 1, 0)$. Then, Schiffer [22] proved that $M(z) := \frac{1}{2}(P_0(z) + P_1(z))$ belongs to $\mathcal{P}(R)$, the set $\mathbb{C}_w \setminus M(R)$ consists of convex domains, and the complementary area $\mathcal{E}(R)$ satisfies $\mathcal{E}(R) = \mathcal{E}_M = \frac{\pi}{2}s$. Further, such function is uniquely determined. By the standard approximation argument we define the Schiffer span s for (R, ζ) for any planar Riemann surface.

5.2.3 The L_1- and L_0-differentials

Let R be an open Riemann surface of genus g $(1 \le g < \infty)$ in a Riemann surface $\widetilde{R}$ such that the boundary ∂R consists of C^ω smooth contours in $\widetilde{R}$; $\partial R = C_1 + \cdots + C_\nu$. Let $\chi = \{A_k, B_k\}_{k=1}^g$ be a canonical homology basis of R mod dividing cycles C_j $(j = 1, \ldots, \nu)$. Let (R, χ) be the marked open Riemann surface, and let $\mathbf{a} = (a_1, \ldots, a_g) \in \mathbb{R}^g \setminus \{(0, \ldots, 0)\}$. Then there exists a unique holomorphic differential $\phi_{\mathbf{a}}^{(1)}$ on R such that

(i) $\phi_{\mathbf{a}}^{(1)}$ is semi-exact on R which has, by definition, no periods along every dividing cycle on R,
(ii) $\mathrm{Re}\,\phi_{\mathbf{a}}^{(1)} = 0$ on each C_j $(j = 1, \ldots, \nu)$,
(iii) $\int_{A_k} \phi_{\mathbf{a}}^{(1)} = a_k \quad (k = 1, \ldots, g)$.

We call $\phi_{\mathbf{a}}^{(1)}$ the L_1 *semi-exact differential on* (R, χ) *normalized by* $\mathbf{a}$, or for simplicity, the L_1*-differential* on (R, χ). In other words, this is the semi-exact canonical differential of the first kind in the sense of Kusunoki [13, Theorem 1] with the normalization (iii). In case where the real part of the Abelian integral $\int \phi_{\mathbf{a}}^{(1)}$

is single-valued on the whole surface R, Re $\int \phi_{\mathbf{a}}^{(1)}$ is the L_1*-principal function* of Sario–Nakai [20, I. 7A].

Shiba ([24, Theorem 4], [26, Theorem 3.5]) showed that, for the given $\phi_{\mathbf{a}}^{(1)}$ on (R, χ), there exist some closed Riemann surface S of the same genus g and some conformal embedding $\iota : R \hookrightarrow S$ such that $\phi_{\mathbf{a}}^{(1)}$ can be holomorphically extended on S and the ideal boundary of R is realized as a set of vertical slits on S with a vanishing total area.

Remark 5.1 Even if the L_1-differential $\phi_{\mathbf{a}}^{(1)}$ on (R, χ) is fixed, the above S and ι are not always uniquely determined, but depend on $(2g - 2 - d)$ real parameters, where d is the degree of the divisor of $\phi_{\mathbf{a}}^{(1)}$ and g is the genus of R (cf. Shiba [24] and Shiba–Shibata [27]).

We can find another holomorphic differential $\phi_{\mathbf{a}}^{(0)}$ instead of $\phi_{\mathbf{a}}^{(1)}$ by changing the L_1-boundary condition (ii) to the L_0-boundary condition: $\operatorname{Im} \phi_{\mathbf{a}}^{(0)} = 0$ on each C_j. Namely, there exists a unique holomorphic differential $\phi_{\mathbf{a}}^{(0)}$ on R such that

(i) $\phi_{\mathbf{a}}^{(0)}$ is semi-exact on R,
(ii) $\operatorname{Im} \phi_{\mathbf{a}}^{(0)} = 0$ on each C_j $(j = 1, \ldots, \nu)$,
(iii) $\int_{A_k} \phi_{\mathbf{a}}^{(0)} = a_k \quad (k = 1, \ldots, g)$.

We call $\phi_{\mathbf{a}}^{(0)}$ the L_0 *semi-exact differential on* (R, χ) *normalized by* $\mathbf{a}$, or for simplicity, the L_0*-differential* on (R, χ).

5.2.4 *The* a*-Span of* (R, χ)

Let (R, χ) be a marked open Riemann surface of genus g $(1 \le g < \infty)$, and let $\mathbf{a} = (a_1, \ldots, a_g) \in \mathbb{R}^g \setminus \{(0, \ldots, 0)\}$. Let $\phi_{\mathbf{a}}^{(1)}$ and $\phi_{\mathbf{a}}^{(0)}$ be the L_1-and L_0-differentials on (R, χ) normalized by $\mathbf{a}$, respectively. Since $\phi_{\mathbf{a}}^{(1)}$ and $\phi_{\mathbf{a}}^{(0)}$ on (R, χ) are uniquely determined, we set

$$\tau_k^{(s)} := \int_{B_k} \phi_{\mathbf{a}}^{(s)} \quad \in \mathbb{C} \qquad (s = 1, 0;\ k = 1, \ldots, g),$$

which is not always in $\mathbb{H}$.

Definition 5.1 (The a-Span of (R, χ) in [7]) For the L_s-differentials $\phi_{\mathbf{a}}^{(s)}$ on (R, χ), $s = 1, 0$, normalized by $\mathbf{a}$, the $\mathbf{a}$*-modulus* is defined by the following complex number

$$\tau_{\mathbf{a}}^{(s)} := \sum_{k=1}^{g} \left(\int_{A_k} \phi_{\mathbf{a}}^{(s)} \int_{B_k} \phi_{\mathbf{a}}^{(s)} \right) = \sum_{k=1}^{g} a_k \tau_k^{(s)} \qquad (s = 1, 0), \tag{5.1}$$

and the **a**-*span of* (R, χ) is defined by the Euclidean length

$$\rho_{\mathbf{a}} := |\tau_{\mathbf{a}}^{(1)} - \tau_{\mathbf{a}}^{(0)}| \quad \in \mathbb{R}. \tag{5.2}$$

Lemma 5.1 *Let the notation be as above. Then we have:*

1. $\operatorname{Re} \tau_{\mathbf{a}}^{(1)} = \operatorname{Re} \tau_{\mathbf{a}}^{(0)}$.
2. $\operatorname{Im} \tau_{\mathbf{a}}^{(1)} \geq \operatorname{Im} \tau_{\mathbf{a}}^{(0)}$. *The equality holds if and only if* $\phi_{\mathbf{a}}^{(1)} = \phi_{\mathbf{a}}^{(0)}$.
3. $\tau_{\mathbf{a}}^{(1)}, \tau_{\mathbf{a}}^{(0)} \in \mathbb{H}$.

Thus, $\tau_{\mathbf{a}}^{(1)} - \tau_{\mathbf{a}}^{(0)}$ *is a pure imaginary number with* $\operatorname{Im}(\tau_{\mathbf{a}}^{(1)} - \tau_{\mathbf{a}}^{(0)}) \geq 0$.

The geometric significance of **a**-span $\rho_{\mathbf{a}}$ and **a**-moduli $\tau_{\mathbf{a}}^{(s)}$ will be shown in Sect. 5.4. They are related to period matrices of closings of (R, χ).

5.3 Variation of the Schiffer Spans and the a-spans

We shall consider a smooth variation of open Riemann surfaces $R(t)$ and state the parameter dependency of the Schiffer spans and the **a**-spans for each $R(t)$.

5.3.1 Smooth Variation of Open Riemann Surfaces

Let $\pi : \widetilde{\mathcal{R}} \to \Delta$ be a holomorphic submersion of a two-dimensional complex manifold $\widetilde{\mathcal{R}}$ into a disk $\Delta = \{t \in \mathbb{C} \mid |t| < r\}$ such that $\widetilde{R}(t) := \pi^{-1}(t)$, $t \in \Delta$, is noncompact. We consider a family $\mathcal{R}$ of open Riemann surfaces $R(t) := (\pi|_{\mathcal{R}})^{-1}(t)$, $t \in \Delta$, of genus g $(0 \leq g < \infty)$ such that $\partial R(t)$ consists of C^{ω} smooth closed curves $C_j(t)$ $(j = 1, \ldots, \nu)$ in $\widetilde{R}(t)$. We identify $\mathcal{R}$ with the *smooth variation* of $R(t)$ in $\widetilde{\mathcal{R}}$. We note that g and ν do not depend on $t \in \Delta$. Then, $\partial\mathcal{R} := \{\partial R(t) \subset \widetilde{R}(t) \mid t \in \Delta\}$ is a hypersurface, C^{ω}-smooth and real three-dimensional in $\widetilde{\mathcal{R}}$.

For a C^2-smooth defining function $\varphi(t, z)$ of $\partial\mathcal{R}$ in $\widetilde{\mathcal{R}}$, we consider the following Levi curvature κ due to Levenberg-Yamaguchi [14, (1.3)]:

$$\kappa(t, z) := \left(\frac{\partial^2 \varphi}{\partial t \partial \bar{t}} \left| \frac{\partial \varphi}{\partial z} \right|^2 - 2\operatorname{Re} \left\{ \frac{\partial^2 \varphi}{\partial \bar{t} \partial z} \frac{\partial \varphi}{\partial t} \frac{\partial \varphi}{\partial \bar{z}} \right\} + \left| \frac{\partial \varphi}{\partial t} \right|^2 \frac{\partial^2 \varphi}{\partial z \partial \bar{z}} \right) \left| \frac{\partial \varphi}{\partial z} \right|^{-3} \tag{5.3}$$

on $\partial\mathcal{R}$. Note that $\kappa(t, z)$ does not depend on the choice of the defining function $\varphi(t, z)$ of $\partial\mathcal{R}$. We see that $\mathcal{R}$ is a smooth variation if and only if there exists a C^{ω} defining function $\varphi(t, z)$ of $\partial\mathcal{R}$ such that $\frac{\partial \varphi}{\partial z} \neq 0$.

5.3.2 Variation of the Schiffer Spans

Let $\mathcal{R}$ be a smooth variation in $\widetilde{\mathcal{R}}$. Assume that there exists a section $\mathbf{a} := \{a(t) \in R(t) \mid t \in \Delta\}$ of $\mathcal{R}$ over Δ, and that there exists a π-local coordinate $\mathcal{U} := \Delta \times \{|z-\zeta| < r\}$ of a neighborhood $\mathcal{V}$ of $\mathbf{a}$ in $\mathcal{R}$ such that $\mathbf{a}$ corresponds to $\Delta \times \{\zeta\}$. Let $t \in \Delta$ be fixed. Then each $R(t)$, $t \in \Delta$, carries the L_i-principal function $p_i(t,z)$ $(i = 1, 0)$ and L_i-constant $\alpha_i(t)$ $(i = 1, 0)$ for $(R(t), \zeta)$. Under the smooth variation $\mathcal{R}$, we established the following variational formulas for $\alpha_i(t)$ for $(R(t), \zeta)$.

Lemma 5.2 ([5])

$$\frac{\partial^2 \alpha_1(t)}{\partial t \partial \bar{t}} = -\left(\frac{1}{\pi}\int_{\partial R(t)} \kappa(t,z)\left|\frac{\partial p_1(t,z)}{\partial z}\right|^2 ds_z + \frac{4}{\pi}\iint_{R(t)} \left|\frac{\partial^2 p_1(t,z)}{\partial \bar{t}\partial z}\right|^2 dxdy\right).$$

$$\frac{\partial^2 \alpha_0(t)}{\partial t \partial \bar{t}} = \frac{1}{\pi}\int_{\partial R(t)} \kappa(t,z)\left|\frac{\partial p_0(t,z)}{\partial z}\right|^2 ds_z + \frac{4}{\pi}\iint_{R(t)} \left|\frac{\partial^2 p_0(t,z)}{\partial \bar{t}\partial z}\right|^2 dxdy$$
$$- \frac{2}{\pi}\mathrm{Im}\left\{\sum_{l=1}^{g} \frac{\partial}{\partial t}\left(\int_{A_l(t)} *dp_0(t,z)\right)\frac{\partial}{\partial \bar{t}}\left(\int_{B_l(t)} *dp_0(t,z)\right)\right\}.$$

Here, $\kappa(t,z)$ is the same as (5.3).

We remark that the above formula for $\alpha_1(t)$ does not concern the genus g of $R(t)$, but that for $\alpha_0(t)$ concerns it. In the case of the deformed *planar* bordered Riemann surface $R(t)$, i.e., $g = 0$, Lemma 5.2 implies the following variational formula of the Schiffer span $s(t) := \alpha_0(t) - \alpha_1(t)$:

$$\frac{\partial^2 s(t)}{\partial t \partial \bar{t}} = \frac{1}{\pi}\int_{\partial R(t)} \kappa(t,z)\left(\left|\frac{\partial p_1(t,z)}{\partial z}\right|^2 + \left|\frac{\partial p_0(t,z)}{\partial z}\right|^2\right) ds_z$$
$$+ \frac{4}{\pi}\iint_{R(t)} \left(\left|\frac{\partial^2 p_1(t,z)}{\partial \bar{t}\partial z}\right|^2 + \left|\frac{\partial^2 p_0(t,z)}{\partial \bar{t}\partial z}\right|^2\right) dxdy. \tag{5.4}$$

In the case of the deforming *planar* bordered Riemann surface $R(t)$, if $\mathcal{R}$ is pseudoconvex, the contrast between the superharmonicity of $\alpha_1(t)$ and the subharmonicity of $\alpha_0(t)$ are unified with the Schiffer span $s(t) = \alpha_0(t) - \alpha_1(t)$ for $(R(t), \zeta)$, which is subharmonic on Δ. We showed in [5] that $s(t)$ is moreover *logarithmically* subharmonic on Δ.

Theorem 5.1 (Lemma 4.1 in [5]; More Generally, Theorem 3 in [6]) *Let $\mathcal{R}$ be a smooth variation of open Riemann surfaces $R(t)$, $t \in \Delta$, of finite genus g $(0 \le g < \infty)$. Assume that there exists a section of $\mathcal{R}$ over Δ.*

(i) *If $\mathcal{R}$ is pseudoconvex in $\widetilde{\mathcal{R}}$, then the L_1-constant $\alpha_1(t)$ for $(R(t), \zeta)$ is superharmonic on Δ;*

(ii) *If* $\mathcal{R}$ *is pseudoconvex in* $\widetilde{\mathcal{R}}$ *and each* $R(t)$, $t \in \Delta$, *is* planar, *then the* L_0*-constant* $\alpha_0(t)$ *for* $(R(t), \zeta)$ *is subharmonic on* Δ;

(iii) *If* $\mathcal{R}$ *is pseudoconvex in* $\widetilde{\mathcal{R}}$ *and each* $R(t)$, $t \in \Delta$, *is* planar, *the Schiffer span* $s(t)$ *for* $(R(t), \zeta)$ *is* logarithmically *subharmonic on* Δ, *namely,* $\log s(t)$ *is subharmonic on* Δ.

To apply these results to Stein manifolds, we need to study the general (non-smooth) variation $\mathcal{R}$ under pseudoconvexity, namely, the variation $\mathcal{R} : t \in \Delta \to R(t)$ is pseudoconvex in $\widetilde{\mathcal{R}}$ such that neither $\partial R(t)$, $t \in \Delta$, is always C^ω smooth nor the variation $\partial\mathcal{R} : t \in \Delta \to \partial R(t)$ is C^ω smooth with $t \in \Delta$. Through the investigation, we applied the *logarithmically* subharmonicity of $s(t)$ to show the simultaneous uniformization of moving *planar* Riemann surfaces $R(t)$ of class O_{AD} (see [5, Main Theorem 1.1]). Here $R \in O_{AD}$ means there exists no non-constant holomorphic function with finite Dirichlet integral on R.

5.3.3 Variation of the **a***-spans*

Let $\mathcal{R}$ be a smooth variation of open Riemann surfaces $R(t)$, $t \in \Delta$, of finite genus g in $\widetilde{\mathcal{R}}$. For $t \in \Delta$, we may take a canonical homology basis $\chi(t) = \{A_k(t), B_k(t)\}_{k=1}^g$ of $R(t)$ modulo dividing cycles $C_j(t)$ $(j = 1, \dots, \nu)$ such that all $A_k(t)$ and $B_k(t)$ move continuously in $\mathcal{R}$ with $t \in \Delta$. Then each $(R(t), \chi(t))$, $t \in \Delta$, carries the **a**-span $\rho_{\mathbf{a}}(t)$ for a given **a** as follows: We always assume that $\mathbf{a} = (a_1, \dots, a_g) \in \mathbb{R}^g \setminus \{(0, \dots, 0)\}$ is fixed once and for all, namely, **a** is a set of constants independent of t. For every $t \in \Delta$, there exist uniquely defined L_s-differentials $\phi_{\mathbf{a}}^{(s)}(t, z)$ on $(R(t), \chi(t))$, $s = 1, 0$. From the normalization condition (iii), we have for all $t \in \Delta$,

$$\int_{A_k(t)} \phi_{\mathbf{a}}^{(s)}(t, z) = a_k \qquad (s = 1, 0;\ k = 1, \dots, g),$$

and we set

$$\tau_k^{(s)}(t) := \int_{B_k(t)} \phi_{\mathbf{a}}^{(s)}(t, z) \qquad (k = 1, \dots, g),$$

so that $\tau_k^{(1)}(t)$ and $\tau_k^{(0)}(t)$ are complex-valued functions on Δ, and $\tau_{\mathbf{a}}^{(1)}(t)$ and $\tau_{\mathbf{a}}^{(0)}(t)$ are also complex-valued functions on Δ;

$$\tau_{\mathbf{a}}^{(s)}(t) = \sum_{k=1}^{g} \left(\int_{A_k(t)} \phi_{\mathbf{a}}^{(s)}(t, z) \int_{B_k(t)} \phi_{\mathbf{a}}^{(s)}(t, z) \right) = \sum_{k=1}^{g} a_k \cdot \tau_k^{(s)}(t), \quad s = 1, 0.$$

For each $t \in \Delta$, the **a**-span $\rho_{\mathbf{a}}(t)$ of $(R(t), \chi(t))$ is the following:

$$\rho_{\mathbf{a}}(t) = \left|\tau_{\mathbf{a}}^{(1)}(t) - \tau_{\mathbf{a}}^{(0)}(t)\right| \quad \geq 0.$$

We established the following variational formulas of the second order for $\operatorname{Im} \tau_{\mathbf{a}}^{(1)}(t)$ and $\operatorname{Im} \tau_{\mathbf{a}}^{(0)}(t)$.

Theorem 5.2 (Theorem 3 in [7]) *Let $\phi_{\mathbf{a}}^{(s)}(t, z) = f_{\mathbf{a}}^{(s)}(t, z)dz$ $(s = 1, 0)$ using the local parameter z of $\overline{R(t)}$. For each $t \in \Delta$,*

$$\frac{\partial^2 \operatorname{Im} \tau_{\mathbf{a}}^{(1)}(t)}{\partial t \partial \bar{t}} = \frac{1}{2} \int_{\partial R(t)} \kappa(t, z) |f_{\mathbf{a}}^{(1)}(t, z)|^2 |dz| + \left\| \frac{\partial \phi_{\mathbf{a}}^{(1)}(t, z)}{\partial \bar{t}} \right\|_{R(t)}^2, \tag{5.5}$$

$$\frac{\partial^2 \operatorname{Im} \tau_{\mathbf{a}}^{(0)}(t)}{\partial t \partial \bar{t}} = -\left(\frac{1}{2} \int_{\partial R(t)} \kappa(t, z) |f_{\mathbf{a}}^{(0)}(t, z)|^2 |dz| + \left\| \frac{\partial \phi_{\mathbf{a}}^{(0)}(t, z)}{\partial \bar{t}} \right\|_{R(t)}^2 \right). \tag{5.6}$$

Here the Dirichlet norm of a holomorphic differential ϕ on R is denoted by $\|\phi\|_R^2$, and $\kappa(t, z)$ is the same as (5.3).

It follows from Lemma 5.1 that $\rho_{\mathbf{a}}(t) = \operatorname{Im} (\tau_{\mathbf{a}}^{(1)}(t) - \tau_{\mathbf{a}}^{(0)}(t))$. Combining (5.5) and (5.6) in Theorem 5.2, we have the following variational formula of the **a**-span $\rho_{\mathbf{a}}(t)$:

$$\frac{\partial^2 \rho_{\mathbf{a}}(t)}{\partial t \partial \bar{t}} = \frac{1}{2} \int_{\partial R(t)} \kappa(t, z) \left\{ |f_{\mathbf{a}}^{(1)}(t, z)|^2 + |f_{\mathbf{a}}^{(0)}(t, z)|^2 \right\} |dz| + \left\| \frac{\partial \phi_{\mathbf{a}}^{(1)}(t, z)}{\partial \bar{t}} \right\|_{R(t)}^2 + \left\| \frac{\partial \phi_{\mathbf{a}}^{(0)}(t, z)}{\partial \bar{t}} \right\|_{R(t)}^2. \tag{5.7}$$

As an application of the precise variational formula (5.7) of the second order, we can prove the rigidity of $\rho_{\mathbf{a}}(t)$ under the pseudoconvexity of the total space: If $\mathcal{R}$ is pseudoconvex in $\widetilde{\mathcal{R}}$, it implies $\kappa(t, z) \geq 0$ from the definition of (5.3). Then, the right hand side of (5.7) is non-negative, namely, $\rho_{\mathbf{a}}(t)$ is subharmonic on Δ. Moreover, if $\rho_{\mathbf{a}}(t)$ is harmonic, the left hand side of (5.7) must be zero for all $t \in \Delta$. Therefore, we see that $\kappa(t, z) = 0$ on $\partial R(t)$ and $\dfrac{\partial \phi_{\mathbf{a}}^{(1)}}{\partial \bar{t}} = \dfrac{\partial \phi_{\mathbf{a}}^{(0)}}{\partial \bar{t}} = 0$ on $R(t)$ for all $t \in \Delta$. Thus, $\partial \mathcal{R}$ is Levi flat and $\phi_{\mathbf{a}}^{(s)}(t, z)$, $s = 1, 0$, are holomorphic for t. Hence, $\tau_{\mathbf{a}}^{(1)}(t)$ and $\tau_{\mathbf{a}}^{(0)}(t)$ are also holomorphic on Δ. On the other hand, by Lemma 5.1, $\tau_{\mathbf{a}}^{(1)}(t) - \tau_{\mathbf{a}}^{(0)}(t)$ is pure imaginary. Therefore, $\tau_{\mathbf{a}}^{(1)}(t) - \tau_{\mathbf{a}}^{(0)}(t)$ must be constant on Δ, so that the **a**-span $\rho_{\mathbf{a}}(t)$ is also so for $t \in \Delta$. We have

Theorem 5.3 (Theorem 2 in [7]) *Let $\mathcal{R}$ be a smooth variation of open Riemann surfaces $(R(t), \chi(t))$, $t \in \Delta$, of finite genus g $(1 \leq g < \infty)$. Let $\mathbf{a} \in \mathbb{R}^g \setminus \{(0, \ldots, 0)\}$. If $\mathcal{R}$ is a pseudoconvex domain in $\widetilde{\mathcal{R}}$, then*

1. *the $\mathbf{a}$-span $\rho_{\mathbf{a}}(t)$ of $(R(t), \chi(t))$ is subharmonic on Δ,*
2. *$\rho_{\mathbf{a}}(t)$ is harmonic on Δ if and only if*
   a. *$\tau_{\mathbf{a}}^{(1)}(t)$ and $\tau_{\mathbf{a}}^{(0)}(t)$ are holomorphic on Δ,*
   b. *$\rho_{\mathbf{a}}(t) \equiv \rho_{\mathbf{a}}(0)$ for all $t \in \Delta$.*

5.4 Geometric Meaning of the a-span's Variation

We shall clarify the geometric meaning of the **a**-span of R which is related to the period matrices of closings of R. Subsequently, we restate the main theorem in terms of the **a**-directional moduli disk.

5.4.1 Closings of a Marked Open Riemann Surface

Let R be an open Riemann surface of genus g $(1 \leq g < \infty)$ and $\chi = \{A_k, B_k\}_{k=1}^g$ be a canonical homology basis of R modulo dividing cycles. For the marked open Riemann surface (R, χ), we consider a triplet $(S, \widetilde{\chi}, \iota)$ consisting of a closed Riemann surface S of the same genus g, a canonical homology basis $\widetilde{\chi} = \{\widetilde{A}_k, \widetilde{B}_k\}_{k=1}^g$ of S, and a conformal embedding ι of R into S such that $\iota(A_k)$ (resp. $\iota(B_k)$) are homologous to the cycles $\widetilde{A}_k$ (resp. $\widetilde{B}_k$) on S for all $k = 1, \ldots, g$. We say that two such triplets $(S, \widetilde{\chi}, \iota)$ and $(S', \widetilde{\chi}', \iota')$ are equivalent if there is a conformal mapping f of S onto S' with $f \circ \iota = \iota'$ on R. Each equivalence class is denoted by $[S, \widetilde{\chi}, \iota]$, called a *closing* of (R, χ) (see [7, 26]). In other words, the closing of (R, χ) is a compact *continuation* of the same genus which preserves the homological type of surfaces (R, χ) and $(S, \widetilde{\chi})$. See [3] for the definition of a continuation (or a prolongation [16, 17], an extension [20]) of R.

Let $C(R, \chi)$ denote the set of closings of (R, χ). To characterize $C(R, \chi)$, we introduce the so-called *Riemann period matrix* T of $(S, \widetilde{\chi})$:

$$C(R, \chi) \ni [S, \widetilde{\chi}, \iota] \quad \mapsto \quad T = T[S, \widetilde{\chi}, \iota] := \left(\tau_{jk}\right)_{1 \leq j,k \leq g} \in \mathfrak{S}_g, \tag{5.8}$$

where $\tau_{jk} = \int_{\widetilde{B}_k} \omega_j$ $(j, k = 1, \ldots, g)$ are the periods of holomorphic differentials ω_j $(j = 1, \ldots, g)$ on S normalized by $\int_{\widetilde{A}_k} \omega_j = \delta_{jk}$ (the Kronecker delta), and $\mathfrak{S}_g$ is the Siegel upper-half space of degree g. Let

$$\mathfrak{M}(R, \chi) := \{T[S, \widetilde{\chi}, \iota] \in \mathfrak{S}_g \mid [S, \widetilde{\chi}, \iota] \in C(R, \chi)\}. \tag{5.9}$$

Oikawa [17] proved that $\mathfrak{M}(R, \chi)$ is compact and connected in the Teichmüller space $\mathfrak{T}_g$. In particular, Shiba [25, Theorem 5] showed that if R is an open Riemann surface of genus one, then the set $\mathcal{M}(R, \chi)$ of moduli of tori that are closings of (R, χ) is a closed disk in the upper half-plane $\mathbb{H}$. Masumoto [15] gave similar information about the case when R is a once-holed torus. In the higher genus cases $(2 \leq g < \infty)$, Schmieder and Shiba [23, Theorem 3] showed that the restriction

$$\mathfrak{M}_j(R, \chi) := \{\tau_{jj}[S, \widetilde{\chi}, \iota] \in \mathbb{H} \mid [S, \widetilde{\chi}, \iota] \in C(R, \chi)\} \tag{5.10}$$

of (5.9) to each diagonal element τ_{jj} of $T[S, \widetilde{\chi}, \iota]$ is a closed disk in $\mathbb{H}$ for each $1 \leq j \leq g$. When we restrict ourselves to the case of $g = 1$, so that $j = 1$, $\mathfrak{M}_1(R, \chi)$ is equal to $\mathcal{M}(R, \chi)$.

As one of our recent results, we prove that the totality $\mathfrak{M}(R, \chi)$ of the period matrices of the closings of (R, χ) satisfy a set of inequalities in $\mathfrak{S}_g$. The details will given in a separate article [9] as a joint work with Shiba.

Remark 5.2 The closing of (R, χ) is not always uniquely determined by the L_1-differential $\phi_{\mathbf{a}}^{(1)}$ on (R, χ), but depend on $(2g-2-d)$ real parameters, where d is the degree of the divisor of $\phi_{\mathbf{a}}^{(1)}$. Since d must be 0 in the case where $g = 1$, the closing induced by $\phi_{\mathbf{a}}^{(1)}$ is uniquely determined, which is an element of $\partial\mathcal{M}(R, \chi)$. On the other hand, in the cases of $g \geq 2$, all but finitely many exceptional ideal boundary components of R are realized as univalent vertical slits on S; the exceptional boundary components are also realized as vertical slits but they are *not univalent*. Therefore, there are infinitely many distinct ways of obtaining S from (R, χ). In such a case, the total order of "*boundary branch points* of Φ with $d\Phi = \phi_{\mathbf{a}}^{(1)}$" does not vanish in the sense of Shiba (see Theorems 4(6) and 6 in [24]).

Now we shall consider a smooth variation $\mathcal{R}$ of marked open Riemann surfaces $(R(t), \chi(t))$, $t \in \Delta$, of finite genus g in $\widetilde{\mathcal{R}}$. Then, one can ask: how does $C(R(t), \chi(t))$ move with t? In the case of $g = 1$, this problem can be formulated in terms of the hyperbolic diameter $\sigma(t)$ of $\mathcal{M}(R(t), \chi(t))$ in $\mathbb{H}$ with respect to the Poincaré metric, which gives a generalization of Schiffer's span for a planar Riemann surface. In particular, corresponding to the parameter dependency of Schiffer spans for a family of planar Riemann surfaces (see Theorem 5.1 (iii)), we proved the following.

Theorem 5.4 (Theorem 1.2 in [8]) *Let $\mathcal{R}$ be a smooth variation of open Riemann surfaces $(R(t), \chi(t))$, $t \in \Delta$, of genus one. If $\mathcal{R}$ is a pseudoconvex domain in $\widetilde{\mathcal{R}}$, then*

(i) *the hyperbolic diameter $\sigma(t)$ of $\mathcal{M}(R(t), \chi(t))$ is subharmonic on Δ.*
(ii) *$\sigma(t)$ is harmonic on Δ if and only if $\mathcal{R}$ is a trivial family $\Delta \times R(0)$.*

In the proof of Theorem 5.4, the unique closing of $(R(t), \chi(t))$ corresponding to each boundary point of $\mathcal{M}(R(t), \chi(t))$ was used to establish the variational formulas [8, Lemma 2.2] of the L_1- and L_0-differentials. However, the method is valid only for Riemann surfaces of genus one. In the genus one case, each boundary point of

$\mathcal{M}(R, \chi)$ corresponds to a *unique* closing of (R, χ). However, in the higher genus cases, even at a boundary point of $\mathfrak{M}_j(R, \chi)$ in (5.10), the closing of (R, χ) is not always uniquely determined. Indeed, even if each of the L_1- and L_0-differentials on (R, χ) is fixed, there can be uncountably many distinct closings of (R, χ) (to be precise, see [24, Theorem 6], [25, Theorem 5], and *Remark* 5.2).

To extend Theorem 5.4 to the cases of higher genus, we needed to introduce a new span of (R, χ) which is defined only by a marked open Riemann surface (R, χ) of finite genus g and $\mathbf{a} \in \mathbb{R}^g \setminus \{\mathbf{0}\}$, without closing of (R, χ), namely, the $\mathbf{a}$-span $\rho_{\mathbf{a}}$ of (R, χ). Then we established a new relation between $\rho_{\mathbf{a}}$ of (R, χ) and the period matrices of closings of (R, χ).

5.4.2 *Geometric Meaning of the* a*-span*

Definition 5.2 (a-directional Moduli disk) Let $T[S, \widetilde{\chi}, \iota]$ be the period matrix of a closing $[S, \widetilde{\chi}, \iota]$ of (R, χ) as above in (5.8). Let $\mathbf{a} = (a_1 \cdots a_g) \in \mathbb{R}^g$ be a $1 \times g$ matrix, and let ${}^t\mathbf{a}$ be the transpose matrix of $\mathbf{a}$. For an arbitrarily fixed matrix $\mathbf{a} \in \mathbb{R}^g \setminus \{\mathbf{0}\}$, the complex number

$$T_{\mathbf{a}}[S, \widetilde{\chi}, \iota] := \mathbf{a} \cdot T[S, \widetilde{\chi}, \iota] \cdot {}^t\mathbf{a} = \sum_{1 \le j,k \le g} a_j a_k \tau_{jk} \tag{5.11}$$

lies in $\mathbb{H}$ by the property of period matrices. Then we can see from [23, Theorem 3] that the set

$$\mathfrak{M}_{\mathbf{a}}(R, \chi) := \{T_{\mathbf{a}} := T_{\mathbf{a}}[S, \widetilde{\chi}, \iota] \in \mathbb{H} \mid [S, \widetilde{\chi}, \iota] \in C(R, \chi)\} \tag{5.12}$$

is a closed disk $\{|T_{\mathbf{a}} - T^*_{\mathbf{a}}| \le \eta_{\mathbf{a}}/2\}$ in $\mathbb{H}$. We call it the **a**-*directional moduli disk.*

We here prepare a key lemma for the restatement of Theorem 5.3. Let $\mathbf{a} = (a_1, \ldots, a_g) \in \mathbb{R}^g \setminus \{(0, \ldots, 0)\}$. Let $\phi^{(1)}_{\mathbf{a}}$ be the L_1-differential on the marked open Riemann surface (R, χ), where $\chi = \{A_k, B_k\}_{k=1}^g$ is a canonical homology basis of $R \bmod \partial R$. There exist some closed Riemann surface $S^{(1)} \supset R$ and some conformal embedding $\iota^{(1)} : R \hookrightarrow S^{(1)}$ such that $\phi^{(1)}_{\mathbf{a}}$ can be holomorphically extended to $S^{(1)}$. Then $\iota^{(1)}$ maps $R \setminus \cup_{k=1}^g (A_k \cup B_k)$ onto a vertical slits domain in $\hat{\mathbb{C}}$ by the L_1-boundary condition (ii) of $\phi^{(1)}_{\mathbf{a}}$. Similarly, the L_0-differential $\phi^{(0)}_{\mathbf{a}}$ on (R, χ) can be holomorphically extended to a closed Riemann surface $S^{(0)}$ such that a conformal embedding $\iota^{(0)} : R \hookrightarrow S^{(0)}$ maps $R \setminus \cup_{k=1}^g (A_k \cup B_k)$ onto a horizontal slits domain in $\hat{\mathbb{C}}$. We may take a canonical homology basis $\widetilde{\chi}^{(i)} = \{\widetilde{A_k}, \widetilde{B_k}\}_{k=1}^g$ of $S^{(i)}$, $i = 1, 0$, with $\iota^{(i)}(A_k) \sim \widetilde{A_k}$ and $\iota^{(i)}(B_k) \sim \widetilde{B_k}$ in $S^{(i)}$ for all $k = 1, \ldots, g$. Then we say that the closing $[S^{(i)}, \widetilde{\chi}^{(i)}, \iota^{(i)}]$ of (R, χ) is the L_i-*closing* of (R, χ) if the closing is induced by the L_i-differential on (R, χ), $i = 1, 0$.

Lemma 5.3 *Let $\tau_{\mathbf{a}}^{(i)}$, $i = 1, 0$, be the* **a**-*modulus induced by $\phi_{\mathbf{a}}^{(i)}$ on (R, χ) as in* (5.1). *Let $T[S^{(i)}, \widetilde{\chi}^{(i)}, \iota^{(i)}]$, $i = 1, 0$, be the period matrix of the L_i-closing $[S^{(i)}, \widetilde{\chi}^{(i)}, \iota^{(i)}]$ of (R, χ) as in* (5.11). *Then, we have*

$$\tau_{\mathbf{a}}^{(i)} = \mathbf{a} \cdot T[S^{(i)}, \widetilde{\chi}^{(i)}, \iota^{(i)}] \cdot {}^t\mathbf{a}, \qquad (i = 1, 0). \tag{5.13}$$

Remark 5.3 In higher genus cases, the right hand side of (5.13) is not free from the ambiguity of gluing the slits (see *Remark* 5.2 in Sect. 5.4.1). But the left hand side of (5.13) is uniquely determined from the definition of $\tau_{\mathbf{a}}^{(i)}$ for each $i = 1, 0$. Therefore, we can establish the variational formulas of $\tau_{\mathbf{a}}^{(1)}(t)$ and $\tau_{\mathbf{a}}^{(0)}(t)$ in Theorem 5.2.

Remark 5.4 It should be noted that, in general, a $(g \times g)$-complex matrix induced by the L_1 semi-exact differentials ϕ_j $(j = 1, \ldots, g)$ on (R, χ), normalized by $\int_{A_k} \phi_j = \delta_{jk}$ (the Kronecker delta), is neither the period matrix of a closed Riemann surface nor a point of the Siegel upper-half space $\mathfrak{S}_g$.

Combining Lemma 5.3 with [23, Theorem 3], we have the following formulation:

Corollary 5.1 *Let $\mathfrak{M}_{\mathbf{a}}(R, \chi)$ be the* **a**-*directional moduli disk $\{T_{\mathbf{a}} := T_{\mathbf{a}}[S, \widetilde{\chi}, \iota] \in \mathbb{H} \mid |T_{\mathbf{a}} - T_{\mathbf{a}}^*| \le \eta_{\mathbf{a}}/2\}$ in $\mathbb{H}$ as seen in* (5.12). *Let $\tau_{\mathbf{a}}^{(s)}$, $s = 1, 0$, be the* **a**-*modulus as seen in* (5.1), *and let $\rho_{\mathbf{a}}$ be the* **a**-*span of (R, χ) as seen in* (5.2). *Then the following holds:*

1. $T_{\mathbf{a}}^* = \frac{1}{2}(\tau_{\mathbf{a}}^{(1)} + \tau_{\mathbf{a}}^{(0)})$,
2. $\eta_{\mathbf{a}} = \rho_{\mathbf{a}}$.

5.4.3 Restatement

Now we ask: how does the **a**-directional moduli disk $\mathfrak{M}_{\mathbf{a}}(R(t), \chi(t))$ vary when $(R(t), \chi(t))$ deforms with the complex parameter t from the several complex variables point of view? For a fixed $\mathbf{a} = (a_1, \ldots, a_g) \in \mathbb{R}^g \setminus \{(0, \ldots, 0)\}$, the set of closings of $(R(t), \chi(t))$, $t \in \Delta$, induces a family of closed disks

$$\mathfrak{M}_{\mathbf{a}}(R(t), \chi(t)) = \{|T_{\mathbf{a}}(t) - T_{\mathbf{a}}^*(t)| \le \eta_{\mathbf{a}}(t)/2\}$$

in $\mathbb{H}$. From Corollary 5.1 based on the key Lemma 5.3, we can restate Theorem 5.3 as follows:

Theorem 5.5 (Restatement of Theorem 5.3) *Assume that $\mathcal{R}$ is a pseudoconvex domain in $\widetilde{\mathcal{R}}$. Then we have*

1. *the diameter $\eta_{\mathbf{a}}(t)$ of $\mathfrak{M}_{\mathbf{a}}(R(t), \chi(t))$ is subharmonic on Δ,*
2. *if $\eta_{\mathbf{a}}(t)$ is harmonic on Δ, then the following hold:*
   a. *the center $T_{\mathbf{a}}^*(t)$ of $\mathfrak{M}_{\mathbf{a}}(R(t), \chi(t))$ is holomorphic on Δ,*
   b. *$\eta_{\mathbf{a}}(t) \equiv \eta_{\mathbf{a}}(0)$ for all $t \in \Delta$.*

In other words, the **a***-directional moduli disk* $\mathfrak{M}_{\mathbf{a}}(R(t), \chi(t))$, $t \in \Delta$, *with the same diameter* $\eta_{\mathbf{a}}(t) \equiv \eta_{\mathbf{a}}(0)$ *moves holomorphically in* $\mathbb{H}$.

Remark 5.5 Theorem 5.3 is a statement purely about open Riemann surfaces, which does not utilize closings of the surfaces. In other words, the **a**-span provides a new approach in capturing the moduli of open Riemann surfaces, without closing them. Indeed, there is a relationship between the **a**-span of open surfaces and the period matrix of their closing, as we see in Lemma 5.3.

In particular, in the case of $\mathbf{a}_j = (0, \ldots, 0, 1, 0, \ldots, 0) \in \mathbb{R}^g$ where 1 is the j-th element, we have $\mathbf{a}_j \cdot T[S, \widetilde{\chi}, \iota] \cdot {}^t\mathbf{a}_j = \tau_{jj}$ from (5.11), which implies that each diagonal element τ_{jj} $(j = 1, \ldots, g)$ of $T[S, \widetilde{\chi}, \iota]$ is a point of $\mathbb{H}$. Thus, in the case of $\mathbf{a}_j$, Theorem 5.5 can be reduced to Theorem 5.4 based on the result of genus one case, so that the following holds:

Theorem 5.6 *Fix* $\mathbf{a}_j = (0, \ldots, 0, 1, 0, \ldots, 0) \in \mathbb{R}^g$ *where* 1 *is the* j*-th element. Let* $\sigma_{\mathbf{a}_j}(t)$ *be the hyperbolic diameter of* $\mathfrak{M}_{\mathbf{a}_j}(R(t), \chi(t))$. *Assume that* $\mathcal{R}$ *is a pseudoconvex domain in* $\widetilde{\mathcal{R}}$. *Then the following holds:*

(1) $\sigma_{\mathbf{a}_j}(t)$ *is subharmonic on* Δ.
(2) If $\sigma_{\mathbf{a}_j}(t)$ *is harmonic on* Δ, *then* $\mathfrak{M}_{\mathbf{a}_j}(R(t), \chi(t)) = \mathfrak{M}_{\mathbf{a}_j}(R(0), \chi(0))$ *for all* $t \in \Delta$. *Namely,* $\mathfrak{M}_{\mathbf{a}_j}(R(t), \chi(t))$ *does not depend on* $t \in \Delta$.

Acknowledgments This article is based on the author's talk at the "Conference in Honor of 65th birthday of Athanase Papadopoulos" held at Galatasaray University, Istanbul, Turkey in June 2022. The author would like to thank the organizers, Professors A. Muhammed Uludağ, Ayberk Zeytin, Özge Ülkem, Sumio Yamada, and Marc Troyanov for the invitation and the warm hospitality at the conference. She would like to expresses her gratitude to Professor Athanase Papadopoulos for his constant encouragement.

References

1. N. A'Campo, L. Ji, A. Papadopoulos, On the early history of moduli and Teichüller spaces, in *Lipman Bers, A Life in Mathematics*, ed. by L. Keen, I. Kra, R.E. Rodriguez (American Mathematical Society, Providence, 2015), pp. 175–262. ISBN 978-1-4704-2056-7. hal-01276761
2. L.V. Ahlfors, A. Beurling, Conformal invariants and function-theoretic null-sets. Acta Math. **83**, 101–129 (1950)
3. L.V. Ahlfors, L. Sario, *Riemann Surfaces*. Princeton Mathematical Series, No. 26 (Princeton University Press, Princeton, 1960), xi+382 pp.
4. S. Bochner, Fortsetzung Riemannscher Flächen. Math. Ann. **98**(1), 406–421 (1928)
5. S. Hamano, Uniformity of holomorphic families of non-homeomorphic planar Riemann surfaces. Ann. Polon. Math. **111**(2), 165–181 (2014)
6. S. Hamano, Log-plurisubharmonicity of metric deformations induced by Schiffer and harmonic spans. Math. Z. **284**(1–2), 491–505 (2016)
7. S. Hamano, On rigidity of pseudoconvex domains fibered by open Riemann surfaces according to directional moduli. Math. Z. **300**(1), 979–993 (2022)
8. S. Hamano, M. Shiba, H. Yamaguchi, Hyperbolic span and pseudoconvexity. Kyoto J. Math. **57**(1), 165–183 (2017)

9. S. Hamano, M. Shiba, The period matrices of an open Riemann surface and its closings in the Siegel upper half-space. (submitted)
10. M. Heins, A problem concerning the continuation of Riemann surfaces, in *Contributions to the Theory of Riemann Surfaces*. Annals of Mathematics Studies, no. 30 (Princeton University Press, Princeton, 1953), pp. 55–62
11. L. Ji, A. Papadopoulos, Historical development of Teichüller theory. Arch. Hist. Exact Sci. **67**(2), 119–147 (2013)
12. J. Kahn, K.M. Pilgrim, D.P. Thurston, Conformal surface embeddings and extremal length. Groups Geom. Dyn. **16**(2), 403–435 (2022)
13. Y. Kusunoki, Theory of Abelian integrals and its applications to conformal mappings. Mem. Coll. Sci. Univ. Kyoto Ser. A. Math. **32**, 235–258 (1959)
14. N. Levenberg, H. Yamaguchi, The metric induced by the Robin function. Mem. Am. Math. Soc. **92**(448), viii+156 pp. (1991)
15. M. Masumoto, Holomorphic mappings of once-holed tori. J. Anal. Math. **129**, 69–90 (2016)
16. A. Mori, A remark on the prolongation of Riemann surfaces of finite genus. J. Math. Soc. Jpn. **4**, 27–30 (1952)
17. K. Oikawa, On the prolongation of an open Riemann surface of finite genus. Kôdai Math. Sem. Rep. **9**, 34–41 (1957)
18. K. Oikawa, On the uniqueness of the prolongation of an open Riemann surface of finite genus. Proc. Am. Math. Soc. **11**, 785–787 (1960)
19. B. Rodin, L. Sario, *Principal Functions*, in collaboration with Mitsuru Nakai (D. Van Nostrand Co., Inc., Princeton-Toronto-London, 1968), xviii+347 pp.
20. L. Sario, M. Nakai, Classification theory of Riemann surfaces. Die Grundlehren der mathematischen Wissenschaften, Band 164 (Springer-Verlag, New York-Berlin, 1970), xx+446 pp.
21. L. Sario, K. Oikawa, Capacity functions. Die Grundlehren der mathematischen Wissenschaften, Band 149 (Springer-Verlag New York, Inc., New York, 1969), xvii+361 pp.
22. M. Schiffer, The span of multiply connected domains. Duke Math. J. **10**, 209–216 (1943)
23. G. Schmieder, M. Shiba, One-parameter variations of the ideal boundary and compact continuations of a Riemann surface. Analysis (Munich) **18**(2), 125–130 (1998)
24. M. Shiba, The Riemann-Hurwitz relation, parallel slit covering map, and continuation of an open Riemann surface of finite genus. Hiroshima Math. J. **14**(2), 371–399 (1984)
25. M. Shiba, The moduli of compact continuations of an open Riemann surface of genus one. Trans. Am. Math. Soc. **301**(1), 299–311 (1987)
26. M. Shiba, Conformal embeddings of an open Riemann surface into another—a counter of univalent function theory. Interdiscip. Inform. Sci. **25**(1), 23–31 (2019)
27. M. Shiba, K. Shibata, Hydrodynamic continuations of an open Riemann surface of finite genus. Complex Variables Theory Appl. **8**(3–4), 205–211 (1987)

Chapter 6
Angle Defect for Super Triangles

Robert Penner

Abstract We prove that the angle defect minus the area of a super hyperbolic triangle is not identically zero and explicitly compute the non-vanishing fermionic correction. This disproves the Angle Defect Theorem for $\mathcal{N} = 1$ super hyperbolic geometry and provides a novel non-trivial additive function of super triangles. The proof techniques involve the orthosymplectic group OSp(1|2) in its action on the super Minkowski space $\mathbb{R}^{2,1|2}$ and brute-force computation.

6.1 Introduction

This chapter continues our ongoing study of triangles in $\mathcal{N} = 1$ super hyperbolic geometry, namely, the geometry of the usual two-real dimensional hyperbolic plane enriched with two real anti-commuting variables. We proved the Law of Cosines for super hyperbolic triangles in [14] as a birthday present for Norbert A'Campo, only to find exactly the classical formula with no corrections; however, we also found that this classical formula relates expressions which do involve so-called *fermionic corrections*. Here as a birthday present for Athanase Papadopoulos, we compute the non-zero correction to the Angle Defect Theorem for super hyperbolic triangles.

The classical hyperbolic Angle Defect Theorem states most elegantly that the sum S of the interior angles of a hyperbolic triangle plus its area A is constant equal to π. That is, $S \geq 0$ is always at most π, the *defect* D is $D = \pi - S \geq 0$, and the theorem is that $A = D$.

It was Lambert who originally proved this result in hyperbolic geometry in 1766 in a manuscript unpublished in his lifetime, which finally appeared in German [17] in 1895, after Johann Bernoulli III, who was in charge of Lambert's estate, considered publishing it and decided against it. A French translation [12] by Athanase and his student Guillaume Théret appeared only recently in 2014.

R. Penner (✉)
Institut des Hautes Études Scientifiques, Le Bois Marie, Bures-sur-Yvette, France
e-mail: rpenner@ihes.fr

A. M. Uludağ, A. Zeytin (eds.), *Essays on Geometry*,
https://doi.org/10.1007/978-3-031-76257-4_6

As was the fashion for centuries, and indeed for millennia, to prove the impossibility of the dubious hyperbolic geometry, theorems were proved within its confines in the hopes of deriving a logical contradiction. This was Lambert's motivation, and his elegant proof comprises all of two statements: Angle defect is additive for triangles, and area is the only such additive function, at least up to an overall scalar. The first assertion is obvious, as is the additivity of area, but the uniqueness part of Lambert's second assertion seems to anticipate developments in measure theory by several centuries.

Contemporary textbook proofs of the Angle Defect Theorem typically involve first computing the area of a triangle with one or more vanishing interior angles and again rely on additivity of area. This approach is not available in super hyperbolic geometry due to the divergence of the fermionic correction to the area of a super hyperbolic triangle with vanishing interior angles, as we prove in the Appendix.

Lambert's proof is likewise unavailable in our context because *area is not the only additive function for super hyperbolic triangles*. This follows from our main result, Theorem 6.4, which explicitly computes the non-zero fermionic correction to the Angle Defect Theorem for super triangles. This furthermore raises the intriguing question of what are the additive functions on super hyperbolic triangles.

The classical Angle Defect Theorem is evidently a direct consequence of the Gauss–Bonnet Theorem in constant curvature. Differential geometric approaches to super geometry are given in [1, 4, 16], and perhaps an argument different from ours could be based on that.

This entire chapter is devoted to the computation of the area and angle defect of a super hyperbolic triangle. We work in the super Minkowski space $\mathbb{R}^{2,1|2}$, which is recalled and discussed in Sect. 6.2, taking the upper sheet $\mathbb{H}$ of the unit hyperboloid as our model for the super hyperbolic plane. As has been noted elsewhere [15], this is the most natural model since in the super analogues of the upper half plane and Poincaré disk, not every geodesic ray is asymptotic to a point of the absolute; in contrast, every super geodesic in $\mathbb{H}$ is indeed asymptotic in each direction to a ray in the super light cone, as we shall recall.

Super geodesics are discussed in Sect. 6.3. The real orthosymplectic group $OSp(1|2)$, which plays the role here of $PSL(2, \mathbb{R})$ or $SL(2, \mathbb{R})$ in the classical case, is recalled and explained in Sect. 6.4. The $OSp(1|2)$-invariant area form Ω on $\mathbb{H}$ is computed in Sect. 6.5 as the pull-back of the usual $PSL(2, \mathbb{R})$-invariant area form in the classical case, and a primitive ω with $d\omega = \Omega$ is derived. Our approach to computing areas is given by Stokes' theorem and hence devolves to computing line integrals of ω on the edges of a super triangle; additivity of area is thus obvious with this formalism.

These calculations are at some points concise and rewarding and at other points routine and tedious. They comprise the bulk of the paper in Sects. 6.6 and 6.7. The final result provides a purely fermionic additive function on super triangles, which is given by the angle defect minus the area. It is obviously not identically zero, but it is complicated and far from illuminating. With a large fermionic correction to the classical formula, this is therefore diametrically opposite to what we found for the Law of Cosines in the earlier work.

Let us emphasize that whereas the classical results we extend involve two-real dimensional geometry, their super analogues include also two real fermions and hence are higher dimensional in a sense. For instance, Theorem 6.2 in [14] provides the further conditions to guarantee the intersection in $\mathbb{H}$ of two super geodesics whose underlying classical geodesics intersect in a point. Exactly what, if anything, our findings might suggest about classical or super scissors congruence in dimensions 2, 3 or 4 remains to be seen. It seems feasible to follow the scheme here and find an analogous correction for the volume of a super hyperbolic 3-simplex.

This chapter is dedicated to Athanase Papadopoulos on the happy occasion of his 65th birthday. He is my dearest friend. We met as post-docs, and since our doctoral theses had substantial overlap, we presumably should have been natural competitors. It was quite the contrary, and from the moment we first shook hands at a conference in Warwick, we have been fast friends.

I had just moved from Princeton to Los Angeles, and he from Orsay to Strasbourg. I more or less immediately joined Athanase in Strasbourg for a semester, and he directly thereafter joined me in Los Angeles for a semester, each of us with our young families in tow. We wrote two papers [6, 7] together at that time on Thurston theory, specifically on pseudo-Ansov mappings and measured foliations. I appreciated Strasbourg more than he enjoyed Los Angeles, and I have often been back to visit him over the last 40 years. Strasbourg has been a second home to me, or perhaps more precisely third.

We had three substantial collaborations after that period: In [8, 9], we showed that the Weil–Petersson Kähler form on Teichmüller space extends naturally to the Thurston two-form on the space of (projective) measured foliations, an early paradigm for tropical geometry as well as for the theory of cluster varieties, which co-opted some of our constructions; in [10], we studied so-called broken hyperbolic structures on surfaces, which have certain controlled discontinuities along geodesics; and in [11], we extended the Hatcher–Thurston theory, among other things, to the setting of non-orientable surfaces, providing moves on pants decompositions of a surface so that finite compositions act transitively. We have long planned to compute the relations among our moves, a project for which I hope we both still have the time, energy and inclination.

We have been through good times and bad times together. As we have taken our separate trajectories through mathematics and through life from our once-similar initial conditions, it has been rewarding to see Athanase develop into the great scholar that he has become. It has furthermore been an abiding joy in my career and in my existence to have counted him among my life-long friends.

6.2 Super Minkowski Space

Let $\hat{\mathbb{R}} = \hat{\mathbb{R}}[0] \otimes \hat{\mathbb{R}}[1]$ be the $\mathbb{Z}/2$-graded module over $\mathbb{R}$ with one central generator $1 \in \mathbb{R} \subseteq \hat{\mathbb{R}}[0]$ of degree zero and countably infinitely many anti-commuting generators $\theta_1, \theta_2, \ldots \in \hat{\mathbb{R}}[1]$ of degree one. An arbitrary $a \in \hat{\mathbb{R}}$ can be written uniquely as

$$a = a_\# + \sum_i a_i \theta_i + \sum_{i<j} a_{ij} \theta_i \theta_j + \sum_{i<j<k} a_{ijk} \theta_i \theta_j \theta_k + \cdots,$$

where $a_\#, a_i, a_{ij}, a_{ijk} \ldots \in \mathbb{R}$. The term $a_\#$ of degree zero is called the *body* of the *super number* $a \in \hat{\mathbb{R}}$. If $a \in \hat{\mathbb{R}}[0]$, then it is said to be an *even* super number or *boson*, while if $a \in \hat{\mathbb{R}}[1]$, then it is said to be an *odd* super number or *fermion*. We adopt the notation throughout that fermions are denoted by lower-case Greek letters.

One allows only finitely many anti-commuting factors in any product and only finitely many summands in any super number, a constraint we shall call *regularity*. Regularity implies nilpotence, i.e., for each $a \in \hat{\mathbb{R}}$, there is some $n \in \mathbb{Z}_{\geq 0}$ with $(a - a_\#)^n = 0$. It follows that if $a_\# \neq 0$, then we may write

$$\begin{aligned} \frac{1}{a} &= \frac{1}{a_\# + (a - a_\#)} = \frac{1}{a_\#} \frac{1}{1 + \frac{a - a_\#}{a_\#}} \\ &= \frac{1}{a_\#}\left[1 - \frac{a - a_\#}{a_\#} + \left(\frac{a - a_\#}{a_\#}\right)^2 - \cdots + (-1)^{n-1}\left(\frac{a - a_\#}{a_\#}\right)^{n-1}\right], \end{aligned}$$

and hence $a \in \hat{\mathbb{R}}$ is invertible if and only if $a_\# \neq 0$. It similarly follows from regularity that the zero divisors in $\hat{\mathbb{R}}$ are given by the ideal generated by $\hat{\mathbb{R}}[1]$.

One analogously extends real-analytic functions on $\mathbb{R}$ to $\hat{\mathbb{R}}$ with Taylor series under appropriate restrictions on the body. For instance for later tacit application, if $b^2 = 0$ for some $b \in \hat{\mathbb{R}}$, then for $a \in \hat{\mathbb{R}}$ with $a_\# \neq 0$, we have $\frac{1}{a+b} = \frac{1}{a}(1 - \frac{b}{a})$, and for $a_\# > 0$, we have $\sqrt{a+b} = \sqrt{a}\,(1 + \frac{b}{2a})$.

The usual order relation $\leq$ on $\mathbb{R}$ induces one on $\hat{\mathbb{R}}$ with $a \leq b$, for $a, b \in \hat{\mathbb{R}}$, if and only if $a_\# \leq b_\#$. Likewise if $a_\# \neq 0$, then the sign $sign(a)$ is defined to be the sign of $a_\#$ and the absolute value is $|a| = sign(a)a$.

Affine $n|m$ dimensional $\hat{\mathbb{R}}$ super space is defined to be

$$\mathbb{R}^{n|m} = \{(x_1, x_2, \ldots, x_n \mid \theta_1, \theta_2, \ldots, \theta_m) \in \hat{\mathbb{R}}^{n+m} : x_i \in \hat{\mathbb{R}}[0], \theta_j \in \hat{\mathbb{R}}[1]\}.$$

One defines n|m *super manifolds* with charts based on affine super space $\mathbb{R}^{n|m}$ in the usual way [1, 4, 16], and a *(Riemannian) super metric* on an $n|m$ super manifold is defined to be a positive definite boson-valued quadratic form on each tangent space $\mathbb{R}^{n|m}$ as usual.

The principal example for us here is *(real) super Minkowski* 2,1|2 *space*

$$\mathbb{R}^{2,1|2} = \{(x_1, x_2, y \mid \phi, \psi) \in \hat{\mathbb{R}}^5 : x_1, x_2, y \in \hat{\mathbb{R}}[0] \text{ and } \phi, \psi \in \hat{\mathbb{R}}[1]\},$$

which supports the boson-valued symmetric bilinear pairing

$$\langle (x_1, x_2, y \mid \phi, \psi), (x_1', x_2', y' \mid \phi', \psi') \rangle = \frac{1}{2}(x_1 x_2' + x_1' x_2) - yy' + \phi\psi' + \phi'\psi$$

with associated quadratic form $x_1 x_2 - y^2 + 2\phi\psi$.

The body of $\mathbb{R}^{2,1|2}$ with this inner product is evidently the classical Minkowski space $\mathbb{R}^{2,1}$ with its (negative) definite restriction to

$$\mathbb{H}' = \{\mathbf{x} = (x_1, x_2, y) \in \mathbb{R}^{2,1} : \langle \mathbf{x}, \mathbf{x} \rangle = 1 \text{ and } x_1 + x_2 > 0\}$$

providing a model of the hyperbolic plane, and we analogously define the *super hyperbolic plane* to be

$$\mathbb{H} = \{\mathbf{x} = (x_1, x_2, y \mid \phi, \psi) \in \mathbb{R}^{2,1|2} : \langle \mathbf{x}, \mathbf{x} \rangle = 1 \text{ and } x_1 + x_2 > 0\} \supseteq \mathbb{H}'$$

with its metric likewise induced from the inner product. Intermediate between the classical and super hyperbolic planes is the *bosonic hyperboloid*

$$\widehat{\mathbb{H}} = \{\mathbf{x} = (x_1, x_2, y|0, 0) \in \mathbb{R}^{2,1|2} : \langle \mathbf{x}, \mathbf{x} \rangle = 1 \text{ and } x_1 + x_2 > 0\},$$

whose coordinates lie in $\hat{\mathbb{R}}[0]$, in contrast to those of $\mathbb{H}'$ lying in $\mathbb{R}$. There are the natural inclusions $\mathbb{H}' \subset \widehat{\mathbb{H}} \subset \mathbb{H}$.

Continuing by analogy, let

$$\mathbb{H}' = \{\mathbf{h} \in \mathbb{R}^{2,1} : \langle \mathbf{h}, \mathbf{h} \rangle = -1\}$$

denote the usual hyperboloid of one sheet and

$$\mathbb{H} = \{\mathbf{h} \in \mathbb{R}^{2,1|2} : \langle \mathbf{h}, \mathbf{h} \rangle = -1\} \supseteq \mathbb{H}'$$

its super analogue. Let L denote the collection of isotropic vectors in $\mathbb{R}^{2,1}$ with

$$L^+ = \{\mathbf{u} = (u_1, u_2, v) \in L : u_1 + u_2 > 0\}$$

the *(open) positive light cone* (whose points are affine duals to horocycles in $\mathbb{H}'$, as in [13]), and let $\mathcal{L}$ denote the collection of isotropic vectors in $\mathbb{R}^{2,1|2}$ with

$$\mathcal{L}^+ = \{\mathbf{u} = (u_1, u_2, v \mid \phi, \psi) \in \mathcal{L} : u_1 + u_2 > 0\}.$$

6.3 Super Geodesics

Super geodesics in $\mathbb{H}$ are parametrized in analogy to the geodesics of general relativity:

Theorem 6.1 (Theorem 1.2 of [2]) *The general form of a super geodesic* $\mathbf{x}(s)$ *in* $\mathbb{H}$ *parametrized by arc length s is given by*

$$\mathbf{x} = \cosh s\, \mathbf{u} + \sinh s\, \mathbf{v},$$

for some $\mathbf{u} \in \mathbb{H}$, $\mathbf{v} \in \mathbb{H}$ *with* $\langle \mathbf{u}, \mathbf{v} \rangle = 0$. *The asymptotes of the corresponding super geodesic are given by the rays in* $\mathcal{L}^+$ *containing the vectors* $\mathbf{e} = \mathbf{u} + \mathbf{v}$, $\mathbf{f} = \mathbf{u} - \mathbf{v}$. *Conversely, points* $\mathbf{e}, \mathbf{f} \in \mathcal{L}^+$ *with* $\langle \mathbf{e}, \mathbf{f} \rangle = 2$ *determine a unique corresponding super geodesic, where* $\mathbf{u} = \frac{1}{2}(\mathbf{e} + \mathbf{f}) \in \mathbb{H}$ *and* $\mathbf{v} = \frac{1}{2}(\mathbf{e} - \mathbf{f}) \in \mathbb{H}$.

For any $\mathbf{u} \in \mathbb{H}$ and $\mathbf{v} \in \mathbb{H}$ with $\langle \mathbf{u}, \mathbf{v} \rangle = 0$, let

$$L_{\mathbf{u},\mathbf{v}} = \{\cosh s\, \mathbf{u} + \sinh s\, \mathbf{v} : s \in \mathbb{R}\}$$

denote the corresponding super geodesic.

Proof This follows directly from the variational principle applied to the functional

$$\int \Big(\sqrt{|\langle \dot{\mathbf{x}}, \dot{\mathbf{x}} \rangle|} + \lambda(\langle \mathbf{x}, \mathbf{x} \rangle - 1)\Big) dt,$$

where the dot stands for the derivative with respect to the parameter s along the curve, with corresponding Euler–Lagrange equations

$$\ddot{\mathbf{x}} = 2\lambda \mathbf{x}, \qquad \langle \mathbf{x}, \mathbf{x} \rangle = 1$$

with s chosen so that $|\langle \dot{\mathbf{x}}, \dot{\mathbf{x}} \rangle| = 1$. Differentiating two times the second equation and combining with the first equation we have

$$\lambda = -\frac{\langle \dot{\mathbf{x}}, \dot{\mathbf{x}} \rangle}{2}.$$

The solution with $\lambda = -1/2$ can be ruled out, and in the remaining case $\lambda = 1/2$, it is expressed as

$$\mathbf{x} = \cosh s\, \mathbf{u} + \sinh s\, \mathbf{v}.$$

The constraints $-\langle \dot{\mathbf{x}}, \dot{\mathbf{x}} \rangle = \langle \mathbf{x}, \mathbf{x} \rangle = 1$ imply the conditions on $\mathbf{u}, \mathbf{v}$.

Corollary 6.1 *For any two distinct points* $\mathbf{P}, \mathbf{Q} \in \mathbb{H}$, *there is a unique super geodesic containing them, and the distance D between them satisfies* $\cosh D = \langle \mathbf{P}, \mathbf{Q} \rangle$.

Proof For existence, note that $\langle \mathbf{P}, \mathbf{Q} \rangle > 1$ for distinct $\mathbf{P}, \mathbf{Q} \in \mathbb{H}$, and define

$$\ell = \sqrt{\frac{\langle \mathbf{P}, \mathbf{Q} \rangle + 1}{\langle \mathbf{P}, \mathbf{Q} \rangle - 1}},$$

so that $\langle \mathbf{P}, \mathbf{Q} \rangle = \frac{\ell^2+1}{\ell^2-1}$. The identity $\cosh^{-1} t = \log_e(t + \sqrt{t^2-1})$ therefore gives

$$e^{\cosh^{-1}\langle \mathbf{P},\mathbf{Q} \rangle} = \langle \mathbf{P}, \mathbf{Q} \rangle + \sqrt{\langle \mathbf{P}, \mathbf{Q} \rangle^2 - 1} = \frac{\ell + 1}{\ell - 1}.$$

We exhibit $L_{\mathbf{u},\mathbf{v}} = L_{\frac{\mathbf{e}+\mathbf{f}}{2}, \frac{\mathbf{e}-\mathbf{f}}{2}}$ containing $\mathbf{P}, \mathbf{Q}$ parametrized as

$$\mathbf{x}(s) = \cosh s\ \mathbf{u} + \sinh s\ \mathbf{v} = \frac{1}{2}[\exp(s)\ \mathbf{e} + \exp(-s)\ \mathbf{f}]$$

by taking

$$\mathbf{e} = \frac{\ell - 1}{2\ell}[(1-\ell)\mathbf{P} + (1+\ell)\mathbf{Q}],$$
$$\mathbf{f} = \frac{\ell + 1}{2\ell}[(1+\ell)\mathbf{P} + (1-\ell)\mathbf{Q}].$$

Direct computation confirms that $\mathbf{e}, \mathbf{f} \in \mathcal{L}^+$ with $\langle \mathbf{e}, \mathbf{f} \rangle = 2$ and

$$\mathbf{x}(0) = \mathbf{P} \text{ and } \mathbf{x}(\cosh^{-1}\langle \mathbf{P}, \mathbf{Q} \rangle) = \mathbf{Q},$$

thus establishing existence as well as that $\cosh D = \langle \mathbf{P}, \mathbf{Q} \rangle$.

For uniqueness, the system of equations

$$\mathbf{P} = \cosh s\ \mathbf{u} + \sinh s\ \mathbf{v},$$
$$\mathbf{Q} = \cosh t\ \mathbf{u} + \sinh t\ \mathbf{v}$$

is tantamount to the linear system

$$\begin{pmatrix} \cosh s\ I & \sinh s\ I \\ \cosh t\ I & \sinh t\ I \end{pmatrix} \begin{pmatrix} \mathbf{u} \\ \mathbf{v} \end{pmatrix} = \begin{pmatrix} \mathbf{P} \\ \mathbf{Q} \end{pmatrix},$$

where I is the 5-by-5 identity matrix, which has a unique solution since the determinant is non-zero for distinct s, t.

Corollary 6.2 *Given distinct* $\mathbf{P}, \mathbf{Q} \in \mathbb{H}$, *the unit tangent vector at* $\mathbf{P}$ *to the line from* $\mathbf{P}$ *to* $\mathbf{Q}$ *is given by* $\frac{\mathbf{Q} - \mathbf{P}\langle \mathbf{P},\mathbf{Q} \rangle}{\sqrt{\langle \mathbf{P},\mathbf{Q} \rangle^2 - 1}}$. □

It follows from Corollary 6.1 that any three non-collinear points of $\mathbb{H}$ define a *super triangle*, namely, three super geodesic segments with disjoint interiors meeting pairwise at the given points. The usual Hyperbolic Law of Cosines holds for super triangles, as proved in [14] and as follows directly from Corollary 6.2 upon computing the cosine of an interior angle based on the usual formula.

6.4 Orthosymplectic Group OSp(1|2)

We next include basic information concerning the orthosymplectic group OSp(1|2), which is among the simplest of Lie super groups and whose body is the classical special linear group SL(2, $\mathbb{R}$). We refer the interested reader to [3, 5] for more information about general Lie super algebras and super groups and to [15] for details about OSp(1|2).

An element $g \in \mathrm{OSp}(1|2)$ can be represented by the 3-by-3 matrix $g = \begin{pmatrix} a & b & \alpha \\ c & d & \beta \\ \gamma & \delta & f \end{pmatrix}$, where a, b, c, d, f are even and $\alpha, \beta, \gamma, \delta$ are odd, with multiplication defined by

$$\begin{pmatrix} a_1 & b_1 & \alpha_1 \\ c_1 & d_1 & \beta_1 \\ \gamma_1 & \delta_1 & f_1 \end{pmatrix}\begin{pmatrix} a_2 & b_2 & \alpha_2 \\ c_2 & d_2 & \beta_2 \\ \gamma_2 & \delta_2 & f_2 \end{pmatrix} = \begin{pmatrix} a_1a_2+b_1c_2-\alpha_1\gamma_2 & a_1b_2+b_1d_2-\alpha_1\delta_2 & a_1\alpha_2+b_1\beta_2+\alpha_1 f_2 \\ c_1a_2+d_1c_2-\beta_1\gamma_2 & c_1b_2+d_1d_2-\beta_1\delta_2 & c_1\alpha_2+d_1\beta_2+\beta_1 f_2 \\ \gamma_1a_2+\delta_1c_2+\delta_1\gamma_2 & \gamma_1b_2+\delta_1d_2+f_1\delta_2 & -\gamma_1\alpha_2-\delta_1\beta_2+f_1 f_2 \end{pmatrix},$$

where g is required to satisfy the two further conditions:

- the *super determinant* or *Berezinian* of g is unity, namely,

$$\mathrm{sdet}\, g = f^{-1} \det\left[\begin{pmatrix} a & b \\ c & d \end{pmatrix} + f^{-1}\begin{pmatrix} \alpha\gamma & \alpha\delta \\ \beta\gamma & \beta\delta \end{pmatrix}\right] = 1;$$

  and
- g is *orthosymplectic*, namely

$$g^{st} J g = J,$$

  where $J = \begin{pmatrix} 0 & -1 & 0 \\ 1 & 0 & 0 \\ 0 & 0 & 1 \end{pmatrix}$ and the *super transpose* $g^{st} = \begin{pmatrix} a & c & \gamma \\ b & d & \delta \\ -\alpha & -\beta & f \end{pmatrix}$.

The extra minus signs in the definition of matrix multiplication come from fermionic anti-commutation and our specification of the ordering of rows and columns. The Berezinian is the analogue of the classical determinant [3, 5, 16] and is characterized by being a multiplicative homomorphism satisfying sdet exp $g = \exp(a + d - f)$, but unlike the classical determinant, it is only defined for invertible super matrices. The functional equation $g^{st} J g = J$ explains the analogy with the classical symplectic group and provides the simple expression

$g^{-1} = J^{-1}g^{st}J = \begin{pmatrix} d & -b & \delta \\ -c & a & -\gamma \\ -\beta & \alpha & f \end{pmatrix}$ for inversion in $\mathrm{OSp}(1|2)$. More explicitly, the two conditions together are equivalent to the constraints

$$\alpha = b\gamma - a\delta, \quad \beta = d\gamma - c\delta, \quad f = 1 + \alpha\beta,$$
$$\gamma = a\beta - c\alpha, \quad \delta = b\beta - d\alpha, \quad f^{-1} = ad - bc.$$

Notice that there is a bosonic Lie super group

$$\widehat{\mathrm{SL}}_2 = \mathrm{SL}(2, \hat{\mathbb{R}}[0]) = \{\begin{pmatrix} a & b & 0 \\ c & d & 0 \\ 0 & 0 & 1 \end{pmatrix} : a, b, c, d \in \hat{\mathbb{R}}[0] \text{ and } ad - bc = 1\}$$

together with canonical inclusions $\mathrm{SL}(2, \mathbb{R}) < \widehat{\mathrm{SL}}_2 < \mathrm{OSp}(1|2)$. Another canonical subspace $\{u(\alpha, \beta) : \alpha, \beta \in \hat{\mathbb{R}}[1]\} \subset \mathrm{OSp}(1|2)$, which is not a subgroup, is parametrized by a pair of fermions α, β and defined by

$$u(\alpha, \beta) = \begin{pmatrix} 1 - \frac{\alpha\beta}{2} & 0 & \alpha \\ 0 & 1 - \frac{\alpha\beta}{2} & \beta \\ \beta & -\alpha & 1 + \alpha\beta \end{pmatrix} \in \mathrm{OSp}(1|2).$$

It is not difficult to prove the following useful

Lemma 6.1 *Any element $g \in \mathrm{OSp}(1|2)$ can be written uniquely as a product*

$$g = \begin{pmatrix} a & b & 0 \\ c & d & 0 \\ 0 & 0 & 1 \end{pmatrix} u(\alpha, \beta) = u(a\alpha + b\beta, c\alpha + d\beta) \begin{pmatrix} a & b & 0 \\ c & d & 0 \\ 0 & 0 & 1 \end{pmatrix},$$

in $\mathrm{OSp}(1|2)$, for some $\begin{pmatrix} a & b & 0 \\ c & d & 0 \\ 0 & 0 & 1 \end{pmatrix} \in \widehat{\mathrm{SL}}_2$ and fermions $\alpha, \beta \in \hat{\mathbb{R}}[1]$. □

Minkowski three-space $\mathbb{R}^{2,1} \approx \mathbb{R}^3$, as the space of binary symmetric bilinear forms, is naturally coordinatized by

$$A = \begin{pmatrix} z - x & y \\ y & z + x \end{pmatrix} = \begin{pmatrix} x_1 & y \\ y & x_2 \end{pmatrix} \in \mathbb{R}^{2|1},$$

and g in the component $\mathrm{SO}_+(1, 2)$ of the identity in $\mathrm{SO}(1, 2)$ acts via orientation-preserving isometry on $A \in \mathbb{R}^{2,1}$ as change of basis via the adjoint

$$g : A \mapsto g.A = g^t A g.$$

This action of $\mathrm{SL}(2, \mathbb{R})$ or $\mathrm{SO}_+(1, 2) \approx \mathrm{PSL}(2, \mathbb{R})$ as the group of isometries of $\mathbb{H}'$ links hyperbolic geometry and elementary number theory.

Likewise, $\mathbb{R}^{2,1|2} \approx \mathbb{R}^{3|2}$ is naturally coordinatized by

$$A = \begin{pmatrix} x_1 & y & \phi \\ y & x_2 & \psi \\ -\phi & -\psi & 0 \end{pmatrix} \in \mathbb{R}^{2,1|2},$$

and $g \in \mathrm{OSp}(1|2)$ acts on A as change of basis again via the adjoint

$$g : A \mapsto g.A = g^{st} A g.$$

One checks using Lemma 6.1 that this action again preserves the inner product on $\mathbb{R}^{2,1|2}$ and hence restricts to an isometric action on $\mathbb{H}$ itself.

The body of this action of $\mathrm{OSp}(1|2)$ on $\mathbb{H}$ is the classical action of orientation-preserving isometries on $\mathbb{H}'$, and this extension is our analogous action of this Lie super group on the super hyperbolic plane. Moreover, the induced action of $\widehat{\mathrm{SL}}_2 < \mathrm{OSp}(1|2)$ preserves the bosonic hyperboloid $\widehat{\mathbb{H}}$, and we therefore find the equivariant tower of isometric actions

$$\begin{array}{ccc} \mathrm{OSp}(1|2) & \circlearrowright & \mathbb{H} \\ \vee & & \cup \\ \widehat{\mathrm{SL}}_2 & \circlearrowright & \widehat{\mathbb{H}} \\ \vee & & \cup \\ \mathrm{SL}(2,\mathbb{R}) & \circlearrowright & \mathbb{H}'. \end{array}$$

6.5 OSp(1|2)-invariant Area Form on $\mathbb{H}$

The classical hyperboloid $\mathbb{H}'$ supports the $\mathrm{SL}(2,\mathbb{R})$-invariant hyperbolic area form given in coordinates by $\Omega'(x_1, x_2, y) = \frac{dx_2 \wedge dx_1}{2y}$, as we shall recall. We promote this to an $\widehat{\mathrm{SL}}_2$-invariant form on $\widehat{\mathbb{H}}$ and then suitably pull-back to derive an $\mathrm{OSp}(1|2)$-invariant two-form on $\mathbb{H}$ itself.

Lemma 6.2 *$\widehat{\mathbb{H}}$ supports the $\widehat{\mathrm{SL}}_2$-invariant two-form $\hat{\Omega} = \frac{\mathrm{d}x_2 \wedge \mathrm{d}x_1}{2y}$.*

Proof We have $\begin{pmatrix} a & b & 0 \\ c & d & 0 \\ 0 & 0 & 1 \end{pmatrix}.(x_1, x_2, y|0,0) = (x_1', x_2', y'|0,0)$, where

$$x_1' = a^2 x_1 + 2acy + c^2 x_2,$$
$$x_2' = b^2 x_1 + 2bdy + d^2 x_2,$$
$$y' = abx_1 + (ad+bc)y + cdx_2.$$

One computes directly that

$$\begin{aligned}\hat{\Omega}' &= \frac{dx_2' \wedge dx_1'}{2y'} \\ &= \frac{(ad-bc)[(ad+bc)(dx_2 \wedge dx_1) + 2dy \wedge (ab\,dx_1 - cd\,dx_2)]}{2[abx_1 + (ad+bc)y + cdx_2]}.\end{aligned}$$

Meanwhile, $x_1x_2 - y^2 = 1$ implies that $dy = \frac{x_1dx_2+x_2dx_1}{2y}$. Substituting this into the numerator finally yields $\hat{\Omega} = \hat{\Omega}'$.

Lemma 6.3 *For any* $\mathbf{x} = (x_1, x_2, y|\phi, \psi) \in \mathbb{H}$, *consider the element* $u^{\mathbf{x}} = u(y\psi - x_2\phi, y\phi - x_1\psi) \in \mathrm{OSp}(1|2)$. *This transformation acts on* $\mathbf{x}$ *to give* $u^{\mathbf{x}}.\mathbf{x} = (1 - \phi\psi)(x_1, x_2, y|0, 0) \in \widehat{\mathbb{H}}$, *and this determines the canonical mapping*

$$\begin{aligned}\mathbb{H} &\to \widehat{\mathbb{H}} \\ (x_1, x_2, y|\phi, \psi) &\mapsto (\hat{x}_1, \hat{x}_2, \hat{y}|0, 0) = (1-\phi\psi)(x_1, x_2, y|0, 0).\end{aligned}$$

□

Theorem 6.2 *The hyperbolic area form on* $\mathbb{H}'$ *pulls-back under the natural mapping* $\mathbb{H} \to \mathbb{H}'$ *to the* $\mathrm{OSp}(1|2)$*-invariant two-form on* $\mathbb{H}$ *given by*

$$\Omega(x_1, x_2, y|\phi, \psi) = \frac{d[(1+\phi\psi)x_2)] \wedge d[(1+\phi\psi)x_1]}{2[(1+\phi\psi)y)]}.$$

Proof This follows from the previous lemma and the factorization in Lemma 6.1.

Lemma 6.4 $\Omega = d\omega$ *on* $\mathbb{H}$ *for the primitive*

$$\omega(x_1, x_2, y|\phi, \psi) = \frac{y(1+\phi\psi)}{2}\, d\log_e\left(\frac{x_1}{x_2}\right).$$

Proof Let $\hat{y} = y(1+\phi\psi)$ and likewise for $\hat{x}_1, \hat{x}_2$. Notice that

$$\hat{x}_1\hat{x}_2 - \hat{y}^2 = (1+2\phi\psi)(x_1x_2 - y^2) = (1+2\phi\psi)(1-2\phi\psi) = 1,$$

so that

$$2\hat{y}\, d\hat{y} = \hat{x}_1 d\hat{x}_2 + x_2 d\hat{x}_1.$$

Thus,

$$\begin{aligned}d\omega &= d\left[\frac{\hat{y}}{2}\,(\frac{d\hat{x}_1}{\hat{x}_1} - \frac{d\hat{x}_2}{\hat{x}_2})\right] \\ &= \frac{\hat{x}_1 d\hat{x}_2 + \hat{x}_2 d\hat{x}_1}{4\hat{y}} \wedge \frac{1}{\hat{x}_1\hat{x}_2}[\hat{x}_2 d\hat{x}_1 - \hat{x}_1 d\hat{x}_2]\end{aligned}$$

$$= \frac{d\hat{x}_2 \wedge d\hat{x}_1}{2\hat{y}}$$

$$= \Omega,$$

and finally $\omega = \frac{\hat{y}}{2}\,\mathrm{dlog}_e(\frac{\hat{x}_1}{\hat{x}_2}) = \frac{\hat{y}}{2}\,\mathrm{dlog}_e(\frac{x_1}{x_2})$.

6.6 Line Integrals

Let $\mathbf{P} = (p_1, p_2, p|\alpha, \beta)$ and $\mathbf{Q} = (q_1, q_2, q|\gamma, \delta)$ be distinct points in $\mathbb{H}$, and define

$$d = \langle \mathbf{P}, \mathbf{Q} \rangle > 1 \text{ and } \ell = \sqrt{\frac{d+1}{d-1}}.$$

It follows that $d = \frac{\ell^2+1}{\ell^2-1}$ and the distance between $\mathbf{P}$ and $\mathbf{Q}$ is given by $D = \cosh^{-1} d$, with $\exp(D) = d + \sqrt{d^2-1} = \frac{\ell+1}{\ell-1}$.

By the proof of Corollary 6.1, the geodesic segment $\overline{\mathbf{PQ}}$ from $\mathbf{P}$ to $\mathbf{Q}$ is parametrized for $t = \exp(s)$ by

$$\mathbf{x}(t) = \frac{1}{2}(t\mathbf{e} + t^{-1}\mathbf{f}), \text{ for } 1 \le t \le \exp(D) = \frac{\ell+1}{\ell-1},$$

where

$$\mathbf{e} = M\,[(1-\ell)\mathbf{P} + (1+\ell)\mathbf{Q}],$$
$$\mathbf{f} = N\,[(1+\ell)\mathbf{P} + (1-\ell)\mathbf{Q}],$$

with

$$M = \frac{\ell - 1}{2\ell} \text{ and } N = \frac{\ell+1}{2\ell},$$

so that $MN = \frac{1}{2(d+1)}$. According to Lemma 6.4, the area primitive on $\overline{\mathbf{PQ}}$ is given in coordinates as

$$\omega(t) = \omega \circ \mathbf{x}(t) = [1 + \phi(t)\psi(t)]\,\bar{\omega}(t),$$

where

$$\bar{\omega}(t) = \frac{1}{4}[tA + t^{-1}B]\,d\log_e \frac{tA_1 + t^{-1}B_1}{tA_2 + t^{-1}B_2},$$

with

$$A = M[(1-\ell)p + (1+\ell)q] \text{ and } B = N[(1+\ell)p + (1-\ell)q],$$
$$A_i = M[(1-\ell)p_i + (1+\ell)q_i] \text{ and } B_i = N[(1+\ell)p_i + (1-\ell)q_i],$$

for $i = 1, 2$, and with

$$2\phi(t) = tM[(1-\ell)\alpha + (1+\ell)\gamma] + t^{-1}N[(1+\ell)\alpha + (1-\ell)\gamma],$$
$$2\psi(t) = tM[(1-\ell)\beta + (1+\ell)\delta] + t^{-1}N[(1+\ell)\beta + (1-\ell)\delta].$$

It is useful here and hereafter to introduce the notation

$$S_i = \frac{(q_i\alpha - p_i\gamma)(q_i\beta - p_i\delta)}{(qp_i - pq_i)^2}, \text{ for } i = 1, 2.$$

Lemma 6.5 *For $i = 1, 2$, we have the identities*

$$A_i B_i = \frac{(qp_i - pq_i)^2}{d^2 - 1}(1 - 2S_i),$$
$$AB_i - BA_i = 2\,\frac{qp_i - pq_i}{\sqrt{d^2 - 1}}.$$

In particular, it follows that A_i and B_i have the same sign provided $(qp_i - pq_i)_\# \neq 0$. In this case, if $p_i, q_i > 0$, then also $A_i, B_i > 0$, and $p_i \leq q_i$ if and only if $\exp(D)\frac{A_i}{B_i} \leq 1$.

Proof For the first identity, compute

$$\begin{aligned}
A_i B_i &= MN[((1-\ell)p_i + (1+\ell)q_i][(1+\ell)p_i + (1-\ell)q_i] \\
&= \frac{1}{2(d+1)}\Big\{(1-\ell^2)(p_i^2 + q_i^2) + 2(1+\ell^2)p_i q_i\Big\} \\
&= \frac{1}{2(d+1)}\Big\{\Big(1 - \frac{d+1}{d-1}\Big)(p_i^2 + q_i^2) + 2\Big(1 + \frac{d+1}{d-1}\Big)p_i q_i\Big\} \\
&= \frac{1}{d^2-1}\Big\{-p_i^2 - q_i^2 + p_i q_i[p_1 q_2 + p_2 q_1 - 2pq + 2(\alpha\delta + \gamma\beta)]\Big\} \\
&= \frac{1}{d^2-1}\Big\{-p_i^2 - q_i^2 + p_i^2 q_1 q_2 + q_i^2 p_1 p_2 - 2p_i q_i[pq - (\alpha\delta + \gamma\beta)]\Big\} \\
&= \frac{1}{d^2-1}\Big\{p_i^2(q^2 - 2\gamma\delta) + q_i^2(p^2 - 2\alpha\beta) - 2p_i q_i[pq - (\alpha\delta + \gamma\beta)]\Big\} \\
&= \frac{1}{d^2-1}\Big\{(qp_i - pq_i)^2 - 2(q_i\alpha - p_i\gamma)(q_i\beta - p_i\delta)\Big\},
\end{aligned}$$

as was asserted. The second identity follows directly from the definitions and elementary algebra.

For the final assertions, let us first note that

$$\exp(D)\frac{A_i}{B_i} = \frac{(1-\ell)p_i + (1+\ell)q_i}{(1+\ell)p_i + (1-\ell)q_i}$$

since $\exp(D)\frac{M}{N} = \frac{\ell+1}{\ell-1}\frac{M}{N} = 1$. Now, $0 < q_i \leq p_i$ implies that $p_i + q_i > \ell(q_i - p_i)$, so $B_i > 0$, whence $A_i > 0$. Furthermore, $\exp(D)\frac{A_i}{B_i} \leq 1$ is equivalent to $(1-\ell)p_i + (1+\ell)q_i \leq (1+\ell)p_i + (1-\ell)q_i$, which is in turn equivalent to $q_i \leq p_i$. In the contrary case, $0 < p_i \leq q_i$ implies that $p_i + q_i > \ell(p_i - q_i)$, so $A_i > 0$, whence $B_i > 0$. Again, $\exp(-D)\frac{B_i}{A_i} \leq 1$ is equivalent to $(1+\ell)p_i + (1-\ell)q_i \leq (1-\ell)p_i + (1+\ell)q_i$, which is in turn equivalent to $p_i \leq q_i$.

Our aim is to explicitly compute $\int_{\mathbf{PQ}} \omega$ in this section in order to apply Stokes' Theorem in the next section, and we first assert that

$$\bar{\omega}(t) = \frac{dt}{2} \sum_{i=1}^{2} (-1)^i \frac{(A + t^{-2}B)B_i}{t^2 A_i + B_i}. \tag{6.1}$$

To see this, compute directly that

$$d\log_e \frac{tA_1 + t^{-1}B_1}{tA_2 + t^{-1}B_2} = \frac{2t^{-1}(B_2A_1 - B_1A_2)}{(tA_1 + t^{-1}B_1)(tA_2 + t^{-1}B_2)}$$

and by partial fractions that

$$\frac{1}{(tA_1 + t^{-1}B_1)(tA_2 + t^{-1}B_2)} = \frac{1}{B_2A_1 - B_1A_2}\left\{\frac{B_2}{t^2A_2 + B_2} - \frac{B_1}{t^2A_1 + B_1}\right\}.$$

Equation 6.1 follows immediately.

Next we compute directly from the definitions that

$$\begin{aligned} 4\phi(t)\psi(t) &= (\alpha\delta + \gamma\beta)[(t^2M^2 + t^{-2}N^2)(1-\ell^2) + 2MN(1+\ell^2)] \\ &\quad + \alpha\beta\,[tM(1-\ell) + t^{-1}N(1+\ell)]^2 \\ &\quad + \gamma\delta\,[tM(1+\ell) + t^{-1}N(1-\ell)]^2 \end{aligned}$$

Collecting like powers of t, we find

$$\phi(t)\psi(t) = Xt^2 + Yt^{-2} + Z, \tag{6.2}$$

where

$$4X = M^2[(1-\ell)^2\alpha\beta + (1+\ell)^2\gamma\delta + (1-\ell^2)(\alpha\delta+\gamma\beta)]$$
$$4Y = N^2[(1+\ell)^2\alpha\beta + (1-\ell)^2\gamma\delta + (1-\ell^2)(\alpha\delta+\gamma\beta)]$$
$$4Z = 2MN[(1-\ell^2)(\alpha\beta+\gamma\delta) + (1+\ell^2)(\alpha\delta+\gamma\beta)]\,.$$

Lemma 6.6 *Suppose that* $0 < p_1 < q_1$ *and* $0 < q_2 < p_2$. *Then we have*

$$\int_{\mathbf{PQ}} \omega = (A_1B_2 - B_1A_2)\left[\frac{AX}{(\ell-1)A_1A_2} - \frac{BY}{(\ell+1)B_1B_2}\right]$$

$$+\sum_{i=1,2}(-1)^i(1-S_i)\left[1+Z-X\frac{B_i}{A_i}-Y\frac{A_i}{B_i}\right]\arctan\left[\frac{qp_i-pq_i}{p_i+q_i}(1-S_i)\right]$$
$$+\pi\, sign(qp_2-pq_2)(1-S_2)\left[1+Z-X\frac{B_2}{A_2}-Y\frac{A_2}{B_2}\right].$$

Proof Setting

$$W(t) = \frac{B_2}{t^2A_2+B_2} - \frac{B_1}{t^2A_1+B_1}$$

and combining Eqs. 6.1 and 6.2, we find that

$$2\omega = \left\{t^2AX + t^0[BX + A(1+Z)] + t^{-2}[AY + B(1+Z)] + t^{-4}BY\right\}W(t)\mathrm{d}t.$$

Together perturbing **P** and **Q** by OSp(1|2) if necessary to arrange that $(pq_i - qp_i)_\# \neq 0$, we have $A_i, B_i > 0$ by Lemma 6.5 and our hypotheses that $p_i, q_i > 0$, for $i = 1, 2$. We may then apply the following indefinite integrals

$$\int\frac{t^2dt}{t^2u+v} = -\sqrt{\frac{v}{u^3}}\arctan\left(\sqrt{\frac{u}{v}}\,t\right) + \frac{t}{u},$$
$$\int\frac{dt}{t^2u+v} = \frac{1}{\sqrt{uv}}\arctan\left(\sqrt{\frac{u}{v}}\,t\right),$$
$$\int\frac{t^{-2}dt}{t^2u+v} = -\sqrt{\frac{u}{v^3}}\arctan\left(\sqrt{\frac{u}{v}}\,t\right) - \frac{1}{vt},$$
$$\int\frac{t^{-4}dt}{t^2u+v} = \sqrt{\frac{u^3}{v^5}}\arctan\left(\sqrt{\frac{u}{v}}\,t\right) - \frac{v-3ut^2}{3v^2t^3}$$

to conclude that

$$
\begin{aligned}
2\int_{\overline{\mathbf{PQ}}}\omega &= AXt\left[\frac{B_2}{A_2}-\frac{B_1}{A_1}\right]+BYt^{-1}\left[\frac{A_1}{B_1}-\frac{A_2}{B_2}\right]\\
&\quad+\frac{AB_2-BA_2}{\sqrt{A_2B_2}}\left[1+Z-X\,\frac{B_2}{A_2}-Y\,\frac{A_2}{B_2}\right]\arctan\left(\sqrt{\frac{A_2}{B_2}}\,t\right)\\
&\quad-\frac{AB_1-BA_1}{\sqrt{A_1B_1}}\left[1+Z-X\,\frac{B_1}{A_1}-Y\,\frac{A_1}{B_1}\right]\arctan\left(\sqrt{\frac{A_1}{B_1}}\,t\right)
\end{aligned}
$$

evaluated between the limits $t=1$ and $t=\exp(D)=\frac{\ell+1}{\ell-1}$.

Now, we compute

$$
\begin{aligned}
\arctan\left(\sqrt{\frac{A_i}{B_i}}\,t\right)\Bigg|_{t=1}^{t=\exp(D)} &= \arctan\left(\sqrt{\frac{A_i}{B_i}}\,\exp(D)\right)-\arctan\left(\sqrt{\frac{A_i}{B_i}}\right)\\
&= \arctan\left\{\frac{\sqrt{\frac{A_i}{B_i}}(\exp(D)-1)}{1+\frac{A_i}{B_i}\exp(D)}\right\}+k_i\,\pi,
\end{aligned}
$$

using standard formulas for the difference of arctangents, our hypotheses on p_i, q_i, and the last part of Lemma 6.5, where $k_i=\begin{cases}0, & \text{if } i=1;\\ 1, & \text{if } i=2.\end{cases}$

Meanwhile, we have

$$
\frac{\frac{\sqrt{A_iB_i}\,[\exp(D)-1]}{A_i\exp(D)+B_i}\exp(-\frac{D}{2})}{\exp(-\frac{D}{2})}=\frac{\sqrt{A_iB_i}\,\sinh(\frac{D}{2})}{A_i\exp(\frac{D}{2})+B_i\exp(-\frac{D}{2})},
$$

and moreover,

$$
\begin{aligned}
A_i\exp(\frac{D}{2})+B_i\exp(-\frac{D}{2}) &= \sqrt{\frac{\ell+1}{\ell-1}}\left(\frac{\ell-1}{2\ell}\right)\Big[(1-\ell)p_i+(1+\ell)q_i\Big]\\
&\quad+\sqrt{\frac{\ell-1}{\ell+1}}\left(\frac{\ell+1}{2\ell}\right)\Big[(1+\ell)p_i+(1-\ell)q_i\Big]\\
&= \frac{\sqrt{\ell^2-1}}{\ell}\,(p_i+q_i)
\end{aligned}
$$

since $\exp(D)=\frac{\ell+1}{\ell-1}$ implies that $\cosh(\frac{D}{2})=\frac{\ell}{\sqrt{\ell^2-1}}$ and $\sinh(\frac{D}{2})=\frac{1}{\sqrt{\ell^2-1}}$.

Finally using the first identity in Lemma 6.5, we have

$$
\begin{aligned}
-k_i\,\pi + \arctan\left(\sqrt{\frac{A_i}{B_i}}\,t\right)\Bigg|_{t=1}^{t=\exp(D)} &= \arctan\left(\frac{2\sqrt{A_i B_i}\,\sinh(\frac{d}{2})\cosh(\frac{d}{2})}{p_i+q_i}\right) \\
&= \arctan\left(\frac{\sinh(D)}{\sqrt{d^2-1}}\,\frac{|qp_i - pq_i|}{p_i+q_i}\,(1-S_i)\right) \\
&= \arctan\left(\frac{|qp_i - pq_i|}{p_i+q_i}\,(1-S_i)\right)
\end{aligned}
$$

since $d = \cosh(D)$.

It follows from Lemma 6.5 that

$$
\frac{AB_i - BA_i}{\sqrt{A_i B_i}} = 2\,\frac{qp_i - pq_i}{|qp_i - pq_i|}\,(1-S_i) = 2\,sign(qp_i - pq_i)\,(1-S_i),
$$

and so

$$
\begin{aligned}
&\frac{AB_i - BA_i}{\sqrt{A_i B_i}}\left[\arctan\left(\sqrt{\frac{A_i}{B_i}}\,t\right)\Bigg|_{t=1}^{t=\exp(D)}\right] \\
&= 2\,(1-S_i)\arctan\left[\frac{qp_i - pq_i}{p_i+q_i}(1-S_i)\right] \\
&+ 2\pi\,k_i\,sign(qp_i - pq_i)\,(1-S_i)
\end{aligned}
$$

since arctangent is an odd function.

6.7 Area of a Super Triangle

Let Δ denote a super triangle with vertices

$$
\begin{aligned}
\mathbf{P} &= (p_1, p_2, p|\alpha, \beta), \\
\mathbf{Q} &= (q_1, q_2, q|\gamma, \delta), \\
\mathbf{R} &= (r_1, r_2, r|\varepsilon, \phi).
\end{aligned}
$$

Lemma 6.7 *Without loss of generality by applying an element of* $\mathrm{OSp}(1|2)$ *and possibly relabeling the vertices, we may assume that we have* $\alpha = \gamma = \varepsilon$, $0 < p_1 < q_1 < r_1$ *and* $0 < r_2 < q_2 < p_2$.

Proof First recall that the classical isometry from $\mathbb{H}'$ to the upper half plane is given by $\Pi : (x_1, x_2, y) \mapsto \frac{i-y}{x_2}$, where $i = \sqrt{-1}$. (This extends to an analogous mapping from $\mathbb{H}$ to the super upper half plane with complex coordinates z, η by

$$(x_1, x_2, y|\phi, \psi) \mapsto z = \frac{i - y - i\phi\psi}{x_2}, \quad \eta = \frac{\psi}{x_2}(1 + iy),$$

but we shall not need this here. Each of these mappings is appropriately equivariant.)

It is thus geometrically clear that by translation in $\widehat{\mathrm{SL}}_2$, we may arrange that $\Pi(\mathbf{P}), \Pi(\mathbf{Q}), \Pi(\mathbf{R})$ each have positive imaginary part and negligible squared real part divided by imaginary part. This implies that $(p_1p_2), (q_1q_2), (r_1r_2)$ are all nearly equal to unity. Now perturb in $\widehat{\mathrm{SL}}_2$ so that these imaginary parts are all distinct and relabel if necessary so that $0 < r_2 < q_2 < p_2$, from which it then follows that also $0 < p_1 < q_1 < r_1$.

Next further perturb the resulting Δ with an element of $\widehat{\mathrm{SL}}_2$ so that

$$t = (qp_1 - pq_1) + (rq_1 - qr_1) + (pr_1 - rp_1)$$

has non-zero body. Given fermions ξ, η, let us suppose that $u(\xi, \eta).\mathbf{P} = (p_1', p_2', p'|\alpha', \beta')$, and likewise respectively γ' and ε' for $\mathbf{Q}$ and $\mathbf{R}$. One computes that

$$\alpha' = p_1\xi + p\eta + \alpha(1 + \tfrac{3}{2}\xi\eta),$$
$$\gamma' = q_1\xi + q\eta + \gamma(1 + \tfrac{3}{2}\xi\eta),$$
$$\varepsilon' = r_1\xi + r\eta + \varepsilon(1 + \tfrac{3}{2}\xi\eta).$$

We therefore seek fermions ξ, η so that in particular

$$0 = \alpha' - \gamma' = (p_1 - q_1)\xi\left[1 + \frac{3}{2}\frac{(\gamma - \alpha)\eta}{p_1 - q_1}\right] + (p - q)\eta + (\alpha - \gamma).$$

Multiplying through by $[1 - \frac{3}{2}\frac{(\gamma-\alpha)\eta}{p_1-q_1}]$ yields

$$(p_1 - q_1)\xi + (p - q)\eta\left[1 - \frac{3}{2}\frac{(\gamma - \alpha)\eta}{p_1 - q_1}\right] = (\gamma - \alpha)\left[1 - \frac{3}{2}\frac{(\gamma - \alpha)\eta}{p_1 - q_1}\right],$$

or equivalently $(p_1 - q_1)\xi + (p - q)\eta = \gamma - \alpha$. We likewise find in particular for $\alpha' = \varepsilon'$ that $(p_1 - r_1)\xi + (p - r)\eta = \varepsilon - \alpha$.

Thus, we find the necessary and sufficient linear system

$$\begin{pmatrix} p_1 - q_1 & p - q \\ p_1 - r_1 & p - r \end{pmatrix}\begin{pmatrix} \xi \\ \eta \end{pmatrix} = \begin{pmatrix} \gamma - \alpha \\ \varepsilon - \alpha \end{pmatrix}$$

with solution

$$\xi = \frac{1}{t}\left[(p-r)(\gamma-\alpha)+(q-p)(\varepsilon-\alpha)\right],$$
$$\eta = \frac{1}{t}\left[(r_1-p_1)(\gamma-\alpha)+(p_1-q_1)(\varepsilon-\alpha)\right],$$

where the determinant

$$\begin{aligned} t &= (p-r)(p_1-q_1)-(p-q)(p_1-r_1) \\ &= (qp_1-pq_1)+(rq_1-qr_1)+(pr_1-rp_1) \end{aligned}$$

has non-zero body by construction.

For the rest of this chapter, we shall assume that the triangle Δ with vertices **P**, **Q**, **R** has been normalized and possibly relabeled to arrange that $\alpha=\gamma=\varepsilon$, $0<p_1<q_1<r_1$ and $0<r_2<q_2<p_2$. In fact, no subsequent fermionic product of degree two will include $\beta\delta$, $\delta\phi$, or $\phi\beta$. Thus $\alpha=\gamma=\varepsilon$ divides all subsequent fermionic expressions, whence the product of any two of them must vanish. We shall refer to this fact as being *due to fermionic degree.*

Attempting to mitigate the inevitable proliferation of notation, set

$$\bar{V} = (A_1B_2-B_1A_2)\left[\frac{AX}{(\ell-1)A_1A_2}-\frac{BY}{(\ell+1)B_1B_2}\right],$$
$$H_i = (Z-X\frac{B_i}{A_i}-Y\frac{A_i}{B_i}),\ \text{for } i=1,2,$$

in the notation of the previous section, so that

$$\begin{aligned} \int_{\overline{\mathbf{PQ}}}\omega &= \bar{V}+\pi\, sign(qp_2-pq_2)\,(1+H_2-S_2) \\ &+\sum_{i=1,2}(-1)^i(1+H_i-S_i)\arctan\left[\frac{qp_i-pq_i}{p_i+q_i}(1-S_i)\right] \end{aligned}$$

since $H_iS_i=0$ due to fermionic degree. This equation, together with its analogues for $\overline{\mathbf{QR}}$ and $\overline{\mathbf{RP}}$, provides a dreadful but explicit closed-form expression $\int_\Delta\Omega = \int_{\overline{\mathbf{PQ}}}\omega+\int_{\overline{\mathbf{QR}}}\omega+\int_{\overline{\mathbf{RP}}}\omega$ for the area of Δ by Stokes' theorem.

Rather than explicate this expression and with our eye on the Angle Defect Formula, let us instead proceed to compute $\cos(\int_{\overline{\mathbf{PQ}}}\omega)$ and $\sin(\int_{\overline{\mathbf{PQ}}}\omega)$. These are tractable since $\cos(a)=1$ and $\sin(a)=a$ for any super number a with $a^2=0$, and since a product of any two factors $\bar{V}$, H_1, H_2, S_1, S_2 must vanish due to fermionic degree.

Theorem 6.3 *We have the identities*

$$\cos \int_{\mathbf{PQ}} \omega = \frac{C+\hat{c}}{F}, \quad \text{and} \quad \sin \int_{\mathbf{PQ}} \omega = \frac{S+\hat{s}}{F},$$

where

$$C = -2[(p+q)^2 + (d+1)(1-pq)] \text{ and } S = (p+q)(p_1q_2 - p_2q_1),$$
$$F = 2(d+1)\sqrt{p_1p_2q_1q_2}, \quad \hat{c} = SV - \bar{c} \text{ and } \hat{s} = -CV - \bar{s},$$

with

$$\begin{aligned}\bar{c} &= 2[(q\alpha - p\gamma)(q\beta - p\delta) + (\alpha+\gamma)(\beta+\delta)] \\ &\quad + (S_1+S_2)(qp_1 - pq_1)(qp_2 - pq_2), \\ \bar{s} &= S_1(p_2+q_2)(qp_1 - pq_1) - S_2(p_1+q_1)(qp_2 - pq_2),\end{aligned}$$

and

$$\begin{aligned}V &= \bar{V} + \pi\, sign(qp_2 - pq_2)(H_2 - S_2) \\ &\quad + \sum_{i=1,2} (-1)^i\,(H_i - S_i)\arctan\left[\frac{qp_i - pq_i}{p_i+q_i}(1-S_i)\right].\end{aligned}$$

Proof We have derived $\int_{\mathbf{PQ}} \omega = (U_2 - U_1) + V + \pi\, sign(qp_2 - pq_2)$ in Lemma 6.6, where

$$U_i = \arctan\left[\frac{qp_i - pq_i}{p_i+q_i}(1-S_i)\right], \text{ for } i = 1, 2.$$

Standard trigonometric identities give

$$\begin{aligned}-\cos \int_{\mathbf{PQ}} \omega &= \cos U_2 \cos U_1 + \sin U_2 \sin U_1 \\ &\quad - V\,[\sin U_2 \cos U_1 - \cos U_2 \sin U_1], \\ -\sin \int_{\mathbf{PQ}} \omega &= [\sin U_2 \cos U_1 - \cos U_2 \sin U_1 \\ &\quad + V\,[\cos U_2 \cos U_1 + \sin U_2 \sin U_1],\end{aligned}$$

where the leading minus signs derive from $\pi\, sign(qp_2 - pq_2)$ and where $V^2 = 0$ follows from considerations of fermionic degree, the latter of which implies that $\cos V = 1$ and $\sin V = V$.

The usual identities cos arctan x $= \frac{1}{\sqrt{1+x^2}}$, sin arctan x $= \frac{x}{\sqrt{1+x^2}}$ imply that

$$\cos U_i = \frac{p_i + q_i}{\sqrt{2\, p_i q_i\, (d+1)}},$$
$$\sin U_i = \frac{(qp_i - pq_i)(1 - S_i)}{\sqrt{2\, p_i q_i\, (d+1)}},$$

for $i = 1, 2$, using the equation

$$(p_i + q_i)^2 + (qp_i - pq_i)^2 = 2[(d+1)p_i q_i + (qp_i - qp_i)^2\, S_i],$$

which relies upon $\langle \mathbf{P}, \mathbf{Q} \rangle = d$ and $\langle \mathbf{P}, \mathbf{P} \rangle = 1 = \langle \mathbf{Q}, \mathbf{Q} \rangle$. A further similar calculation gives

$$\begin{aligned} &(p_1 + q_1)(p_2 + q_2) + (qp_1 - pq_1)(qp_2 - pq_2) \\ &= 2[(p+q)^2 + (d+1)(1-pq) - (q\alpha - p\gamma)(q\beta - p\delta) - (\alpha + \gamma)(\beta + \delta)], \end{aligned}$$

and a more elementary direct calculation gives

$$(p+q)(q_1 p_2 - q_2 p_1) = (p_1 + q_1)(qp_2 - pq_2) - (p_2 + q_2)(qp_1 - pq_1).$$

It remains only to collect the fermionic terms.

Adopting for $\overline{\mathbf{QR}}$ and $\overline{\mathbf{RP}}$ the analogous respective alphabetic notation introduced for $\overline{\mathbf{PQ}}$, Corollary 6.3 provides the expressions

$$\cos \int_{\overline{\mathbf{QR}}} \omega = \frac{D + \hat{d}}{G} \quad \text{and} \quad \sin \int_{\overline{\mathbf{QR}}} \omega = \frac{T + \hat{t}}{G},$$

$$\cos \int_{\overline{\mathbf{RP}}} \omega = \frac{E + \hat{e}}{H} \quad \text{and} \quad \sin \int_{\overline{\mathbf{RP}}} \omega = \frac{U + \hat{u}}{H},$$

where for example

$$\begin{aligned} &D = -2[(q+r)^2 + (e+1)(1-qr)] \ \text{ and } \ T = (q+r)(q_1 r_2 - q_2 r_1), \\ &E = -2[(r+p)^2 + (f+1)(1-rp)] \ \text{ and } \ U = (r+p)(r_1 p_2 - r_2 p_1), \\ &G = 2(e+1)\sqrt{q_1 q_2 r_1 r_2} \ \text{ and } \ H = 2(f+1)\sqrt{p_1 p_2 r_1 r_2}, \end{aligned}$$

with $e = \langle \mathbf{Q}, \mathbf{R} \rangle$, $f = \langle \mathbf{R}, \mathbf{P} \rangle$.

Corollary 6.3 *We have the identity*

$$\cos \int_{\triangle} \Omega = P + R,$$

where

$$\begin{aligned}(FGH)\, P &= \hat{c}(DE - TU) - \hat{s}(DU + ET)\\ &\quad + \hat{d}(CE - SU) - \hat{t}(CU + ES)\\ &\quad + \hat{e}(CD - ST) - \hat{u}(CT + DS),\\ (FGH)\, R &= CDE - CTU - DSU - EST,\end{aligned}$$

with $FGH = 8(d+1)(e+1)(f+1)\, p_1 p_2 q_1 q_2 r_1 r_2$.

Proof This follows directly from Corollary 6.3 and its analogues for $\overline{\mathbf{QR}}$ and $\overline{\mathbf{RP}}$, using standard trigonometric formulas.

6.8 Angle Defect

Lemma 6.8 *The cosine of the sum of the interior angles of* $\triangle$ *is given by*

$$L = 1 - \frac{1 + 2def - d^2 - e^2 - f^2}{(d+1)(e+1)(f+1)},$$

where we have $d = \langle \mathbf{P}, \mathbf{Q} \rangle$, $e = \langle \mathbf{Q}, \mathbf{R} \rangle$, *and* $f = \langle \mathbf{R}, \mathbf{P} \rangle$ *as before.*

Proof Let x, y, z be the respective cosines of the interior angles of $\triangle$ at $\mathbf{P}$, $\mathbf{Q}$, $\mathbf{R}$, so for example

$$x = -\langle \frac{\mathbf{Q} - \mathbf{P}d}{\sqrt{d^2 - 1}}, \frac{\mathbf{R} - \mathbf{P}f}{\sqrt{f^2 - 1}} \rangle = \frac{df - e}{\sqrt{d^2 - 1}\sqrt{f^2 - 1}}$$

by Corollary 6.2. The usual formula for the sum of arccosines gives

$$\begin{aligned}&\arccos x + \arccos y + \arccos z\\ &\quad = \arccos\Big[xyz - z\sqrt{(1 - x^2)(1 - y^2)}\\ &\qquad - y\sqrt{(1 - x^2)(1 - z^2)} - x\sqrt{(1 - y^2)(1 - z^2)}\Big],\end{aligned}$$

with for example

$$x\sqrt{(1-y^2)(1-z^2)} = (df-e)\,\frac{(1+2def-d^2-e^2-f^2)}{(d^2-1)(e^2-1)(f^2-1)}$$

by direct calculation, noting that $1+2def \geq d^2+e^2+f^2$ by the Hyperbolic Law of Cosines for $\triangle$.

Thus, we find

$$\begin{aligned}(d^2-1)(e^2-1)(f^2-1)L &= (ef-d)(df-e)(de-f)\\ &\quad + (1+2def-d^2-e^2-f^2)(d+e+f-ef-df-ed).\end{aligned}$$

Dividing this equation by the common factor $(d-1)(e-1)(f-1)$ of its two sides finally yields the asserted identity.

In the notation of the previous section, define

$$R = \frac{CDE-CTU-DSU-EST}{FGH}. \tag{6.3}$$

Let us also define $I = p_1q_2+p_2q_1$, $J = q_1r_2+q_2r_1$, and $K = r_1p_2+r_2p_1$, with $i = \alpha\delta+\gamma\beta = d-\frac{1}{2}I+pq$, $j = \gamma\varepsilon+\delta\phi = e-\frac{1}{2}J+qr$ and $k = \varepsilon\beta+\alpha\phi = f-\frac{1}{2}K+rp$, and with $\Theta = \frac{\alpha\beta}{p^2+1}+\frac{\gamma\delta}{q^2+1}+\frac{\varepsilon\phi}{r^2+1}$.

Lemma 6.9 *We have $L+R=Q$ with*

$$\begin{aligned}
&8(d+1)(e+1)(f+1)Q\\
&=-4\left[\frac{\varepsilon\phi I^2}{r^2+1}+\frac{\alpha\beta J^2}{p^2+1}+\frac{\gamma\delta K^2}{q^2+1}\right]\\
&\quad+8[Ii+Jj+Kk]-2[iJK+jIK+kIJ]\\
&\quad+2\Theta[I^2(r^2-1)+J^2(p^2-1)+K^2(q^2-1)]\\
&\quad-4\Theta[IJ(1+rp)+JK(1+pq)+IK(1+qr)]\\
&\quad-8\Theta[I(r^2+1)(1-pq)+J(p^2+1)(1-qr)+K(q^2+1)(1-rp)]\\
&\quad-2\left[\frac{KJi(pq+1)}{(p^2+1)(q^2+1)}+\frac{KIj(qr+1)}{(q^2+1)(r^2+1)}+\frac{IJk(rp+1)}{(p^2+1)(r^2+1)}\right]\\
&\quad+4(p+q)(q+r)(r+p)\\
&\left[\frac{Ji+iJ}{(q^2+1)(r+p)}+\frac{Jk+Kj}{(r^2+1)(p+q)}+\frac{Ki+iK}{(p^2+1)(q+r)}\right]
\end{aligned}$$

$$+4\frac{(p+q)(q+r)(r+p)}{(p^2+1)(q^2+1)(r^2+1)}$$

$$\left[Ii\frac{(1-pq)(r^2+1)}{p+q}+Jj\frac{(1-qr)(p^2+1)}{q+r}+Kk\frac{(1-rp)(q^2+1)}{r+p}\right]$$

$$+8(p+q)(q+r)(r+p)$$

$$\left[\frac{I\left(\frac{\alpha\beta}{p^2+1}+\frac{\gamma\delta}{q^2+1}\right)-i}{p+q}+\frac{J\left(\frac{\gamma\delta}{q^2+1}+\frac{\varepsilon\phi}{r^2+1}\right)-j}{q+r}+\frac{K\left(\frac{\alpha\beta}{p^2+1}+\frac{\varepsilon\phi}{r^2+1}\right)-k}{r+p}\right].$$

One critical conclusion, which is tantamount to the classical Angle Defect Theorem, is that the body $Q_\# = 0$ vanishes.

Proof First note for example that

$$IJ = q_1q_2K + p_1q_2^2r_1 + p_2q_1^2r_2,$$

so

$$\begin{aligned}(p_1q_2 - p_2q_1)(q_1r_2 - q_2r_1) &= q_1q_2K - p_1q_2^2 - p_2q_1^2r_2\\ &= 2q_1q_2K - IJ,\end{aligned}$$

and likewise

$$\begin{aligned}(q_1r_2 - q_2r_1)(r_1p_2 - r_2p_1) &= 2r_1r_2I - JK,\\ (p_1q_2 - p_2q_1)(r_1p_2 - r_2p_1) &= 2p_1p_2J - IK.\end{aligned}$$

It follows that

$$\begin{aligned}IJK &= q_1q_2K^2 + (p_1q_2^2r_1 + p_2q_1^2r_2)K\\ &= q_1q_2K^2 + p_1p_2(q_1^2r_2^2 + q_2^2r_1^2) + r_1r_2(p_1^2q_2^2 + p_2^2q_1^2)\\ &= q_1q_2K^2 + p_1p_2[(q_1r_2 + q_2r_1)^2 - 2q_1q_2r_1r_2]\\ &+ q_1q_2[(p_1q_2 + p_2q_1)^2 - 2p_1p_2r_1r_2]\\ &= q_1q_2K^2 + p_1p_2J^2 + r_1r_2I^2 - 4p_1p_2q_1q_2r_1r_2,\end{aligned}$$

and moreover that

$$\begin{aligned}(p_1q_2 - q_1p_2)(q_1r_2 - q_2r_1)K &= q_1q_2K - p_1p_2J - r_1r_2I + 4p_1p_2q_1q_2r_2r_2,\\ (p_1q_2 - q_1p_2)(r_1p_2 - r_2p_1)J &= p_1p_2J - q_1q_2K - r_1r_2I + 4p_1p_2q_1q_2r_2r_2,\\ (q_1r_2 - q_2r_1)(r_1p_2 - r_2p_1)I &= r_1r_2I - p_1p_2J - q_1q_2K + 4p_1p_2q_1q_2r_2r_2.\end{aligned}$$

Turning first to L and setting $L' = 8(d+1)(e+1)(f+1)L$, we find from Lemma 6.8 that

$$L' = -def + d + e + f + de + ef + df + d^2 + e^2 + f^2.$$

A moderate computation from the definitions using the expression above for IJK and collecting like terms, we find that

$$\begin{aligned}
L' &= 4p_1p_2q_1q_2r_1r_2\\
&+ 8[p^2q^2r^2 + p^2q^2 + q^2r^2 + r^2p^2 + pqr(p+q+r) - pq - qr - rp]\\
&+ I^2[2 - r_1r_2] + 4I[1 - pqr^2 - r(p+q) - 2pq]\\
&+ J^2[2 - p_1p_2] + 4J[1 - p^2qr - p(q+r) - 2qr]\\
&+ K^2[2 - q_1q_2] + 4K[1 - pq^2r - q(r+p) - 2rp]\\
&+ 2[IJ(1+rp) + JK(1+pq) + IK(1+qr)]\\
&- 2iJK + 4i[J + K + Jrp + Kqr] + 8i[1 - pqr^2 + I - 2pq - r(p+q)]\\
&- 2jIK + 4j[I + K + Irp + Kpq] + 8j[1 - p^2qr + J - 2qr - p(r+q)]\\
&- 2kIJ + 4k[I + J + Iqr + Jpq] + 8k[1 - pq^2r + K - 2rp - q(r+p)].
\end{aligned}$$

Turning now to the calculation of R and using the previous formulas, we find that

$$\begin{aligned}
-CTU &= 2(q+r)(r+p)(p^2+1)(q^2+1)[2r_1r_2I - JK]\\
&+ 2i(q+r)(r+p)(1-pq)[2r_1r_2I - JK]\\
&+ (q+r)(r+p)(1-pq)[r_1r_2I^2 - q_1q_2K^2 - p_1p_2J^2 + 4p_1p_2q_1q_2r_1r_2],
\end{aligned}$$

$$\begin{aligned}
-DSU &= 2(p+q)(r+p)(q^2+1)(r^2+1)[2p_1p_2J - IK]\\
&+ 2j(p+q)(r+p)(1-qr)[2p_1p_2J - IK]\\
&+ (p+q)(r+p)(1-qr)[p_1p_2J^2 - r_1r_2I^2 - q_1q_2K^2 + 4p_1p_2q_1q_2r_1r_2],
\end{aligned}$$

$$\begin{aligned}
-EST &= 2(p+q)(q+r)(r^2+1)(P^2+1)[2q_1q_2K - IJ]\\
&+ 2k(p+q)(q+r)(1-rp)[2q_1q_2K - IJ]\\
&+ (p+q)(q+r)(1-rp)[q_1q_2K^2 - p_1p_2J^2 - r_1r_2I^2 + 4p_1p_2q_1q_2r_1r_2],
\end{aligned}$$

$$\begin{aligned}
-CDE &= 8(p^2+1)^2(q^2+1)^2(r^2+1)^2\\
&+ (1-pq)(1-qr)(1-rp)[IJK + 2IJk + 2JKi + 2IKj]
\end{aligned}$$

$$+2(p^2+1)(q^2+1)(1-qr)(1-rp)[JK+2kJ+2jK]$$
$$+2(q^2+1)(r^2+1)(1-pq)(1-rp)[IK+2iK+2kI]$$
$$+2(p^2+1)(r^2+1)(1-pq)(1-qr)[IJ+2iJ+2jI]$$
$$+4(p^2+1)^2(q^2+1)(r^2+1)(1-qr)[J+2j]$$
$$+4(p^2+1)(q^2+1)^2(r^2+1)(1-rp)[K+2k]$$
$$+4(p^2+1)(q^2+1)(r^2+1)^2(1-pq)[I+2i].$$

It is now purely a matter of using the earlier expression for IJK and collecting like terms to derive the asserted expression for Q. The vanishing of the body of Q is equivalent to the classical Angle Defect Theorem and gives an internal consistency check.

Combining the previous Theorem with Corollary 6.3, we finally have:

Theorem 6.4 (The Angle Defect Theorem for Super Triangles) *The angle defect $\mathcal{D}$ minus the area $\mathcal{A}$ of the super triangle $\triangle$ is given by the fermionic correction*

$$\mathcal{D}-\mathcal{A}=\frac{P+Q}{\sqrt{1-R^2}},$$

where P and Q are respectively given explicitly in Corollary 6.3 and Lemma 6.9, and R is expressed in Eq. 6.3. This fermionic correction is not identically zero.

Proof We have derived in the referenced results that

$$\cos\mathcal{A}=R+P \ \text{ and } \ \cos\mathcal{S}=Q-R,$$

where $\mathcal{S}$ is the sum of the interior angles of $\triangle$. The standard formulas $\arccos x = \frac{\pi}{2}-\sum_{n=0}^{\infty}\frac{(2n)!}{2^{2n}(n!)^2}\frac{x^{2n+1}}{2n+1}$ and $\frac{\mathrm{d}}{\mathrm{d}x}\arccos x=-\frac{1}{\sqrt{1-x^2}}$ therefore give

$$\mathcal{A}=\arccos R-\frac{P}{\sqrt{1-R^2}}, \ \text{ and } \ \mathcal{S}=\pi-\arccos R-\frac{Q}{\sqrt{1-R^2}}$$

since $P^2=Q^2=0$. The asserted formula therefore follows since by definition $\mathcal{D}-\mathcal{A}=(\pi-\mathcal{S})-\mathcal{A}$.

It is a triviality that our fermionic correction is not identically zero on super triangles. Indeed, suppose $\mathcal{D}-\mathcal{A}$ does vanish for some super triangle $\triangle$, so in particular $\mathcal{A}_\#=\mathcal{A}$ by the classical Angle Defect Theorem. Modify $\triangle$ by translating its two fermionic coordinates by respective odd constants ϕ and ψ with $\phi\psi\neq 0$.

The resulting defect $\mathcal{D}' = \mathcal{D}$ is unchanged, and yet the resulting area $\mathcal{A}' = \mathcal{A}(1 + \phi\psi) = \mathcal{A}_{\#}(1 + \phi\psi)$ according to the formula for the area primitive. Thus, our fermionic correction $\mathcal{D}' - \mathcal{A}' = \mathcal{A}_{\#}\,\phi\psi$ cannot then also vanish for the modified triangle.

Appendix: Super Ideal Triangles

In this appendix, we show that the fermionic part of the area of a super ideal triangle diverges. To this end given $\mathbf{e}', \mathbf{f}', \mathbf{g}' \in \mathcal{L}^+$, define

$$\mathbf{e} = \sqrt{\frac{2\langle \mathbf{f}', \mathbf{g}'\rangle}{\langle \mathbf{e}', \mathbf{f}'\rangle\,\langle \mathbf{g}', \mathbf{e}'\rangle}}\,\mathbf{e}', \quad \mathbf{f} = \sqrt{\frac{2\langle \mathbf{g}', \mathbf{e}'\rangle}{\langle \mathbf{e}', \mathbf{f}'\rangle\,\langle \mathbf{f}', \mathbf{g}'\rangle}}\,\mathbf{f}', \quad \mathbf{g} = \sqrt{\frac{2\langle \mathbf{e}', \mathbf{f}'\rangle}{\langle \mathbf{f}', \mathbf{f}'\rangle\,\langle \mathbf{g}', \mathbf{e}'\rangle}}\,\mathbf{g}',$$

so that $\langle \mathbf{e}, \mathbf{f}\rangle = \langle \mathbf{f}, \mathbf{g}\rangle = \langle \mathbf{g}, \mathbf{e}\rangle = 2$, and introduce coordinates

$$\begin{aligned}
\mathbf{e} &= (p_1, p_2, p|\alpha, \beta),\\
\mathbf{f} &= (q_1, q_2, q|\gamma, \delta),\\
\mathbf{g} &= (r_1, r_2, r|\varepsilon, \phi).
\end{aligned}$$

These define respective super geodesics

$$\begin{aligned}
E &= \{\frac{1}{2}(t\mathbf{f} + t^{-1}\mathbf{g}) \in \mathbb{H} : 0 < t < \infty\},\\
F &= \{\frac{1}{2}(t\mathbf{g} + t^{-1}\mathbf{e}) \in \mathbb{H} : 0 < t < \infty\},\\
G &= \{\frac{1}{2}(t\mathbf{e} + t^{-1}\mathbf{f}) \in \mathbb{H} : 0 < t < \infty\},
\end{aligned}$$

as in Theorem 6.1, bounding a super ideal triangle.

The OSp(1|2)-invariant two-form Ω on $\mathbb{H}$ admits the same primitive ω as in Sect. 6.5, and we begin by computing $\int_G \omega$. The calculations in Sect. 6.6 apply directly to conclude that

$$2\omega = \left\{t^2 pX + t^0[qX + p(1+Z)] + t^{-2}[pY + q(1+Z)] + t^{-4}qY\right\} U(t)\,\mathrm{d}t,$$

where $4X = \alpha\beta$, $4Y = \gamma\delta$, $4Z = \alpha\delta + \gamma\beta$, and $U(t) = \frac{q_2}{t^2 p_2 + q_2} - \frac{q_1}{t^2 p_1 + q_1}$, and moreover that

$$\begin{aligned}2\int_G \omega &= pXt\left[\frac{q_2}{p_2}-\frac{q_1}{p_1}\right]+qYt^{-1}\left[\frac{p_1}{q_1}-\frac{p_2}{q_2}\right]\\&+\frac{pq_2-qp_2}{\sqrt{p_2q_2}}\left[1+Z-X\,\frac{q_2}{p_2}-Y\,\frac{p_2}{q_2}\right]\arctan\left(\sqrt{\frac{p_2}{q_2}}\,t\right)\\&-\frac{pq_1-qp_1}{\sqrt{p_1q_1}}\left[1+Z-X\,\frac{q_1}{p_1}-Y\,\frac{p_1}{q_1}\right]\arctan\left(\sqrt{\frac{p_1}{q_1}}\,t\right),\end{aligned}$$

but in this instance evaluated between the limits $t=0$ and $t=\infty$.

The latter two summands are bounded, and the former two summands give

$$pXt\left[\frac{q_2}{p_2}-\frac{q_1}{p_1}\right]+qYt^{-1}\left[\frac{p_1}{q_1}-\frac{p_2}{q_2}\right]=(p_1q_2-p_2q_1)\left[\frac{Xt}{p}+\frac{Y}{qt}\right].$$

This expression diverges for the specified limits provided only that we have $p_1q_2-p_2q_1\neq 0\neq qXt^2+pY$, so in particular $\alpha\beta\neq 0\neq\gamma\delta$.

There can be no cancellation from $\int_E \omega$ and $\int_F \omega$, since these involve different fermionic products, and it follows that the fermionic part of the area of an ideal triangle indeed diverges. One can readily trace through the remainder of the calculations in the main text to confirm that the body of this area equals π.

References

1. B. DeWitt, *Supermanifolds*, 2nd edn. Cambridge Monographs on Mathematical Physics (Cambridge University Press, Cambridge, 1992)
2. Y. Huang, R. Penner, A. Zeitlin, Super McShane identity J. Differ. Geom. 509–551 (1223)
3. V. Kac, Lie superagebras. Adv. Math. **26**, 8–96 (1977)
4. E. Kessler, *Supergeometry, Super Riemann Surfaces and the Superconformal Action Functional*. Lecture Notes in Mathematics, vol. 2230 (Springer-Nature, Berlin, 2019)
5. Yu.I. Manin, *Topics in Noncommutative Geometry* (Princeton University Press, Princeton, 1991)
6. A. Papadopoulos, R. Penner, A characterization of pseudo-Anosov foliations. Pac. J. Math. **130**, 359–377 (1987)
7. A. Papadopoulos, R. Penner, Enumerating pseudo-Anosov conjugacy classes. Pac. J. Math. **142**, 159–173 (1990)
8. A. Papadopoulos, R. Penner, La forme symplectique de Weil-Petersson et le bord de Thurston de l'espace de Teichmüller. Comptes Rendus Acad. Sci. Paris **312** Série I, 871–874 (1991)
9. A. Papadopoulos, R. Penner, The Weil-Petersson symplectic structure at Thurston's boundary. Trans. Am. Math. Soc. **335**, 891–904 (1993)
10. A. Papadopoulos, R. Penner, Broken hyperbolic structures. Ann. Glob. Anal. Geom. **27**, 53–77 (2005)
11. A. Papadopoulos, R. Penner, Hyperbolic metrics, measured foliations and pants decompositions for non-orientable surfaces. Asian J. Math. **20**, 157–182 (2016)
12. A. Papadopoulos, G. Théret, *La théorie des parallèles de Johann Heinrich Lambert*. Critical edition with French translation and mathematical and historical commentaries, ed. Blanchard, coll. Sciences dans l'Histoire, Paris, 214 pp. (2014)

13. R. Penner, *Decorated Teichmüller Theory*. QGM Masters Class Series, volume 1 (European Math Society, Berlin, 2012)
14. R. Penner, Super hyperbolic law of cosines: same formula with different content, in *Essays in Geometry. Dedicated to Norbert A'Campo*, vol. 34, ed. by A. Papadopoulos (European Mathematical Society, Berlin, 2023), pp. 73–96
15. R. Penner, A. Zeitlin, Decorated super-Teichmüller space. J. Differ. Geom. **111**, 527–566 (2019)
16. A. Rogers, *Supermanifolds: Theory and Applications* (World Scientific, Singapore, 2007)
17. P. Stäckel, F. Engel, Die Theorie der Parallellinien von Euklid bis auf Gauss, eine Urkundensammlung zur Vorgeschichte der nicht-euklidischen Geometrie (B. G. Teubner, Leipzig, 1895)

Chapter 7
Lipschitz and Quasiconformal Mappings in Cartography

Hideki Miyachi and Ken'ichi Ohshika

Dedicated to our best friend Athanase Papadopoulos on the occasion of his sixty-fifth birthday

Abstract We study the best Lipschitz constant of a geographical map known as the Delisle map, a projection of a region of the sphere onto a cone. This map was used by Euler for drawing the map of the Russian empire. Euler considered that this map is the most appropriate, given the extent of the region to draw. We provide numerical evidence to show that the Delisle map is indeed the best among several maps from the sphere to a cone, for the region considered by Euler. The maps we consider are natural maps from the sphere to a cone which are also used in geography. We also discuss the quasiconformal dilatations of these maps.

7.1 Introduction

We study several maps from a spherical surface to a Euclidean surface, from the point of view of their Lipschitz and conformal distortions. These maps are projections from the sphere to a cone, which are used in geography. The study by geometers of the distortion of geographical maps is an old problem. Ptolemy, in his *Geography* [11], was a prominent example of a mathematician-geographer who was involved in the problem of finding geographical maps of least distortion. Closer to us, Euler was thoroughly involved in geography, and he wrote several memoirs on this subject, see [3–5]. In his memoir [5], Euler described a projection from a

H. Miyachi
School of Mathematics and Physics, College of Science and Engineering, Kanazawa University, Kakuma-machi, Kanazawa, Ishikawa, Japan
e-mail: miyachi@se.kanazawa-u.ac.jp

K. Ohshika (✉)
Department of Mathematics, Gakushuin University, Mejiro, Toshima-ku, Tokyo, Japan
e-mail: ohshika@math.gakushuin.ac.jp

A. M. Uludağ, A. Zeytin (eds.), *Essays on Geometry*,
https://doi.org/10.1007/978-3-031-76257-4_7

region of the sphere, representing the Earth, contained between two parallels onto a cone intersecting the sphere along these two parallels, and he analysed the metrical distortion of this geographical map. Several other prominent mathematicians studied geographical maps and their distortions; we mention Lagrange, Gauss, Chebyshev, Bonnet, Darboux, and there are many others. See the various chapters of the book [2] for overviews of some aspects of ancient and more recent mathematical geography. Among the recent mathematicians who studied problems arising in geography, we mention Milnor who, in his 1969 paper titled *A problem in cartography* [8] (see also his comment on the paper contained in [9]), formulated a general problem on geographical maps of least distortion and studied more specifically the distortion of a map called the *azimuthal equidistant projection*.

In this chapter, we study geographical maps which are closely related to the map studied by Euler in [5]. To be more concrete, we shall consider several naturally defined projections from the sphere into a cone and we shall show that indeed, the map studied by Euler is, in a precise way, the best map among them. The practical advantage of considering geographical maps whose range lies in a cone rather than the Euclidean plane is that several projections can be naturally defined from (a subset of) the sphere into a cone, and then, one can easily unroll this cone and obtain the image of the map on a plane, that is, what is usually called a geographical map.

We shall also compute the quasiconformal dilatation of the best map from a region of the sphere bounded by two parallels (which is the domain of all the maps we consider) into a cone. Let us recall here that the study of quasiconformal maps originates in geography. Teichmüller, in §33 of his paper *Extremale quasikonforme Abbildungen und quadratische Differentiale* [14], makes a few historical comments on the history of quasiconformal mappings and he writes there that the origin of this theory lies in geography, more precisely, in the art of drawing geographical maps. He mentions in this respect the name of the French mathematician and geographer Nicolas-Auguste Tissot (1824–1897). Indeed, the latter worked extensively on the quasiconformal dilatation of geographical maps (although he did not use the word "quasiconformal"). Assuming that the map is differentiable, Tissot introduced at each point of its image an ellipse (which was later called by geographers the "Tissot indicatrix"), which is the image of a circle of unit radius in the tangent space at the source of the map. This ellipse is a visible representation of the quasiconformal dilatation of the map at the given point, see [15]. In the last section of the present chapter, we shall draw the Tissot indicatrices of the geographical maps which we study in the previous sections and we shall see that the map used by Euler is the closest to conformal among them, except for the best quasi-conformal map, in the setting of his regions (the Tissot indicatrices are closest to circles of equal radius).

This chapter grew out of a discussion with Athanase Papadopoulos in Centro di Ricerca Matematica Ennio de Giorgi in Pisa. We would like to express our heartfelt gratitude to Athanase who motivated this work and taught us Euler's work on cartography.

7.2 Formulation of the Problem

In this section, we set some terminology, and we formulate the problems and the results.

Let S be the unit sphere in the 3-dimensional Euclidean space $\mathbb{E}^3$. We regard S as a model of the Earth. We take L_1, L_2 to be two parallels on S with latitudes θ_1 and θ_2, not necessarily contained in the same hemisphere. We denote by A the region bounded by L_1 and L_2 and we denote by $\mathcal{N}$ and $\mathcal{S}$ the centers of these parallels. We regard these points as the North and the South Poles of the Earth. In this way, L_1 and L_2 become parallels (in the usual geographical sense). We shall call the line joining $\mathcal{N}$ and $\mathcal{S}$ the axis of the Earth. The great circle centered at $\mathcal{N}$ (or $\mathcal{S}$) is the equator. Let C be a cone in $\mathbb{E}^3$ intersecting S along $L_1 \cup L_2$. We denote the annular region in C bounded by $L_1 \cup L_2$ by B, and the apex angle of C by θ. Note that we do not exclude the case where L_1 and L_2 are symmetric with respect to the equator of the sphere. In this case, the cone C is a cylinder, and its apex is at infinity.

We consider a class of embeddings $f : D \to C$ from a domain D on S containing A and satisfying the following properties.

1. $f(A) = B$
2. f is the identity on $L_1 \cup L_2$.
3. f sends parallels contained in A to circles parallel to L_1 and L_2 on C.
4. f sends any intersection with A of a meridian of the sphere to a straight line on C orthogonal to the images of the parallels.

We shall study the following five natural maps:

(a) The projection of S to C from the center of S. This is called, by geographers, the *Murdoch second projection*, see [6, p. 184]. It was introduced in geography by the geographer Patrick Murdoch in 1758, see [10]. The domain of definition D of this map is the complement of the South Pole if $|\theta_2| > |\theta_1|$, the complement of the North Pole if $|\theta_2| < |\theta_1|$, and the complement of both poles if $|\theta_1| = |\theta_2|$.

(b) The *horizontal projection* of S to C: Any point in A is sent to the intersection with C of the horizontal ray joining a point on the axis of the Earth to the point A. The domain of definition D of this map is the complement $S \setminus \{\mathcal{N} \cup \mathcal{S}\}$ of the North and the South Poles of the sphere.

(c) The *conical equidistant map from S to C*. This also carries the name *the Delisle projection* and it was studied by Euler in his memoir [5] titled *On Delisle's geographical projection used for a general map of the Russian empire.*[1] It is also sometimes called the *third Murdoch projection*, see [6, p. 181] and [10]. This map, besides satisfying the properties from (1) to (4) above, has the property that on each meridian, distances are preserved up to a constant. The domain of definition D of this map is also the complement $S \setminus \{\mathcal{N} \cup \mathcal{S}\}$ of the North and the South Poles of S.

[1] Euler's memoir is written in Latin, and an English translation is provided in [5].

(d) The projection from the sphere to the cone C orthogonally to C. We call this map the *orthogonal projection*. In this case, D is again the complement $S \setminus \{\mathcal{N} \cup \mathcal{S}\}$ of the North and the South Poles.
(e) A best quasiconformal map from A to B, that is, a homeomorphism minimizing the global quasiconformal constant. This map is defined only on A, and hence $D = A$.

Warning In geography, the terminology is not as universally established as it is in mathematics. Thus, although the names we have given to the above maps are used by geographers (in particular in the references we give in the bibliography), the same maps have different names in some geography treatises.

Euler, in his memoir [5], considers only the map (c), and our initial aim was to study this map. However, since each of the five maps has its own merit, we shall discuss the four of them. Comparing the properties of all these maps in terms of Lipschitz and quasiconformal dilatations is worth trying, but our aim was most of all to check whether Euler was right in preferring the Delisle map for the cartography of the Russian empire, and indeed we verify this numerically in the sections below.

We shall use modern terminology[2] including a definition of the bi-Lipschitz constant of a map given by Milnor in his paper on geography [8], and we shall also use the notion of quasiconformal mapping, which Euler did not have.

We note that by general theorems on quasiconformal mappings dating back to Teichmüller, the quasiconformal dilatation of the best quasiconformal map referred to in (e) is constant on A. Therefore, this map can be regarded as an analogue of the Mercator map (which is angle-preserving) in the setting of projections to cones. The Mercator map, since it is conformal, is also a best quasiconformal map from the sphere to a cylinder.

Let us now be more precise on the notions of dilatation and distortion we use.

We fix the annulus A and we let f be one of the above five maps. Following Euler in [5], we try to compute the "worst error" of f on A and also a wider annulus A', both contained in D. We define such a notion more precisely, using modern terminology. In fact, there are several possible notions which can be used for the measure of the deviation of a map from being distance-preserving (up to a constant), locally and globally, and we shall recall some of them.

For $D \supset A$, let $f : D \to C$ be a C^1-embedding. Let $A' \subset D$ be another annulus bounded by parallels whose middle parallel coincides with that of A. For a point $x \in D$, we define two constants, M and m, where

$$M(x) := \sup_{v \in T_x S \setminus \{0\}} |df(v)|/|v|$$

and

[2] The terminology is "modern", compared to the one used in Euler's times.

$$m(x) := \inf_{v \in T_x S \setminus \{0\}} |df(v)|/|v|.$$

The (quasiconformal) dilatation of f at x is defined to be

$$K_f(x) := \frac{M(x)}{m(x)}.$$

In the theory of quasiconformal C^1 mappings, the central notion is the maximal dilatation of f on A', defined as

$$K_f(A') := \sup_{x \in A'} K(x).$$

Milnor, in his paper [8], gives a definition of the *maximum scale*, *minimum scale* and *distortion* of a geographic map. We can recover the *maximum scale*, in the sense of Milnor as

$$\sigma_2(f, A') = \sup_{x \in A'} M(x),$$

the *minimum scale* as

$$\sigma_1(f, A') = \inf_{x \in A'} m(x),$$

and the distortion as

$$\delta_{A'}(f) = \log\left(\frac{\sigma_2(f, A')}{\sigma_1(f, A')}\right).$$

We also define the *infinitesimal bi-Lipschitz constant* $\sigma(x)$ of the map f to be $\sigma(x) := \max\{M(x), m(x)^{-1}\}$. We introduce the notation

$$L_{A'}(f) := \sup_{x \in A'} \sigma(x)$$

and

$$\ell_{A'}(f) := \inf_{x \in A'} \sigma(x).$$

We call

$$v_{A'}(f) = \log\left(\frac{L_{A'}(f)}{\ell_{A'}(f)}\right)$$

the *variation of the infinitesimal bi-Lipschitz constant*, which we also call the *metrical distortion*.

Euler's original task was to draw a geographical map[3] of the Russian empire, and he used the Delisle projection for this map, in the *Russian Atlas* he edited, published in Saint Petersburg in 1745 [1].[4] It was natural for him to compute the metrical distortion of this map under the condition that A' is the smallest band containing this empire. Our aim in this chapter is to give numerical values for the metrical distortion of the Delisle projection of the Russian empire and to compare it with that of other natural projections.

In the rest of this chapter, we address the problem of computing $M(x)$, $m(x)$ and $\sigma(x)$ for $x \in D$ for each of the five maps in the case when $A' = A$. In Sect. 7.3, we obtain formulae for the values $L_A(f), \ell_A(f), v_A(f), \sigma_1(f, A), \sigma_2(f, A), \delta_A(f)$ associated with the first map.

In Sect. 7.4, we give numerical computations in the case when the annuli A and A' are the ones that Euler used, i.e., when A is the band bounded by two parallels corresponding to $47°30'$ and $62°30'$ and A' is the band bounded by two parallels corresponding to $40°$ and $70°$.

We also give numerical values for the numbers $\delta_{A'}$ and $v_{A'}$ for the four maps (for the fifth cannot be defined in A'), and show that f_3, that is, Delisle's map, gives the smallest values for $\delta_{A'}$ and $v_{A'}$ among the four maps.

Section 7.5 contains figures that indicating the Tissot ellipses of the various maps we defined, from which we can see that the Delisle map is also a very reasonable choice for the region considered.

7.3 The Formulae for the Five Maps

We first set some more notation, and we introduce coordinates for the regions on the sphere and the cone contained by the parallels L_1 and L_2.

We let A be the region bounded by L_1, L_2, where for $i = 1, 2$, L_i is the parallel at latitude θ_i. The equator is regarded as the parallel with longitude of angle 0. Angles are expressed in radians. The apex angle of the cone C is equal to $\theta_1 + \theta_2$. We denote by B the annular region on the cone between L_1 and L_2. We use the polar coordinates (ϕ, ψ) on the sphere S, setting the coordinates of the equator to be $\phi = 0$ (Fig. 7.1), and the coordinates (α, β) on C, where α denotes the direction

[3] We shall sometimes omit the adjective "geographical" when talking about a geographical map, when the context is clear. Let us note that the terminology "map" in geometry has its source in geography. The term "projection" is also used for a geographical map, and we shall sometimes adopt this terminology.

[4] The map of the whole Russian Empire edited by Euler is the last map in the *Russian atlas*. The other maps are maps of regions of Russia, and for these smaller regions, Euler used other projections. Euler started working on this *Russian Atlas* with Joseph-Nicolas Delisle, who was the leading astronomer in Saint Petersburg, and then, after Delisle left for France, Euler became the head of this project.

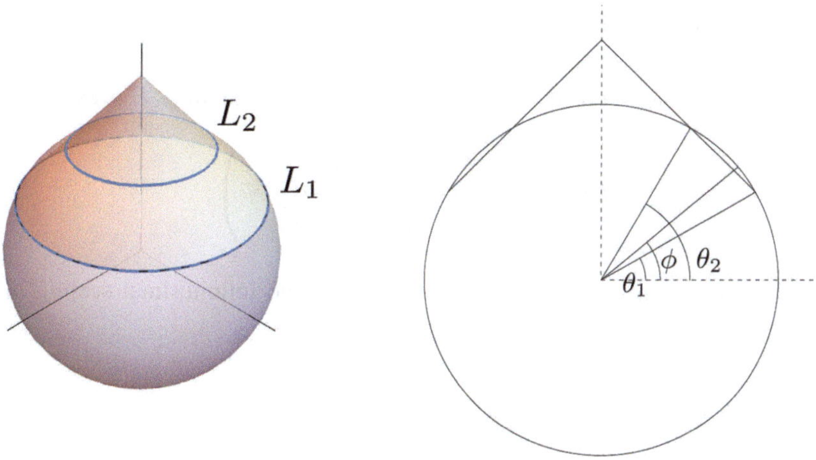

Fig. 7.1 The notation for the coordinates on the sphere S : the angles θ_1 and θ_2 of the two parallels L_1 and L_2 and the angle coordinate ϕ

Fig. 7.2 The coordinates α, β on the cone C

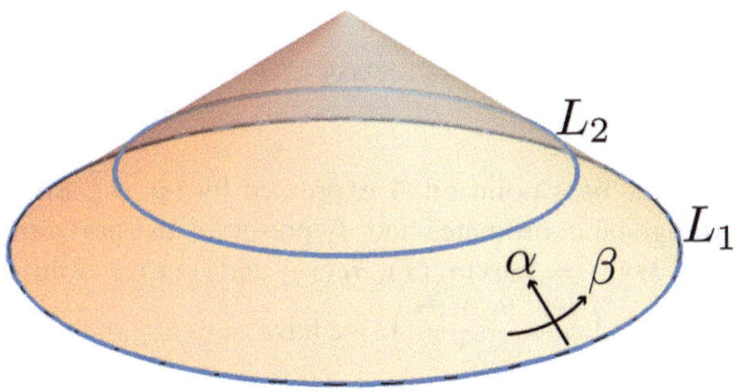

of cones with length parameter with $\alpha = 0$ set to be L_1, and β the direction of the horizontal circles with angle parameter (Fig. 7.2).

The coordinates of A and B are concretely given by

$$\begin{array}{ccc} [\theta_1, \theta_2] \times [0, 2\pi] & \longrightarrow & A \\ (\phi, \psi) & \mapsto & (\cos\psi\cos\phi, \sin\psi\cos\phi, \sin\phi) \end{array}$$

and

$$\begin{array}{ccc} \left[0, 2\sin\left(\frac{\theta_2-\theta_1}{2}\right)\right] \times [0, 2\pi] & \longrightarrow & B \\ (\alpha, \beta) & \mapsto & \left(T(\alpha)\cos\beta, T(\alpha)\sin\beta, \alpha\cos\left(\frac{\theta_1+\theta_2}{2}\right)\right), \end{array}$$

where $T(\alpha) = \cos(\theta_1) - \alpha\sin\left(\frac{\theta_1+\theta_2}{2}\right)$.

We now compute the dilatations and the distortions.

7.3.1 The Projection from the Center of S

Let f_1 be the projection from the center of S to C. Then the map f_1 is expressed as

$$f_1(\phi, \psi) = \left(\frac{\sin(\phi - \theta_1)}{\cos\left(\phi - \frac{\theta_1+\theta_2}{2}\right)}, \psi \right),$$

using the coordinate systems (ϕ, ψ) and (α, β). The infinitesimal scaling in the vertical direction is

$$\sigma_v(\phi, \psi) = \frac{\cos\left(\frac{\theta_2-\theta_1}{2}\right)}{\cos^2\left(\phi - \frac{\theta_2-\theta_1}{2}\right)}.$$

The infinitesimal scaling in the horizontal direction is

$$\sigma_h(\phi, \psi) = \frac{\cos\left(\frac{\theta_2-\theta_1}{2}\right)}{\cos\left(\phi - \frac{\theta_1+\theta_2}{2}\right)}.$$

Let x be a point on A expressed by (ϕ, ψ). Since both (ϕ, ψ) and (α, β) are orthogonal coordinates and f_1 preserves the horizontal and vertical directions, we have $M(x) = \max\{\sigma_v(x), \sigma_h(x)\}$ and $m(x) = \min\{\sigma_v(x), \sigma_h(x)\}$. Since $\sigma_h(x) = \sigma_v(x) \cos\left(\phi - \dfrac{\theta_1+\theta_2}{2}\right)$, we have $\sigma_h(x) \leq \sigma_v(x)$, and equality holds at the point where $\phi = \dfrac{\theta_1+\theta_2}{2}$. Therefore,

$$M(x) = \sigma_v(x) = \frac{\cos\left(\frac{\theta_2-\theta_1}{2}\right)}{\cos^2\left(\phi - \frac{\theta_2-\theta_1}{2}\right)},$$

$$m(x) = \sigma_h(x) = \frac{\cos\left(\frac{\theta_2-\theta_1}{2}\right)}{\cos\left(\phi - \frac{\theta_1+\theta_2}{2}\right)},$$

$$K_{f_1}(x) = \frac{\sigma_v(x)}{\sigma_h(x)} = \frac{1}{\cos\left(\phi - \frac{\theta_1+\theta_2}{2}\right)},$$

and

$$\sigma(x) = \max\{\sigma_v(x), \sigma_h(x)^{-1}\}.$$

This implies that

$$K_{f_1}(A) = \frac{1}{\cos\left(\frac{\theta_2-\theta_1}{2}\right)}.$$

We can also easily see that $\sigma_v(x)$ on A takes the maximum value

$$\sigma_2 = \frac{1}{\cos\left(\frac{\theta_2-\theta_1}{2}\right)}$$

on L_1 and L_2, and $\sigma_h(x)$ on A takes the minimum value

$$\sigma_1 = \cos\left(\frac{\theta_2-\theta_1}{2}\right)$$

at the middle parallel. Thus we have

$$\delta_A(f_1) = -2\log\cos\left(\frac{\theta_2-\theta_1}{2}\right).$$

Now, $\sigma(x)$ takes the maximum value on three parallels, with $\phi = \theta_1, \frac{\theta_1+\theta_2}{2}, \theta_2$. The maximum $L_A(f_1)$ is $\cos^{-1}\left(\frac{\theta_2-\theta_2}{2}\right)$. The function $\sigma(x)$ takes a minimum value on two parallels, where $\cos^3\left(\phi - \frac{\theta_1+\theta_2}{2}\right) = \cos^2\left(\frac{\theta_2-\theta_1}{2}\right)$, and the minimum is $\ell_A(f_1) = \cos^{-1/3}\left(\frac{\theta_2-\theta_2}{2}\right)$. Therefore we have

$$v_A(f_1) = \log\left(\frac{L_A(f_1)}{\ell_A(f_1)}\right) = -\frac{2}{3}\log\left(\cos\left(\frac{\theta_2-\theta_1}{2}\right)\right).$$

7.3.2 *The Horizontal Projection*

Using the same coordinate systems (ϕ, ψ) and (α, β), we can express the horizontal projection $f_2\colon A \to B$ as

$$f_2(\phi, \psi) = \left(\frac{\sin(\phi) - \sin(\theta_1)}{\cos\left(\frac{\theta_1+\theta_2}{2}\right)}, \psi\right).$$

We can calculate σ_v, σ_h as before using this expression. Then we have

$$\sigma_v(\phi, \psi) = \frac{\cos(\phi)}{\cos\left(\frac{\theta_1+\theta_2}{2}\right)},$$

and

$$\sigma_h(\phi, \psi) = \frac{\cos(\frac{\theta_2-\theta_1}{2}) - \sin(\phi)\sin(\frac{\theta_1+\theta_2}{2})}{\cos(\phi)\cos(\frac{\theta_1+\theta_2}{2})}.$$

7.3.3 The Conical Equidistant Map

Using again the coordinates (ϕ, ψ) and (α, β), the conical equidistant map f_3 is expressed as

$$f_3(\phi, \psi) = \left(\frac{2\sin\left(\frac{\theta_2-\theta_1}{2}\right)}{\theta_2-\theta_1}(\phi-\theta_1), \psi\right).$$

This implies that

$$\sigma_v(\phi, \psi) = \frac{2\sin\left(\frac{\theta_2-\theta_1}{2}\right)}{\theta_2-\theta_1}$$

and

$$\sigma_h(\phi, \psi) = \frac{\cos(\theta_1) - \frac{2\sin(\frac{\theta_1+\theta_2}{2})\sin(\frac{\theta_2-\theta_1}{2})(\phi-\theta_1)}{(\theta_2-\theta_1)}}{\cos(\phi)}.$$

7.3.4 The Orthogonal Projection

The orthogonal projection f_4 is expressed as

$$f_4(\phi, \psi) = \left(\sin(\phi - \frac{\theta_1+\theta_2}{2}) + \sin(\frac{\theta_2-\theta_1}{2}), \psi\right).$$

Then we can calculate and get

$$\sigma_v(\phi, \psi) = \cos(\phi - \frac{\theta_1+\theta_2}{2})$$

and

$$\sigma_h(\phi, \psi) = \frac{\cos(\theta_1) + \cos(\theta_2) - 2\sin(\phi - \frac{\theta_1+\theta_2}{2})\sin(\frac{\theta_1+\theta_2}{2})}{2\cos(\phi)}.$$

7.3.5 The Best Quasiconformal Map

To compute the best quasiconformal map, we need to obtain the isothermal coordinates of the annuli A and B. For this, we first consider the intrinsic forms of the metrics induced from $\mathbb{R}^3$ on A and B. Using the coordinates of A and B discussed at the beginning of this section, the induced metrics ds_A^2 and ds_B^2 on A and B are given by

$$\begin{aligned} ds_A^2 &= d\phi^2 + \cos^2(\phi)\, d\psi^2; \\ ds_B^2 &= d\alpha^2 + T(\alpha)^2 d\beta^2. \end{aligned}$$

The isothermal coordinates of A and B are then

$$\begin{cases} [\theta_1, \theta_2] \times [0, 2\pi] \ni (\phi, \psi) \mapsto \Phi_A(\phi, \psi) = (T_1(\phi), \psi)\,; \\ [0, 2\sin(\theta_1 + \theta_2)/2] \times [0, 2\pi] \ni (\alpha, \beta) \mapsto \Phi_B(\alpha, \beta) = (T_2(\alpha), \beta) \end{cases} \tag{7.1}$$

respectively, where

$$T_1(\phi) = \int_{\theta_1}^{\phi} \frac{dt}{\cos t} = \frac{1}{2}\left(\log\frac{1+\sin(\phi)}{1-\sin(\phi)} - \log\frac{1+\sin(\theta_1)}{1-\sin(\theta_1)}\right);$$

$$T_2(\alpha) = \int_0^{\alpha} \frac{dt}{T(t)} = \frac{1}{\sin\left(\frac{\theta_1+\theta_2}{2}\right)}\left(\log\frac{\cos\theta_1}{\cos\theta_1 - \alpha\sin\left(\frac{\theta_1+\theta_2}{2}\right)}\right),$$

and the metrics ds_A^2 and ds_B^2 on A and B are presented as conformal metrics

$$\begin{aligned} ds_A^2 &= \cos(T_1^{-1}(x))^2(dx^2 + dy^2); \\ ds_B^2 &= T(T_2^{-1}(u))^2(du^2 + dv^2) \end{aligned}$$

under the isothermal coordinates (7.1). Indeed, the map Φ_A is nothing but Mercator's map. For the case of the annulus B, Let $(\alpha_0, \beta_0) \in [0, 2\sin((\theta_2 - \theta_1)/2)] \times [0, 2\pi]]$ and $(u_0, v_0) = \Phi_B(\alpha_0, \beta_0)$. For a tangent vector $V = \xi \frac{\partial}{\partial \alpha} + \eta \frac{\partial}{\partial \beta}$ at (α_0, β_0) in (α, β)-plane, the pull-back metric satisfies

$$
\begin{aligned}
&\Phi_B^*(T(T_2^{-1}(u_0))(du^2 + dv^2))(V, V) \\
&= T(T_2^{-1}(T_2(\alpha_0)))(du^2 + dv^2)((\Phi_B)_*(v), (\Phi_B)_*(v)) \\
&= T(\alpha_0)^2(du^2 + dv^2)\left(\xi \frac{\partial T_2}{\partial \alpha}(\alpha_0)\frac{\partial}{\partial u} + \eta \frac{\partial}{\partial v}, \xi \frac{\partial T_2}{\partial \alpha}(\alpha_0)\frac{\partial}{\partial u} + \eta \frac{\partial}{\partial v}\right) \\
&= T(\alpha_0)^2(du^2 + dv^2)\left(\xi \frac{1}{T(\alpha_0)}\frac{\partial}{\partial u} + \eta \frac{\partial}{\partial v}, \xi \frac{1}{T(\alpha_0)}\frac{\partial}{\partial u} + \eta \frac{\partial}{\partial v}\right) \\
&= T(\alpha_0)^2 \frac{\xi^2 + T(\alpha_0)^2\eta^2}{T(\alpha_0)^2} = ds_B^2(V, V).
\end{aligned}
$$

In terms of the isothermal coordinates Φ_A and Φ_B, both A and B are conformally equivalent to the following flat annuli:

$$
\begin{aligned}
A &= [0, T_1(\theta_2)] \times \mathbb{R}/(x, y) \sim (x, y + 2\pi); \\
B &= [0, T_2(2\sin((\theta_2 - \theta_1)/2))] \times \mathbb{R}/(u, v) \sim (u, v + 2\pi).
\end{aligned}
$$

Therefore, their moduli are equal to

$$
\mathrm{Mod}(A) = \frac{T_1(\theta_2)}{2\pi} = \frac{1}{4\pi} \log\left(\frac{1 + \sin(\theta_2)}{1 - \sin(\theta_2)} \frac{1 - \sin(\theta_1)}{1 + \sin(\theta_1)}\right);
$$

$$
\mathrm{Mod}(B) = \frac{T_2(\sin((\theta_2 - \theta_1)/2))}{2\pi} = \left(2\pi \sin\left(\frac{\theta_2 + \theta_1}{2}\right)\right)^{-1} \log \frac{\cos(\theta_1)}{\cos(\theta_2)}.
$$

Since we are considering the best quasiconformal map, by the Teichmüller theory its dilatation $K(x)$ is constant, and

$$
K_{f_5}(A) = \frac{\mathrm{Mod}(B)}{\mathrm{Mod}(A)}.
$$

We consider Euclidean annuli conformal to A and B, and develop them onto rectangles in such a way that the horizontal directions are parallel to the real axis and the vertical directions are parallel to the imaginary axis. Then the best quasiconformal map from A to B is given by an affine map

$$
\begin{array}{ccc}
[0, T_1(\theta_2)] \times [0, 2\pi] & \longrightarrow & [0, T_2(\sin((\theta_2 - \theta_1)/2)] \times [0, 2\pi] \\
x + iy & \mapsto & u + iv = Kx + iy
\end{array}
$$

In fact, the best quasiconformal map is unique up to post-composing a conformal automorphism of the target. Hence, under these coordinates, we have an expression

$$f_5(\phi, \psi) = (T_2^{-1}(KT_1(\phi)), \psi)$$

where

$$T_2^{-1}(KT_1(\phi)) = \frac{\cos(\theta_1)}{\sin\left(\dfrac{\theta_1+\theta_2}{2}\right)}\left(1 - \left(\frac{1+\sin(\theta_1)}{1-\sin(\theta_1)}\frac{1-\sin(\phi)}{1+\sin(\phi)}\right)^{\frac{K}{2}\sin\left(\frac{\theta_1+\theta_2}{2}\right)}\right)$$

and $K = K_{f_5}(A)$. Hence,

$$\sigma_h(\phi, \psi) = \frac{K(\theta_1, \theta_2)}{\cos(\phi)}\left|\cos(\theta_1) - T_2^{-1}(K_{f_5}(A)T_1(\phi))\sin\left(\frac{\theta_1+\theta_2}{2}\right)\right|;$$

$$\sigma_v(\phi, \psi) = \frac{1}{\cos(\phi)}\left|\cos(\theta_1) - T_2^{-1}(K_{f_5}(A))T_1(\phi))\sin\left(\frac{\theta_1+\theta_2}{2}\right)\right|.$$

7.4 Numerical Computations for the Various Maps

7.4.1 Comparison of Maps on A

We compare the graphs of $\sigma(x)$ and the values of $L(f)/\ell(f)$ for the five maps we are considering.

In Fig. 7.3, we have drawn the graph of $\sigma(x)$ for the five maps $f_1, \dots, f_5$, where we have set $\theta_1 = \pi/6$ and $\theta_2 = \pi/3$. In Fig. 7.4, we have drawn the graph of $\sigma(x)$ for the four maps f_1, f_3, f_4 and f_5, eliminating f_2, which is apparently the worst.

We observe that in this setting, f_4, and even f_1, may give a smaller variation of infinitesimal distortions than f_3 used by Euler. Indeed, if we take the same L_1 and L_2 as Euler, i.e. in the case when $A = \{(\phi, \psi) \in S \mid 47°30' \leq \phi \leq 52°30'\}$, we have $v_A(f_1) = 0.00572796$, $v_A(f_2) = 0.191678$, $v_A(f_3) = 0.00576879$, $v_A(f_4) = 0.00463332$, and $v_A(f_5) = 0.0071773$. Thus f_4 gives the best value for v_A. On the other hand, if we look at the distortion δ_A, we see that, approximately, $\delta_A(f_1) = 0.0171839$, $\delta_A(f_2) = 0.380566$, $\delta_A(f_3) = 0.00862621$, $\delta_A(f_4) = 0.00866925$ and $\delta_A(f_5) = 0.0115244$. Thus, Delisle's map f_3 is slightly better than the orthogonal projection f_4 in this respect. As for the maximal dilatation, f_5 gives the best value by definition. The approximate values are $K_{f_1}(A) = 1.00863$, $K_{f_2}(A) = 1.24218$, $K_{f_3}(A) = 1.00579$, $K_{f_4}(A) = 1.00867$, and $K_{f_5}(A) = 1.0029$, again for Euler's A.

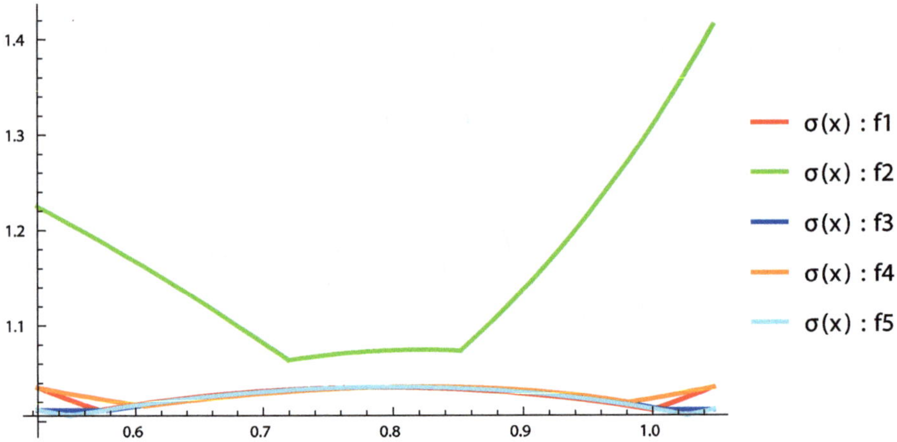

Fig. 7.3 The comparison of $\sigma(x)$ for the five maps $f_1, \ldots, f_5$

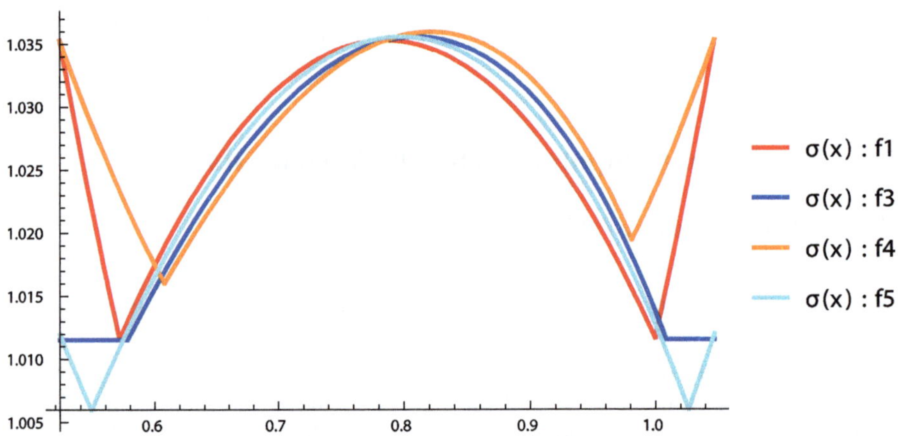

Fig. 7.4 An enlarged graph of $\sigma(x)$ for the four maps f_1, f_3, f_4 and f_5

7.4.2 Euler's Problem

We now turn to the situation where we fix a region A' and choose A which may be different from A' containing it, with the central parallels being the same. We recall that in Euler's memoir, A' was the smallest band containing the entire Russian empire. According to [6], to draw the map of the Russian empire, A was taken to be $\{(\phi, \psi) \in S \mid 47°30' \leq \phi \leq 62°30'\}$, whereas A' was taken to be the smallest one covering the Russian empire, i.e., $A' = \{(\phi, \psi) \in S \mid 40° \leq \phi \leq 70°\}$. Then we have the following graphs, one for f_1, f_2, f_3 and f_4 and the other eliminating f_2 which has much worse distortion than the others (Figs. 7.5 and 7.6).

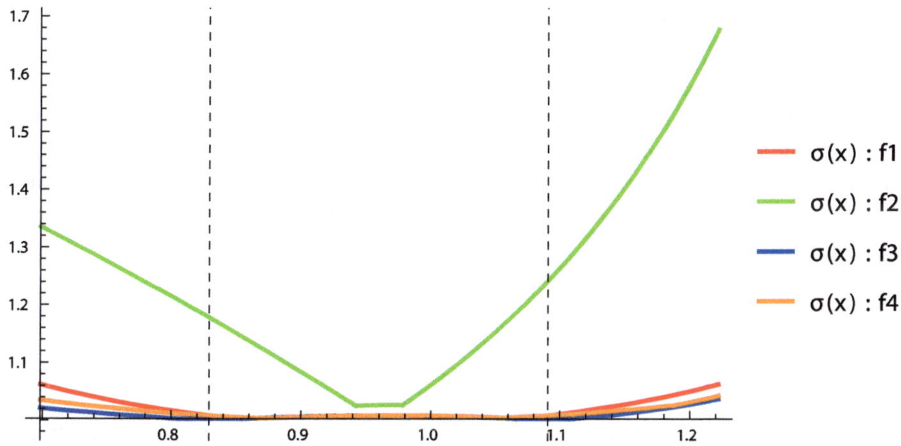

Fig. 7.5 The comparison of $\sigma(x)$ for the four maps $f_1, \ldots, f_4$ in the wider region A'. The dotted lines correspond to L_1 and L_2

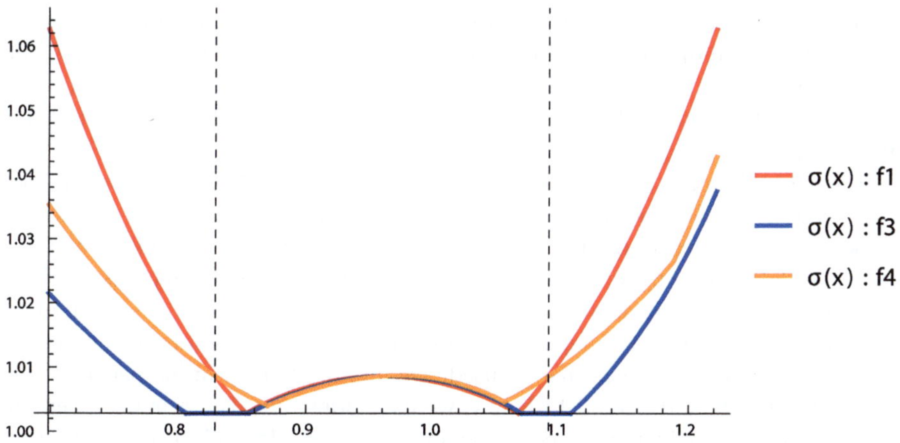

Fig. 7.6 The comparison of $\sigma(x)$ for the three maps f_1, f_3 and f_4 in the wider region A'

We have numerically calculated both $v_{A'}$ and $\delta_{A'}$. Indeed we have $v_{A'}(f_1) = 0.0578805$, $v_{A'}(f_2) = 0.491829$, $v_{A'}(f_3) = 0.0339056$ and $v_{A'}(f_4) = 0.0378697$. Thus even for $v_{A'}$, Delisle's map becomes the best. As for $\delta_{A'}$, we have $\delta_{A'}(f_1) = 0.0693365$, $\delta_{A'}(f_2) = 0.806371$, $\delta_{A'}(f_3) = 0.0453892$, $\delta_{A'}(f_4) = 0.0765739$. We can see that the difference between f_3 and f_4 gets larger than in the case when $A' = A$. As for maximal dilatations, we have approximate values $K_{f_1}(A') = 1.03528$, $K_{f_2}(A') = 1.89518$, $K_{f_3}(A') = 1.04042$ and $K_{f_4}(A') = 1.07958$. Here f_1 has a better value, but f_3 is the second best, and the difference between the values for f_1 and f_3 is small. But Euler, in the choice of the map that would give the best representation of the Russian empire, was interested in the metrical distortion, rather than the conformal one. Therefore it was quite reasonable for him to choose Delisle's map rather than the other projections in this setting.

Since the best quasiconformal map f_5 is defined only on A, we cannot compare this with the others on A'. Instead, we can consider the best quasiconformal map from A', by letting L_1 and L_2 be parallels with latitudes 40° and 70° respectively. For this new f_5, we have $v_{A'}(f_5) = 0.0292036, \delta_{A'}(f_5) = 0.0474334$, and $K_{f_5}(A') = 1.01223$. This map gives better $v_{A'}$ and $K_{A'}$. Still Delisle's map has better distortion.

7.5 Tissot Ellipses

A *Tissot ellipse*, also called a *Tissot indicatrix*, on a geographical map is an ellipse, drawn on this map, which is the image of a unit circle in the tangent bundle by the mapping. Mathematically it represents the quasiconformal dilatation of the map at the given point. In fact, the Tissot ellipse contains more than the quasiconformal information, since the latter (in the usual definition) is represented by the shape of the ellipse (the directions and the ratio of the big axis over the small axis), but the information on the Tissot ellipse involves also the direction and the size of this ellipse, and therefore it also represents the infinitesimal Lipschitz constant at the given point.

The name Tissot refers to the French geographer Nicolas-Auguste Tissot (1824–1897) who introduced this device. See [12] or [13] for a historical review on Tissot's work. As we recalled in the introduction, Teichmüller, in his paper [14], refers to the work of Tissot, as the origin of the theory of quasiconformal mappings. Herbert Grötzsch, one of the main founders of quasiconformal theory, in several figures illustrating quasiconformal mappings in his papers, drew the Tissot indicatrices on the domain of the mappings that he used; see e.g. the figures in his paper [7]. In Figs. 7.7, 7.8, 7.9, 7.10, and 7.11 below, we have drawn the Tissot indicatrices for the maps f_1 to f_5 which we used in the above sections. In each of the four cases from f_1 to f_4, we have taken the annuli A' bounded by the latitudes 40° and 70° and A between 47°30′ and 62°30′. For f_5, which is defined only on A, we assumed that both A and A' are the latter one, bounded by the latitudes between 47°30′ and 62°30′. In all cases, except for the map f_2 (7.8), the map is almost conformal (the Tissot ellipses are close to circles), and is almost distance-preserving (all the Tissot ellipses have almost the same size). In particular, for the map f_3 (Fig. 7.9), this is due to the choice of the two latitudes bounding the annulus, which, as Euler took them, are close to the lower point and highest point of the Russian Empire.

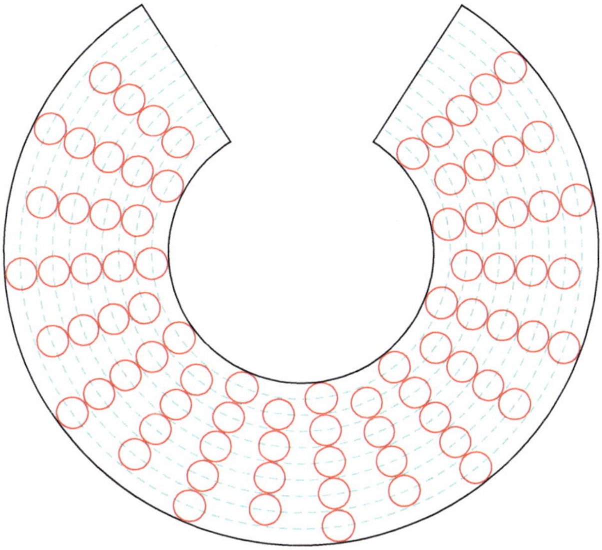

Fig. 7.7 Tissot ellipses for f_1

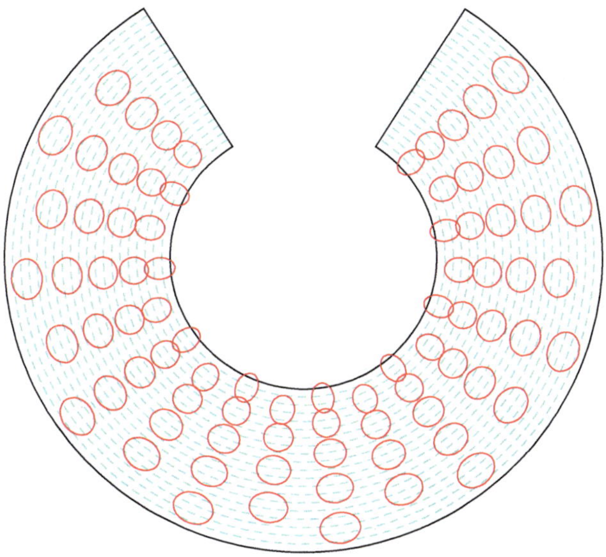

Fig. 7.8 Tissot ellipses for f_2

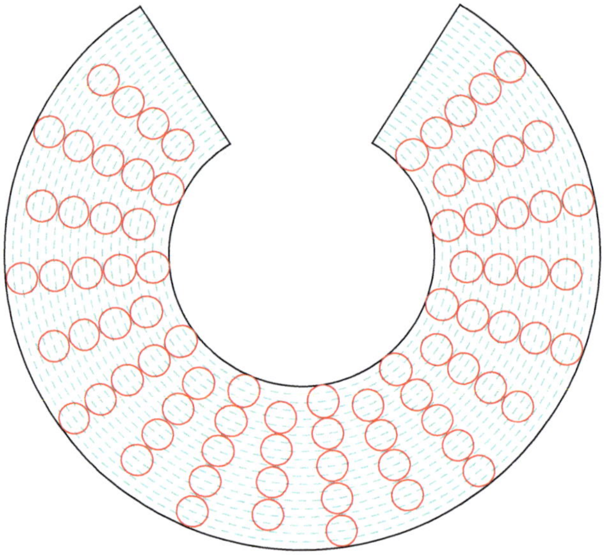

Fig. 7.9 Tissot ellipses for f_3

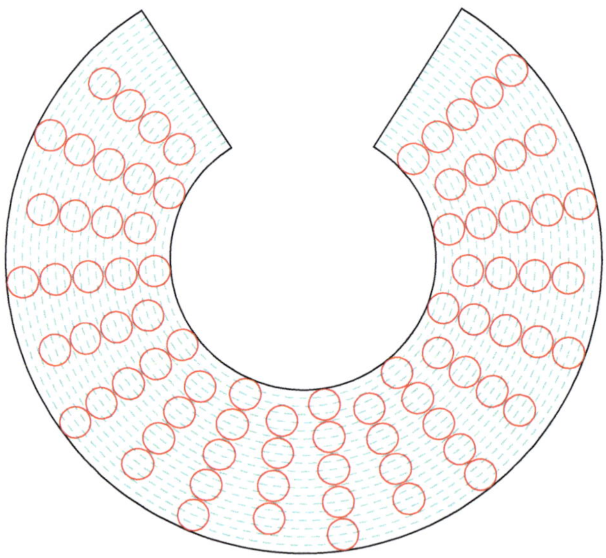

Fig. 7.10 Tissot ellipses for f_4

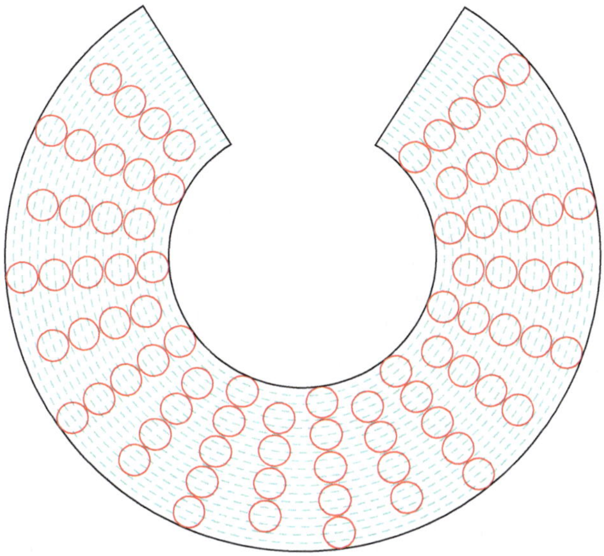

Fig. 7.11 Tissot ellipses for f_5

References

1. Atlas russicus mappa una generali et undeviginti specialibus vastissimum imperium russicum cum adiacentibus regionibus secundum leges geographicas et recentissimas obsevariones delineatum exhibens. Cura et opera Academiae Imperialis Scientiarum Petropolitanae (1745)
2. R. Caddeo, A. Papadopoulos (eds.), *Mathematical Geography in the Eighteenth Century: Euler, Lagrange and Lambert* (Springer, Cham, 2022)
3. L. Euler, De repraesentatione superficiei sphaericae super plano, Acta Academiae Scientiarum Imperialis Petropolitanae 1777, 1778, pp. 107–132, Opera Omnia, Series 1, Volume 28, 248–275. English translation by R. Caddeo, On the representation of the spherical surface on the plane, in *Mathematical Geography in the Eighteenth Century: Euler, Lagrange and Lambert*, ed. by R. Caddeo, A. Papadopoulos (Springer, Cham, 2022) pp. 243–278
4. L.Euler, De proiectione geographica superficiei sphaericae, Acta Academiae Scientiarum Imperialis Petropolitanae 1777, 1778, pp. 133–142, Opera Omnia, Series 1, Volume 28, 276–287. English translation by R. Caddeo, On the geographical projection of the surface of the sphere, in *Mathematical Geography in the Eighteenth Century: Euler, Lagrange and Lambert*, ed. by R. Caddeo, A. Papadopoulos (Springer, Cham, 2022), pp. 279–294
5. L. Euler, De proiectione geographica De Lisliana in mappa generali imperii russici usitata, Acta Academiae Scientiarum Imperialis Petropolitanae 1777, 1778, Opera Omnia, Series 1, Volume 28, 143–153. English translation by R. Caddeo, On Delisle's geographical projection used for a general map of the Russian Empire, in *Mathematical Geography in the Eighteenth Century: Euler, Lagrange and Lambert*, ed. by R. Caddeo, A. Papadopoulos (Springer, Cham, 2022), pp. 295–306
6. A.A.Ch. Germain, Traité des projections des cartes géographiques: représentation plane de la sphère et du sphéroïde (Arthus Bertrand, Paris, 1866)

7. H. Grötzsch, über die Verzerrung bei nichtkonformen schlichten Abbildungen mehrfach zusammenhängender schlichter Bereiche. Leipz. Ber. **82**, 69–80 (1930). English translation by M. Karbe, On the distortion of non-conformal schlicht mappings of multiply-connected schlicht regions, in *Handbook of Teichmüller Theory*, vol. VII. IRMA Lectures in Mathematics and Theoretical Physics, vol. 30, Chapter 16 (European Mathematical Society (EMS), Zürich, 2020), pp. 375–385
8. J. Milnor, A problem in cartography. Am. Math. Mon. **76**, 1101–1112 (1969)
9. J. Milnor, John Milnor Collected Papers: Volume I: Geometry. Publish or Perish, Inc., 1994, New edition (American Mathematical Society, Providence, 2007)
10. P. Murdoch, Of the best form of geographical maps. Philos. Trans. **50**, 553–568 (1757–1758)
11. Claudius Ptolemy, Traité de géographie, translated from the Greek into French by N. Halma (Eberhart, Paris, 1828)
12. A. Papadopoulos, Nicolas-Auguste Tissot: a link between cartography and quasiconformal theory. Arch. Hist. Exact Sci. **71**(4), 319–336 (2017)
13. A. Papadopoulos, A note on Nicolas-Auguste Tissot: at the origin of quasiconformal mappings, in *Handbook of Teichmüller Theory*, ed. by A. Papadopoulos, vol. VII. IRMA Lect. Math. Theor. Phys, vol. 30 (European Mathematical Society, Berlin, 2020), pp. 289–299
14. O. Teichmüller, Extremale quasikonforme Abbildungen und quadratische Differentiale. Abh. Preuß. Akad. Wiss., Math.-Naturw. Kl. 1940, No. 22, 1–197 (1940). English translation by G. Théret: Extremal quasiconformal mappings and quadratic differentials, in *Handbook of Teichmüller Theory* ed. by A. Papadopoulos, vol. V. IRMA Lectures in Mathematics and Theoretical Physics, vol. 26 (European Mathematical Society, Zürich, 2016), pp. 321–483
15. N.-A. Tissot, *Mémoire sur la représentation des surfaces et les projections des cartes géographiques* (Gauthier-Villars, Paris, 1881)

Chapter 8
Spherical Representations of the Group of Isometries of Semi-homogeneous Trees

Massimo A. Picardello

This chapter is dedicated to Athanase Papadopoulos, whose enthusiasm in exploring the links between geometry, geography, history, music and art and whose readiness to share their beauty have been significant for my mathematical research and for our friendship.

Abstract We consider the group $\mathcal{G}$ of isometries of a semi-homogeneous tree $T = T_{q_+,q_-}$ with valencies $q_+ + 1$ and $q_- + 1$ and its two orbits, respectively V_+ and V_-, on the set of vertices. We make use of the action of $\mathcal{G}$ to equip each of $V_\pm$ with a convolution product, hence with a notion of positive definite functions. The ℓ^1-functions radial around a root vertex v_0 in, say, V_+ form an abelian convolution algebra. We study its multiplicative functionals, called spherical functions, that are eigenfunctions of the nearest-neighbor isotropic transition operator (the *Laplace operator* on T), and determine which of them are positive definite. Each positive definite function gives rise to a unitary representation of $\mathcal{G}$; in this way, we produce a series of unitary spherical representations. For $q_+ < q_-$, the representation whose spherical function has eigenvalue 0 is square-integrable.

8.1 Introduction

Let T be a homogeneous tree, that is, a graph without loops and with the same number $q + 1$ of neighbors for each vertex; q is called the homogeneity degree. The isotropic nearest-neighbor transition operator μ_1 on its set V of vertices (called *Laplace operator*) has been studied since a long time (see [12] for references), and is related to the Poisson boundary Ω of T (a topological boundary that yields a compactification of V) and its Poisson kernel K. The setup of a homogeneous tree

M. A. Picardello (✉)
Dipartimento di Matematica, Università di Roma "Tor Vergata", Roma, Italy
e-mail: picard@mat.uniroma2.it

A. M. Uludağ, A. Zeytin (eds.), *Essays on Geometry*,
https://doi.org/10.1007/978-3-031-76257-4_8

has close ties to harmonic analysis, because functions on V are endowed with a convolution product induced by the group of isometries $\mathcal{G} = \operatorname{Aut} T$ (or a simply transitive subgroup thereof), and μ_1 is a convolution operator. Appropriate estimates for the norms of convolution operators [14] can then be used to determine the spectrum of μ_1 on $\ell^p(V)$ for $1 \leqslant p < \infty$ [12, Chapter 3, Theorem 3.3]. Choose a root vertex v_0 and denote by $\mathcal{G}_{v_0}$ its stability subgroup in $\mathcal{G}$. Then, with the convolution product induced by $\mathcal{G}$, the algebra of two-sided $\mathcal{G}_{v_0}$-invariant function on $\mathcal{G}$ (or, in an equivalent realization, the algebra of functions on V radial around v_0) is commutative; that is, $(\mathcal{G}, \mathcal{G}_{v_0})$ is a Gelfand pair (see [1, 12] and references therein). Its spherical functions are the eigenfunctions of μ_1 radial around v_0 and normalized to 1 at v_0: the spherical function at the eigenvalue γ, usually denoted by $\phi(\,\cdot\,, v_0 \,|\, \gamma)$ in the literature, was computed in [11] (see also [2]).

It was shown in [12] that the spectrum of μ_1 on $\ell^p(V)$ is the closure of the set $\{\gamma \in \mathbb{C} \colon \phi(\,\cdot\,, v_0 \,|\, \gamma) \in \ell^r(V) \text{ for every } r > p\}$. Moreover, since μ_1 is invariant under $\mathcal{G}$, its eigenspaces give rise to spaces of representations of $\mathcal{G}$, called *spherical representations*. The representation at the eigenvalue γ is unitary if γ belongs to the ℓ^2-spectrum of μ_1, and unitarizable if $\phi(\,\cdot\,, v_0 \,|\, \gamma)$ is positive definite.

A recent article [7] considers trees that are *semi-homogeneous*, i.e., with two alternating homogeneity degrees q_+ and q_-. The semi-homogeneous Laplace operator μ_1 on functions on V is again as the average operator on neighbors, and gives rise to a nearest-neighbor transition operator. Its spherical functions are obtained as generalized Poisson transforms, by means of an explicit realization of the Poisson kernel for each eigenvalue γ, derived by computing the first visit probability associated to the random walk generated by μ_1.

On the other hand, the semi-homogeneous and the homogeneous settings are considerably different. Indeed, the group $\mathcal{G}$ has two orbits V_+ and V_- on V, corresponding respectively to the homogeneity degrees q_+ and q_-, hence it is not transitive and does not give rise to a convolution product. For the same reason, the natural definition of positive definite function does not make sense. As a consequence, we do not have the convolution estimates of [14] that have been used in the literature to compute the spectrum of μ_1, and it is much harder to build an analytic series of spherical representations of $\mathcal{G}$ by means of spherical functions.

Indeed, the basic problem raised by the existence of two orbits under $\mathcal{G}$ is the lack of a convolution product, hence of a suitable convolution algebra $\mathcal{R}$ of radial functions (see also [5]). In the homogeneous setting, the spectrum of μ_1 is obtained from the multiplicative functionals on the Banach algebra generated by $\mathcal{R}$ in the ℓ^1-norm. Here we need a new approach. The operator μ_1, being a nearest-neighbor transition operator, induces jumps only between vertices of different homogeneities. An idea developed in [7] with the goal of computing the ℓ^p-spectrum of μ_1 focuses attention onto its square μ_1^2, that preserves the homogeneity, hence acts on each of the two orbits, and is transitive on each of them. On the other hand, μ_1^2 is related to the step-2 isotropic transition operator μ_2; indeed, up to a scale factor, it coincides with μ_2 plus a positive multiple of the identity.

The notion of adjacency induced by μ_2 on V_+, that corresponds to distance 2 in V, transforms V_+ into the polygonal graphs Γ_+, studied in [16], that consists of infinitely many complete polygons with $q_- + 1$ vertices attached in a tree-like fashion, with $q_+ + 1$ polygons joining at each vertex; by exchanging q_+ and q_-, a similar description holds for the polygonal graph Γ_- that corresponds to the action on V_-. Both these actions are transitive and we obtain on, say, V_+ a nice convolution algebra of functions radial in this graph around a fixed reference vertex $v_0 \in V_+$. For every p, the ℓ^p-spectrum of μ_2 on $V_\pm$ is known [16]. By making use of this, the ℓ^p-spectrum of μ_1 on a semi-homogeneous tree has been computed in [7].

An unusual fact occurs in the semi-homogeneous setting [7]: there is a spherical function that belongs to $\ell^2(V)$, and also to $\ell^p(V)$ for some $p < 2$. Indeed, the spherical function at the eigenvalue 0 belongs to $\ell^p(V)$ for every $p > 1 + \dfrac{\ln q_+}{\ln q_-}$, that is smaller than 2 if and only if $q_+ < q_-$. Every other bounded spherical function $\phi(\,\cdot\,, v_0 \mid \gamma)$ belongs to $\ell^p(V)$ for some $p > 2$ depending on γ.

The lack of transitivity of $\mathcal{G}$ gives rise to some interesting questions. Here again, μ_1 is $\mathcal{G}$-invariant, i.e., it commutes with $\mathcal{G}$, so its eigenspaces are also invariant under $\mathcal{G}$, hence they give rise to representations of this group, namely, the spherical representations. The representation space at the eigenvalue γ is the span of all translates of $\phi(\,\cdot\,, v_0 \mid \gamma)$. In the homogeneous setting, this representation is unitary or unitarizable if and only if $\phi(\,\cdot\,, v_0 \mid \gamma)$ is positive definite (thereby providing the necessary Hilbert norm via the GNS construction) [11, 12]. In the semi-homogeneous setting, the notion of positive definite makes sense for functions defined on the group $\mathcal{G}$, but not for functions on V, that cannot be lifted to $\mathcal{G}$ since the group action is not transitive. Even though there are only two orbits, it is not known how to equip the space of functions on V with a suitable definition of positive definite function, or at least a Hilbert norm on the linear span of all translates of $\phi(\,\cdot\,, v_0 \mid \gamma)$. Therefore it is not clear which spherical representations are unitary.

For the same reason, as we have mentioned, $\mathcal{G}$ does not induce a convolution product on functions on V. On the other hand, both V_+ and on V_- are homogeneous spaces for $\mathcal{G}$, indeed $V_\pm = \mathcal{G}/\mathcal{G}_{v_\pm}$ for any choice of reference vertices $v_\pm \in V_\pm$, one for each homogeneity. Given two functions supported on the same of these two homogeneous spaces $V_\pm$, we shall show that $\mathcal{G}$ allows to define their convolution product, but only if one of the two functions is bi-$\mathrm{Aut}_{v_\pm}$-invariant.

More precisely, let us restrict attention to functions on V_+ and denote again by v_0 the reference vertex in V_+. If $v \in V$ and $\lambda \in \mathcal{G}$ is such that $v = \lambda v_0$, the convolution product induced by $\mathcal{G}$ on its orbit $\mathcal{G}/\mathcal{G}_{v_0} \approx V_+$ should be defined as $f * g(v) = \int_{\mathcal{G}} f(\tau^{-1}\lambda v_0)\, g(\tau v_0)\, d\tau$, where $d\tau$ is the Haar measure of the (unimodular) group $\mathcal{G}$. On the other hand, the result is invariant on right cosets only if f is two-sided invariant, that is, radial around v_0 (see Sect. 8.3 for more details). In particular, we obtain a radial convolution algebra on V_+ (and another on V_-). This actually leads to two different convolution products, one for each orbit (see [3]). It is not clear how to define the convolution of functions not supported on a single orbit in such a way that it coincides with the usual definition in the setting of homogeneous trees. Moreover, an analogous of Haagerup's convolution estimate [14], that is the main tool for computing the spectra of the homogeneous Laplacian in the well-established approach of [12, Chapter 3], is complicated in this setting [15].

On a homogeneous tree, the spherical representation π_γ of a discrete subgroup $G \subset \mathcal{G}$ isomorphic to a free product, acting simply transitively on a homogeneous tree, is known to be irreducible for $\gamma \in \operatorname{sp}_{\ell^1}(\mu_1)$, by [12, Chapter 5] and [10], hence π_γ is irreducible as a representation of $\mathcal{G}$. The argument, later extended to non-isotropic nearest-neighbor transition operators [13], proceeds by proving that the projector onto a cyclic vector is the weak limit, as $n \to \infty$, of the average of $\pi_\gamma(\tau)$ over all the elements $\tau \in \mathcal{G}$ that move v_0 to vertices of length n. For a semi-homogeneous tree, the lack of transitivity makes it unhandy to reproduce the same argument. Nevertheless, instead of an irreducible representation, we shall obtain a representation reducible as the direct sum of two components, one for each orbit.

In this short note, we study unitary representation of $\mathcal{G}$ by means of its action on $V_+ \approx \mathcal{G}/\mathcal{G}_{v_0}$ (or the analogous action on V_-) and the convolution induced by this action on the radial space $\ell^1_\#(V_+)$, that turns out to be the abelian convolution algebra generated by the Laplacian. The group $\mathcal{G}$ can be factorized as $\mathcal{G} = \mathcal{F}\mathcal{G}_{v_0}$, where $\mathcal{F}$ is any discrete subgroup acting simply transitively on V_+, associated to an appropriate choice of generators. The convolution product is much simpler when considered on $\mathcal{F}$; there is a suitable choice of $\mathcal{F}$ whose Cayley graph is the polygonal graph Γ_+ associated to V_+ [16]. A representation theory for $\mathcal{F}$ could be obtained in the same way as here, perhaps with more readable statements and proofs, but we consider more general and more elegant to avoid a presentation that relies explicitly on a choice of a simply transitive subgroups, that is, of suitable generators, and we work with the full group $\mathcal{G}$ that is naturally associated to the whole tree (observe that any isometry of T restricts to isometries of $V_\pm$, and conversely, every isometry of V_+ (or V_-) extends uniquely to an isometry of T). Actually, the dependence on a choice of generators leads to the fact that the choice of $\mathcal{F}$ is not unique; for instance, the choice of $\mathcal{F}$ in [16] is the free product of $q_+ + 1$ copies of $\mathbb{Z}_{q_-+1}$, but another choice is $\mathbb{Z}_{q_++1} * \mathbb{Z}_{q_-+1}$, that is a discrete group acting simply transitively onto the dual graph of Γ_+. We do not make use of any such simply transitive subgroup, but, for an easier understanding, we encourage the reader to rephrase our proofs in terms of any specific $\mathcal{F}$, and compare our results with those of [16], that yield the ℓ^p-spectrum of the Laplacian μ_2 on Γ_+. For the ℓ^p-spectrum of the Laplacian on a semi-homogeneous tree, see [7]; here, however, we do not study spectra, only positive definite functions and group representations.

Our presentation proceeds as follows. We make use of the action of $\mathcal{G}$ on V_+ and its natural involution to introduce positive definite functions on V_+. Moreover, we study the multiplicative functionals on the algebra $\ell^1_\#(V_+)$, that correspond to the normalized bounded radial eigenfunctions of the Laplacian *(spherical functions)*, prove that the spherical functions are positive definite if and only if they are real-valued and bounded, and conclude that the spherical function $\phi(\,\cdot\,, v_0 \,|\, \gamma)$ that corresponds to a real eigenvalue γ in the ℓ^1-spectrum of the Laplacian gives rise, via the GNS construction, to two unitary representations of $\mathcal{G}$, one for each parity, that are irreducible on the respective GNS completions $\mathcal{H}^\pm_\gamma$ of the span of translates of $\phi(\,\cdot\,, v_0 \,|\, \gamma)$, whereas they decompose as direct sum of two irreducible representations on $\mathcal{H}_\gamma = \mathcal{H}_\gamma^+ \oplus \mathcal{H}_\gamma^-$. If $q_+ < q_-$, the spherical function at the eigenvalue 0 belongs to $\ell^2(V_+)$ and is positive definite, and the representation π_0 is square-integrable, that is, it is a unitary subrepresentation of the regular representation of $\mathcal{G}$. It is worth noting that, for $q_+ < q_-$, a discrete subrepresentation of the regular representation of the free product of $q_+ + 1$ copies of $\mathbb{Z}_{q_-+1}$ (the discrete group acting simply transitively on Γ_+ introduced above) was found in [9, 17].

The author acknowledges many enlightning conversations with Enrico Casadio-Tarabusi. An alternative approach to the results of this article will appear in a forthcoming joint paper.

8.2 Semi-homogeneous Trees and Their Laplace Operator

A *tree* T is a connected, countably infinite, locally finite graph without loops. The nodes of T are called *vertices*; the set of all vertices is denoted by V. Two distinct vertices v, v' are *adjacent*, or *neighbors*, if they belong to the same edge: we shall write $v \sim v'$. Let us fix a reference vertex v_0. The number of neighbors of v is q_v+1, where q_v is the *homogeneity degree*, that is, the number of outward neighbors of v with respect to v_0 if $v \neq v_0$. Let us fix a *parity*, that is, an alternating function $\epsilon\colon V \to \{\pm 1\}$. The level sets of ϵ are denoted by V_+ and V_-.

A *semi-homogeneous* tree $T = T_{q_+,q_-}$ has two alternating homogeneity degrees q_+ on V_+ and q_- on V_-. If $q_+ = q_-$ then the tree is *homogeneous*. We assume $q_+, q_- > 1$ and, without loss of generality, we choose $v_0 \in V_+$, that is,

$$q_{v_0} = q_+. \tag{8.1}$$

Let us specialize our attention to the example of the well-known homogeneous setting. Since it has no loops, a homogeneous tree is the Cayley graph of the free product $\mathfrak{F}$ of $q + 1$ copies of the two-element group $\mathbb{Z}_2$. Indeed, this free product embeds into the group $\mathcal{G}$ of *isometries* of T, the invertible self-maps of V that preserve adjacency: the generators of the factors are isometries that reverse the edges that contain v_0 (see, for instance, [12, Chapter 3, Section V]), and we have the factorization $\mathcal{G} = \mathfrak{F}\,\mathcal{G}_{v_0}$. In particular, $\mathfrak{F}$ acts simply transitively on T and induces a convolution product on functions on V. Therefore $\mathcal{G}$ is transitive on V if T is homogeneous. On the space of functions on $\mathcal{G}$ that are two-sided-invariant under the stability subgroup $\mathcal{G}_{v_0} \subset \mathcal{G}$ of v_0, the convolution operation induced by the group $\mathcal{G}$ is a lifting of the convolution induced by $\mathfrak{F}$ (for homogeneous trees, see [6, Appendix] and [4, Subsection 3.1.1]).

On the other hand, in the strictly semi-homogeneous setting $q_+ \neq q_-$, there are two orbits of $\mathcal{G}$ on V, namely V_+ and V_-. In both settings, the Laplace operator applied to a function f on V is defined as follows: at each vertex v it yields the average of the values of f at the neighbors of v. The number of neighbors depends on the parity of the vertex. That is,

$$\mu_1 f(v) = \frac{1}{q_{\epsilon(v)} + 1} \sum_{w \sim v} f(w). \tag{8.2}$$

Remark 8.1 Following [7], let us consider the space $\mathfrak{R}$ of summation operators with kernel, acting on $\ell^1(V)$, that is, of the type $Rf(v) = \sum_{w\in V} r(v,w) f(w)$, whose kernel $r(v,w)$ is of finite range (that is, it vanishes if $\operatorname{dist}(v,w) > N$ for some $N \in \mathbb{N}$ depending on r). Then μ_1 belongs to $\mathfrak{R}$. More generally, for $n \in \mathbb{N}$ we set

$$\mu_n f(v) = \frac{1}{|\{w\colon \operatorname{dist}(w,v) = n\}|} \sum_{\operatorname{dist}(w,v)=n} f(w). \tag{8.3}$$

An elementary computation shows that, for every $n > 0$,

$$\mu_1 \mu_n f(v) = \mu_n \mu_1 f(v) = \frac{1}{q_v + 1}\, \mu_{n-1} + \frac{q_v}{q_v + 1}\, \mu_{n+1}\,. \tag{8.4}$$

The right-hand side of (8.4) consists of terms of the same parity in n. Hence:

Proposition 8.1 *For every $n \in \mathbb{N}$, there exists a (unique) polynomial Q_n of degree n such that $\mu_n = Q_n(\mu_1)$. If n is even (respectively, odd), then Q_n is even (respectively, odd).*

Let us fix a reference vertex $v_0 \in V_+$ and write $|v| = \operatorname{dist}(v, v_0)$. It follows from (8.4) that the vector space $\mathfrak{R}$ is a commutative algebra generated by μ_1, and on every ray $[v_0, v_1, \dots)$ each γ-eigenfunction of μ_1 radial around v_0 satisfies for every $n \in \mathbb{N}$ the recurrence relation

$$\begin{aligned} &\gamma\, f(v_0) = f(v_1) \qquad \text{if } |v_1| = 1;\\ &\gamma\, f(v_n) = \frac{1}{q_{v_n} + 1}\, f(v_{n-1}) + \frac{q_{v_n}}{q_{v_n} + 1}\, f(v_{n+1}) \qquad \text{if } |v_n| = n > 0. \end{aligned} \tag{8.5}$$

In particular, for each eigenvalue γ there is exactly one γ-eigenfunction f radial around any given vertex v_0 and satisfying the initial condition $f(v_0) = 1$. If $\gamma = 0$, this eigenfunction must vanish on each v with $|v| = 1$ by the first identity in (8.5), henceafter on all of V_- by the second identity; it is now easy to compute this eigenfunction, that we denote by $\psi(v, v_0 \,|\, 0)$, in the usual case $v_0 \in V_+$:

$$\psi(v, v_0 \,|\, 0) = \begin{cases} (-1)^{\frac{|v|}{2}} q_-^{-\frac{|v|}{2}} & \text{if } |v| \text{ is even},\\ 0 & \text{if } |v| \text{ is odd}. \end{cases} \tag{8.6}$$

The usual terminology for this eigenfunction defined on all of V is $\phi(v, v_0 \,|\, 0)$ [7]: notation here is different because $\phi(v, v_0 \,|\, 0)$ will denote the nomalized radial γ-eigenfunction of μ_2 acting on V_+, that is a function defined on.ly on V_+.

Let

$$p_{\text{crit}} = \frac{\ln(q_+ q_-)}{\ln q_-}. \tag{8.7}$$

If $C(n) = \{v\colon |v| = n\}$, it is immediately seen that, for every n, $|C(2n)| = (q_+ q_-)^n$. This yields the following ℓ^p-behavior:

$$\psi(\,\cdot\,, v_0 \mid 0) \in \ell^p(V) \text{ if and only if } p > p_{\text{crit}}. \tag{8.8}$$

In particular, $\psi(\,\cdot\,, v_0 \mid 0) \in \bigcap_{p > p_{\text{crit}}} \ell^p(V)$. Note that $p_{\text{crit}} = 2$ in the homogeneous setup $q_+ = q_-$, and $p_{\text{crit}} > 2$ if and only if $q_+ > q_-$. Therefore

$$\psi(\,\cdot\,, v_0 \mid 0) \in \ell^2(V) \iff q_+ < q_-. \tag{8.9}$$

As a consequence, if (and only if) $q_+ < q_-$, then 0 belongs not only to $\text{sp}(\mu_1;\, \ell^2(V))$, but also to the discrete spectrum of μ_1 as an operator on ℓ^2 and more generally on ℓ^p for every $p \geqslant 2$.

Definition 8.1 For $\gamma \in \mathbb{C}$, the γ-eigenfunction of μ_1 radial around v_0 with value 1 at v_0 is called the *spherical function* on V with eigenvalue γ.

8.3 The Convolution Algebra of Radial Functions and Its Multiplicative Functionals

Given a locally compact group G and a compact subgroup K, the pair (G, K) is a *Gelfand pair* if the convolution algebra $L^1(K\backslash G/K)$ is commutative. Here we are interested in the set-up where $G = \mathcal{G}$ is the group of isometriess of an infinite homogeneous or semi-homogeneous tree.

The isotropy subgroup $\mathcal{G}_v$ of $\mathcal{G}$ at any $v \in V$ is compact, and $\mathcal{G}/\mathcal{G}_v$ is discrete. If T is homogeneous, $\mathcal{G}$ acts transitively upon V, and $\mathcal{G}/\mathcal{G}_v$ is in bijection with V. Moreover, $\mathcal{G}_v$ acts transitively upon the circle C_n of vertices at any distance n from v, or equivalently, the action of $\mathcal{G}$ is doubly transitive. On the other hand, if T is strictly semi-homogeneous (i.e.,semi-homogeneous but not homogeneous), $\mathcal{G}$ has the two orbits $V_\pm \subsetneq V$, and for any $v_+ \in V_+$, $v_- \in V_-$, $n \in \mathbb{N}$, $K_{v_\pm}$ acts transitively upon the circle of vertices in $V_\pm$ at distance n from $v_\pm$, respectively. Therefore $\mathcal{G}_{v_\pm}$ in bijection with $V_\pm$.

From now on, we shall restrict attention to the strictly semi-homogeneous setting, and to the orbit V_+, with reference vertex v_0.

By this bijection, summable functions on the discrete space V_+ lift to summable functions on $\mathcal{G}$ (with respect to its Haar measure). We now show under which conditions the convolution product on $\mathcal{G}$ gives rise to a convolution product on $\ell^1(V_+)$, and in particular on finitely supported functions therein; we shall often restrict attention to this space. Note that the liftings from $\mathcal{G}/\mathcal{G}_{v_0}$ to $\mathcal{G}$ identify right $\mathcal{G}_{v_0}$-invariant functions on $\mathcal{G}$ with functions on V_+. Let us write $K = \mathcal{G}_{v_0}$: then the two-sided K invariant functions on $\mathcal{G}$ are *radial* functions on V_+, in the sense that they depend only on the distance from v_0. We denote the corresponding ℓ^1 space by $\ell^1_\#(V_+)$. We shall show that $(\mathcal{G}, \mathcal{G}_{v_0})$ is a Gelfand pair, that is, $\ell^1_\#(V_+)$ is an abelian convolution algebra. All this also works word by word for V_-. In the special case of homogeneous trees, the convolution induced by $\mathcal{G}$ was studied in [4, 6]; some preliminary facts about convolution in the semi-homogeneous settings are in [3, 5],

Consider a function $f : \mathcal{G} \mapsto \mathbb{C}$, and the Haar measure μ on $\mathcal{G}$ normalized on $\mathcal{G}_{v_0}$. The usual definition of convolution for functions on $\mathcal{G}$ is $u_1 * u_2(\tau) = \int_{\mathcal{G}} u_1(\lambda^{-1}\tau)\, u_2(\lambda)\, d\mu(\lambda)$; this product is associative. Let $\lambda \mapsto \widetilde{\lambda}$ be the canonical projection of $\mathcal{G}$ to $\mathcal{G}/\mathcal{G}_{v_0}$, $\widetilde{\mu}$ the quotient measure of μ (that is, the counting measure) on $\mathcal{G}/\mathcal{G}_{v_0}$. Let us assume that u_2 is right-$\mathcal{G}_{v_0}$-invariant and u_1 bi-$\mathcal{G}_{v_0}$-invariant. Then on $\ell^1(V_+)$ the convolution can be defined as a product between functions u_1 on $\mathcal{G}_{v_0}\backslash\mathcal{G}/\mathcal{G}_{v_0}$ and u_2 on $\mathcal{G}/\mathcal{G}_{v_0}$, as follows:

$$
\begin{aligned}
u_1 * u_2(\widetilde{\tau}) &= \int_{\mathcal{G}_{v_0}} \int_{\mathcal{G}_{v_0}} u_1(\kappa^{-1}\lambda^{-1}\tau)\, u_2(\lambda\kappa)\, d\mu(\kappa)\, d\widetilde{\mu}(\widetilde{\lambda}) \\
&= \int_{\mathcal{G}_{v_0}} u_1\left(\lambda^{-1}\tau\right)\, u_2(\lambda)\, d\widetilde{\mu}(\widetilde{\lambda}) = \int_{\mathcal{G}_{v_0}} u_1\left(\widetilde{\lambda^{-1}\tau}\right)\, u_2(\widetilde{\lambda})\, d\widetilde{\mu}(\widetilde{\lambda}).
\end{aligned}
$$

From now on, we shall use the term *radial* function on V_+ (around v_0) instead of bi-$\mathcal{G}_{v_0}$-invariant function on $\mathcal{G}$. Because of the bijections introduced above, the last identity defines a convolution product between functions on V_+ and radial functions on V_+. We shall now show what happens if also u_1 is radial.

Let us regard functions f and g on $\mathcal{G}$ as functions on V_+, with f radial; by identifying f with a function on $\mathbb{N}$, their convolution on V_+ becomes

$$
f * g(v) = \sum_{v' \in V} f\big(\operatorname{dist}(v, w)\big)\, g(w). \tag{8.10}
$$

In view of this definition, for every n it is equivalent write $\mu_n * g$ instead of $\mu_n g$, by identifying the operator μ_n of (8.3) with the function $\mu_n \delta_{v_0}$, and we shall often do so in the sequel.

Every $\tau \in \mathcal{G}/\mathcal{G}_{v_0}$ extends to an operator on functions on V_+ by the rule

$$
\tau f(v) = f(\tau^{-1} v). \tag{8.11}
$$

Definition 8.2 For every $v \in V_+$ let us choose and fix $\tau[v] \in \mathcal{G}$ such that $\tau[v]v_0 = v$; the choice of $\tau[v]$ is determined only up to elements of its right coset modulo $\mathcal{G}_{v_0}$.

Then

$$|\tau[v]^{-1}v_0| = |\tau[v]v_0| = |v|; \tag{8.12}$$

moreover, the definition (8.10) of convolution of f and g with f radial is equivalent to

$$f * g(v) = \langle \tau[v]\, f,\ \overline{g}\rangle, \tag{8.13}$$

where $\langle\cdot,\ \cdot\rangle$ denotes the inner product in $\ell^2(V_+)$ (with respect to the counting measure). Of course, this implies

$$f * g(v_0) = \langle f,\ \overline{g}\rangle.$$

and, for $f, g, h \in \ell^1_\#(V_+)$,

$$\langle f,\ g * h\rangle = f * (g * h)(v_0) = (f * g) * h(v_0) = \langle f * g, h\rangle. \tag{8.14}$$

Therefore, if f and g are both bi-$\mathcal{G}_{v_0}$-invariant, i.e., radial around v_0, and $v \in V_+$,

$$f * g(v) = \sum_{w\in V_+} f\big(\mathrm{dist}(v,\ w)\big)\, g\big(\mathrm{dist}(w,\ v_0)\big). \tag{8.15}$$

It follows easily that the convolution of radial functions is radial, hence $\ell^1_\#(V_+)$ is a convolution algebra. This algebra is the closure in the ℓ^1 norm of the algebra $\mathfrak{R}_\#(V_+)$ of radial finitely supported functions, that, by (8.4), is the span of the convolution powers of μ_1, hence it is commutative:

$$f * g = g * f \tag{8.16}$$

for all radial functions f, g such that the convolution makes sense.

For all functions f, g on V_+ with f radial, and $\tau \in \mathcal{G}_{v_0}$, one has $\tau(f * g) = f * \tau g$, where τg is defined by $\tau g(v) = g(\tau^{-1}v)$. Indeed, for $v \in V_+$,

$$\begin{aligned}
\tau(f * g)(v) &= \sum_{v'\in V_+} g(v')\, f\big(\mathrm{dist}(\tau^{-1}v,\ v')\big)\\
&= \sum_{v''\in V_+} g(\tau^{-1}v'')\, f\big(\mathrm{dist}(\tau^{-1}v,\ \tau^{-1}v'')\big)\\
&= \sum_{v''} g(\tau^{-1}v'')\, f\big(\mathrm{dist}(v,\ v'')\big) = f * \tau g(v).
\end{aligned} \tag{8.17}$$

Denote by $\mathcal{E}$ the radialization operator around v_0 on finitely supported functions, that is, $\mathcal{E}g(v) = \dfrac{1}{\#C(|v|)}\sum_{w\in C(|v|)} g(w)$, where $C(|v|)$ denotes the circle $\{w\colon |w| = |v|\}$. Then, by (8.17), for all $f, g \in \ell^1(V)$,

$$\mathcal{E}(f * g) = f * \mathcal{E}g \quad \text{if } f \text{ is radial.} \tag{8.18}$$

In particular, the operator μ_n of (8.3), lifted to functions on V, becomes the convolution operator by $\mu_n = \mathcal{E}\delta_v$ for $|v| = n$, whose value $\mu_n(v)$ is $p_n(v_0, v)$, where p_n is the n-step isotropic transition probability induced by the operator. Therefore, by (8.10) (see also [7]),

$$\mu_2 * \mu_{2n} = \frac{1}{(q_+ + 1)q_-}\big(\mu_{2n-2} + (q_- - 1)\mu_{2n} + q_+q_-\mu_{2n+2}\big). \tag{8.19}$$

It follows that the radial convolution algebra $\ell^1_\#(V_+)$ is generated by μ_2, hence it is abelian.

By (8.19) we obtain

Proposition 8.2 *For every $n \in \mathbb{N}$, there exists a (unique) polynomial P_n of degree n such that $\mu_{2n} = P_n(\mu_2)$.*

Definition 8.3 For $\gamma \in \mathbb{C}$, the γ-eigenfunction of μ_2 defined on V_+ and radial around v_0 with value 1 at v_0 is called *spherical function on V_+* with eigenvalue γ and is denoted by $\phi(\,\cdot\,, v_0 \mid \gamma)$. The eigenvalue property implies that $\phi(v, v_0 \mid \gamma)$ takes the value γ if $\operatorname{dist}(v, v_0) = 2$, and by (8.19) it is uniquely determined by γ, and is real-valued if and only if $\gamma \in \mathbb{R}$.

Lemma 8.1 *On a semi-homogeneous tree, the following properties of a function $\phi \not\equiv 0$ on V_+ are equivalent:*

- (*i*) *ϕ is a spherical function on V_+;*
- (*ii*) *for all $v, w \in V_+$, $\mathcal{E}\big(\tau[v]\phi\big)(w) = \phi(v)\phi(w)$;*
- (*iii*) *the functional $L_\phi(f) = \langle f, \phi\rangle$ is multiplicative on the convolution algebra $\ell^1_\#(V_+)$.*

Proof Let $u, v \in V_+$ and $|w| = 2n$. By Proposition 8.2, there exists a polynomial P_n of degree $2n$ such that $\mu_{2n} = P_n(\mu_2)$. Then μ_n acts on ϕ and $\mu_{2n} * \phi = P_n(\mu_2)\phi = P_n(\gamma)\phi$. By (8.18), $\mathcal{E}(\tau[v]\phi)(w) = \langle \tau[v]\phi, \mu_{2n}\rangle = \phi * \mu_{2n}(v^{-1}) = P_n(\gamma)\phi(v^{-1}) = P_n(\gamma)\phi(v)$. On the other hand, since ϕ is radial, $\phi(w) = \langle \mu_{2n}, \phi\rangle = \mu_{2n} * \phi(v_0) = P_n(\gamma)\phi(v_0) = P_n(\gamma)$. Therefore (i) implies (ii).

Now let ϕ be a non-zero function on V_+ that satisfies (ii) and choose v such that $\phi(v) \neq 0$. If (ii) holds, for every $w \in V_+$ we have $\phi(w) = \mathcal{E}(\tau[v]\phi)(w)/\phi(v)$. Therefore ϕ is radial; moreover, for all radial f, g on V_+, by (8.16), (8.13) and (8.12),

$$\begin{aligned}
L_\phi(f * g) &= \langle f * g, \overline{\phi}\rangle = \langle f, \overline{g * \phi}\rangle = \langle f, \overline{\phi * g}\rangle \\
&= \big\langle \langle \tau[\,\cdot\,] f, g\rangle, \phi(\,\cdot\,)\big\rangle = \sum_v f(v) \sum_u \big(\tau[v]^{-1}\phi\big)(u) g(u) \\
&= \sum_v f(v) \mathcal{E}\big(\tau[v]^{-1}\phi\big)(u) g(u) = \sum_{v,w \in V_+} f(v)\, g(w)\, \phi(v)\, \phi(w) \\
&= L_\phi(f)\, L_\phi(g),
\end{aligned} \tag{8.20}$$

since ϕ is radial. Thus (ii) implies (iii).

If (iii) holds, $f \in \ell^1_\#(V_+)$ and ϕ_2 denotes the value of the radial function ϕ on vertices of length 2, then

$$L_\phi(\mu_2 * f) = \langle \phi,\ \mu_2\rangle\, \langle \phi,\ \overline{f}\rangle = \phi_2\, \langle \phi,\ \overline{f}\rangle.$$

On the other hand, by (8.14),

$$L_\phi(\mu_2 * f) = \langle \phi,\ \mu_2 * \overline{f}\rangle = \langle \phi * \mu_2,\ \overline{f}\rangle.$$

Since this holds for each radial function f, it follows that ϕ is an eigenfunction of μ_2 (with eigenvalue ϕ_2). Moreover, $\phi(v_0) \neq 0$, because a radial eigenfunction of μ_2 that vanish at v_0 must vanish everywhere. Now,

$$\phi(v_0) = \phi * \delta_{v_0}(v_0) = \phi * (\delta_{v_0} * \delta_{v_0})(v_0) = \phi(v_0)^2, \tag{8.21}$$

hence $\phi(v_0) = 1$, and (i) follows.

Corollary 8.1 *If ϕ is a spherical function, then $\phi(v_0) = 1$, $\mu_2 * \phi = \gamma\phi$ with $\gamma = \phi(v)$ for $|v| = 2$, and $\mu_{2n} * \phi = P_n(\gamma)\, \phi$, where P_n is the polynomial of Proposition 8.2.*

Proof By (8.21), $\phi(v_0) = L_\phi(\delta_{v_0}) = 1$. Moreover, $\mu_2 * \phi = \mathcal{E}(\tau[v]\, \phi)$ for any v with $|v| = 2$. Therefore, by Lemma 8.1 (ii), $\mu_2 * \phi = \phi(v)\, \phi$ for every such v. The remainder of the statement follows from Proposition 8.2.

Definition 8.4 $\ell^1(V_+)$ and $\ell^1_\#(V_+)$ are involutive algebras equipped with the involution $f^*(v) = f^*(\tau[v]v_0) = \overline{f\big(\tau[v]^{-1}v_0\big)}$.

8.4 Positive Definite Spherical Functions on V_+

It is well known [8, Theorem 14.3.5] that a function is positive definite for $\mathcal{G}$ if it induces a (continuous) positive functional on the involutive algebra $L^1(\mathcal{G})$, or equivalently if and only if it is a matrix coefficient of a unitary representation π_ϕ of $\mathcal{G}$, acting on a vector space $\mathcal{H}$ equipped with an inner product, of the type $\phi(\tau) = \langle \pi_\phi(\tau)h, h\rangle$ for some $h \in \mathcal{H}$. Here we look for functions on $V_+ \approx \mathcal{G}/\mathcal{G}_{v_0}$, that is, right-$\mathcal{G}_{v_0}$-invariant on $\mathcal{G}$, and we take advantage of the fact that $\mathcal{G}/\mathcal{G}_{v_0} \approx \mathcal{F}$ is now a group in order to transfer the usual theory of positive definite functions on groups to this homogeneous space.

A continuous positive definite function on $\mathcal{G}$ is right-$\mathcal{G}_{v_0}$-invariant if and only if it induces a positive continuous functional on the involutive algebra $L^1(\mathcal{G}/\mathcal{G}_{v_0})$, and we take advantage of the fact that, $L^1(\mathcal{G}/\mathcal{G}_{v_0}) \approx \ell^1(V_+)$. Such a function will be called *positive definite on* V_+.

By [8, Theorem 14.3.5], a function on $\mathcal{G}$, is positive definite on V_+ if and only if it is a matrix coefficient of a unitary representation π_ϕ of $\mathcal{G}$, acting on a vector space $\mathcal{H}$ equipped with an inner product, of the type

$$\phi(v) = \langle \pi_\phi(\tau[v])h, h\rangle \qquad \text{for some } h \in \mathcal{H} \text{ invariant under } \pi_\phi(\mathcal{G}_{v_0}). \tag{8.22}$$

Let ϕ be a positive definite function on V_+ and $\mathcal{V}_\phi$ the linear span of all translates $\tau\phi\colon \tau \in \mathcal{G}$. Since ϕ is right-$\mathcal{G}_{v_0}$-invariant, $\mathcal{V}_\phi = \{\tau[v]\phi\colon v \in V_+\}$. Then ϕ induces a positive semi-definite inner product on $\mathcal{V}_\phi$ by the rule

$$\langle \tau[v]\phi,\ \tau[w]\phi\rangle_\phi = \phi\big(\tau[v]^{-1}w\big) \qquad \text{for every } v, w \in V_+. \tag{8.23}$$

From now on, we shall denote by $\langle\,\cdot\,,\ \cdot\,\rangle_\phi$ this inner product and with $\langle\,\cdot\,,\ \cdot\,\rangle$ the ℓ^2-inner product. By (8.23), for every $v \in V_+$,

$$\|\tau[v]\phi\|_\phi^2 = \phi(v_0). \tag{8.24}$$

Denote by $\mathcal{N}_\phi$ the subspace of $\mathcal{V}_\phi$ of all function f such that $\langle f,\ f\rangle_\phi = 0$, and $\mathcal{H}_\phi = \mathcal{V}_\phi/\mathcal{N}_\phi$. Then $\|f\|_\phi = \sqrt{\langle f,\ f\rangle_\phi}$ is a Hilbert space norm on $\mathcal{H}_\phi$ (the so-called *GNS-norm*). Let us normalize h in (8.22) so that $\phi(v_0) = 1$; then (8.24) becomes $\|\pi_\phi(\tau[v])\phi\|_\phi = \|\tau[v]\phi\|_\phi = 1$ for every v. More generally, the GNS-representation generated by a normalized definite function ϕ (that is, with $\phi(v_0) = 1$) is unitary in the GNS-norm.

Let now $\phi = \phi(\,\cdot\,, v_0\,|\,\gamma)$ and write $\mathcal{V}_\gamma$ instead of $\mathcal{V}_\phi$, i.e., the linear span of the functions $\{v \mapsto \phi(\tau v, v_0\,|\,\gamma)\colon \tau \in \mathcal{G}\} = \{v \mapsto \phi(\tau[w]v, v_0\,|\,\gamma)\colon w \in V_+\}$. Then $\mathcal{V}_\gamma$ is invariant under $\mathcal{G}$, and the action of $\mathcal{G}$ gives rise to an (algebraic) representation π_γ of $\mathcal{G}$ on $\mathcal{V}_\gamma$. If ϕ is positive definite, then π_γ extends to a topological representation of $\mathcal{G}$ on the Hilbert space given by the GNS-norm closure $\overline{\mathcal{H}}_\gamma$ of $\mathcal{H}_\gamma = \mathcal{V}_\gamma/\mathcal{N}_\gamma$, called the spherical representation at the eigenvalue γ. Since $\phi(v_0) = 1$, it follows from (8.24) that

$$\|\pi_\gamma(\tau[v])\phi\|_\phi = \|\tau[v]\phi\|_\phi = 1 \qquad \text{for every } v \in V_+, \tag{8.25}$$

and, more generally, π_γ is a unitary representation of $\mathcal{G}$ on $\overline{\mathcal{H}}_\gamma$.

We emphasize the fact that π_γ acts by left translation on $\overline{\mathcal{H}_\gamma}$.

Corollary 8.2 *A spherical function is positive definite on V_+, that is, it defines a continuous positive functional on the involutive algebra $\ell^1(V_+)$, if and only if it is bounded and real-valued.*

Proof Let us first assume that a spherical function ϕ be bounded and real-valued, and show that L_ϕ is a continuous positive functional, i.e., ϕ is positive definite with respect to $\mathcal{G}$. Observe that ϕ is radial, hence, by (8.12),

$$\phi\big(\tau[v]^{-1}v_0\big) = \phi(v). \tag{8.26}$$

By Lemma 8.1, or more directly by (8.20), since ϕ is real-valued,

$$\begin{aligned}
L_\phi(f^* * f) = L_\phi(f^*)\, L_\phi(f) &= \Big(\sum_{v\in V_+} \overline{f\big(\tau[v]^{-1}v_0\big)}\,\phi(v)\Big)\Big(\sum_{w\in V_+} f(w)\phi(w)\Big)\\
&= \Big(\sum_{v\in V_+} \overline{f(v)}\,\phi(\tau[v]v_0)\Big)\Big(\sum_{w\in V_+} f(w)\phi(w)\Big)\\
&= \Big(\sum_{w\in V_+} \overline{f(v)}\,\phi(v)\Big)\Big(\sum_{w\in V_+} f(w)\phi(w)\Big)\\
&= \langle \overline{f},\, \phi\rangle\,\langle f,\, \phi\rangle = |\langle f,\, \phi\rangle|^2 \geqslant 0,
\end{aligned}$$

hence L_ϕ is a positive functional on the involutive algebra $\ell^1_\#(V_+)$. Now let $h \in \ell^1(V_+)$. By (8.18), $L_\phi(h) = \langle h,\, \phi\rangle = \phi{*}h(v_0) = \phi{*}\mathcal{E}h(v_0) = \langle \mathcal{E}h,\, \phi\rangle = L_\phi(\mathcal{E}h)$. Therefore, by [12, Chapter 3, Lemmas 1.2 and 1.3], L_ϕ is also a positive functional on the involutive algebra $\ell^1(V_+)$.

Conversely, let ϕ be a positive definite spherical function on V_+. Then ϕ is bounded (by its value at v_0), and $\phi\big(\tau[v]^{-1}v_0\big) = \overline{\phi(\tau[v]v_0)}$, as in [8, Subsection 13.4.3]. Moreover, ϕ is radial, hence, it follows from (8.26) that $\phi(v) = \phi\big(\tau[v]^{-1}v_0\big) = \overline{\phi}(\tau[v]v_0) = \overline{\phi}(v)$ for every $v \in V_+$. Hence ϕ is real-valued.

Theorem 8.1 *If a spherical function $\phi = \phi(\,\cdot\,, v_0\,|\,\gamma)$ on V_+ belongs to $\ell^2(V_+)$, then ϕ is positive definite, and for every function $f_v = \tau[v]\phi$ in $\mathcal{H}_\gamma$, its ℓ^2-norm and its GNS-norm (defined by the inner product (8.23)) are related by $\langle f_v, \phi\rangle_{\ell^2(V_+)} = \|\phi\|^2_{\ell^2(V_+)}\,\langle f_v, \phi\rangle_\phi$ for every $f_v \in \mathcal{H}_\gamma$ of the type $f_v = \tau[v]\phi = \pi_\gamma(\tau[v])\phi$. Moreover, the unitary representation π_γ of $\mathcal{G}$ acting by translations on $\overline{\mathcal{H}_\gamma}$ is unitarily equivalent to the quasi-regular representation of $\mathcal{G}$ on $\ell^2(V_+)$, or equivalently to the regular representation of $\mathcal{F}$ on $\ell^2(V_+)$, by regarding V_+ as a group isomorphic to $\mathcal{F}$.*

Proof The GNS-representation π_γ associated to $\phi(\,\cdot\,, v_0\,|\,\gamma)$ is the left regular representation of $\mathcal{G}$ on the subspace of functions on $\mathcal{G}/\mathcal{G}_{v_0} \equiv V_+$ generated by translates of ϕ, that is, $\pi_\phi(\tau[v])h(w) = h(\tau[v]^{-1}w)$ for every $v, w \in V_+$ and for every function h on V_+ that is a linear combination of translates of ϕ; if $\phi \in \ell^2(V_+) = L^2(\mathcal{G}/\mathcal{G}_{v_0}) = \ell^2(\mathcal{F})$, this is a unitary representation in the $\ell^2(V_+)$-norm. We have seen that π_ϕ is also a unitary representation in the norm of $\ell^2(V_+) \approx L^2(\mathcal{G}/\mathcal{G}_{v_0})$, and we shall show that these two realizations of π_ϕ (the first by translations on the space $\mathcal{H}_\phi$ generated by the translates of ϕ, and the second on the space $\ell^2(V_+)$ generated by the translates of δ_{v_0}) are unitarily equivalent. By Lemma 8.1 (ii),

$$\langle \pi_\gamma(\tau[v])\phi, \phi\rangle = \langle \mathcal{E}(\pi(\tau[v])\phi), \phi\rangle = \phi(v)\, \|\phi\|^2_{\ell^2(V_+)}. \tag{8.27}$$

Therefore the function ϕ is a (positive) multiple of a matrix coefficient of a unitary representation, hence it is positive definite by Definition 8.4.

Moreover, by Corollary 8.2, ϕ is real-valued. Then, by (8.23) and (8.11), for every $v \in V_+$

$$\langle \pi_\gamma(\tau[v])\phi, \phi\rangle_\phi = \langle \phi\big(\tau[v]^{-1}\,\cdot\big), \phi(\,\cdot\,)\rangle_\phi = \langle \tau[v]\phi, \phi\rangle_\phi = \phi(v), \tag{8.28}$$

since the GNS representation π_ϕ acts by left translation (this confirms the assumption that ϕ is a matrix coefficient of π_γ, but tells us also which matrix coefficient it is). By comparing (8.27) and (8.28) we obtain the identity in the statement. On the other hand, by (8.25), $\|f_v\|_\phi = \|\phi\|_\phi = 1 = \|\delta_{v_0}\|_2 = \|\delta_v\|_2$. Therefore the map $\delta_v \mapsto f_v = \pi_\gamma(\tau[v])\phi$ is an isometry of $\ell^2(V_+)$ onto $\mathcal{H}_\gamma$ and intertwines the translation actions of π_γ on these two spaces.

Remark 8.2 If $\phi = \phi(\cdot, v_0 \mid \gamma) \in \ell^2(V_+)$, then $\gamma \in \mathrm{sp}_{\ell^1(V_+)}\, \mu_2$.

Indeed, by Corollary 8.2, a spherical function on V_+ is positive definite if and only if it is real-valued and bounded. On the other hand, it is known [7] that a spherical function on V_+ is real valued and bounded if (and only if) its eigenvalue belongs to $\mathrm{sp}_{\ell^1(V_+)}\, \mu_2$.

8.5 Spherical Representations of $\mathcal{G}$

On the basis of the previous Sections, the results of [12] now open the way to the spherical representation theory of $\mathcal{G}$:

Theorem 8.2 *Let V be the set of vertices of a strictly semi-homogeneous tree. Let us denote by S_1 the spectrum of μ_1 on $\ell^1(V)$, and by S_2 its spectrum on $\ell^2(V)$.*

- (i) *$\phi(\,\cdot\,, v_0 \mid \gamma)$ is real-valued and bounded if and only if $\gamma \in S_1 \cap \mathbb{R}$. For these eigenvalues, the spherical function is positive definite and the GNS representation π_γ is a unitary representation on the GNS Hilbert space $\overline{\mathcal{H}_\gamma}$.*
- (ii) *The Plancherel measure associated to the regular representation ρ of $\mathcal{G}$ on $\ell^2(V_+)$ is absolutely continuous with respect to Lebesgue measure on $S_2 \subsetneq \mathbb{R}$,*

except for an atom at the eigenvalue 0 *if (and only if)* $q_+ < q_-$. *For these eigenvalues* γ, *the representation* π_γ *is weakly contained in the regular representation* ρ.

(*iii*) *If* $\gamma = 0$ *and* $q_+ < q_-$, *then* $\phi(\,\cdot\,, v_0 \mid 0) \in \ell^2(V)$ *and, on the norm-closure of* $\mathcal{V}_0$, π_0 *is a square-integrable representation, that is, a subrepresentation of the regular representation of* $\mathcal{G}$ *on* $\ell^2(V_+)$.

(*iv*) *All the representations* π_γ *on* $\overline{\mathcal{H}}_\gamma$ *for* $0 \neq \gamma \in \operatorname{Int} S_1$ *are irreducible.*

(*v*) *The representation* π_γ *can be defined also on spaces of functions on* V_- *and on* V, *where it becomes the regular representation* ρ *given by left translations. Since the action of* $\mathcal{G}$ *preserves* $V_\pm$, *the representations* ρ *and every* π_γ, *when regarded on spaces of functions on* V, *decompose as direct sum of their restrictions to* V_+ *and* V_-. *In particular, for* $\gamma \in \operatorname{Int} S_1$, π_γ *is the direct sum of two irreducible subrepresentations.*

Proof For part (*i*) it is enough to show that $\phi(\,\cdot\,, v_0 \mid \gamma)$ is real-valued and bounded if and only if $\gamma \in S_1$, and that the norm $\|\cdot\|_\phi$ is the ℓ^2-norm if and only if $\gamma \in S_2$: since γ is the value of $\phi(\,\cdot\,, v_0 \mid \gamma)$ on the neighbors of v_0 and ϕ is real-valued if and only if $\gamma \in \mathbb{R}$, this follows from Corollary 8.2. See also [7, Theorem 5.8].

For part (*ii*), we observe, as in [7] (see also [5]), that the set V_+, equipped with its natural adjacency relation, i.e., regarded as the Cayley graph of the step-2 isotropic operator μ_2, becomes a polygonal graph in the sense of [16], therefore the regular representation of $\mathcal{G}$ on $\ell^2(V_+)$ can be realized by the action of $\mathcal{G}$ on this polygonal graph. In this setting, the Plancherel measure m_ρ was computed in [9, 17]. An accurate comparison of our current setting with that of [9, 16, 17] shows that m_ρ is absolutely continuous with respect to Lebesgue measure, with a continuous Radon-Nykodim derivative, except for an atom at $\gamma = 0$ if $q_+ < q_-$, and ρ decomposes on $\ell^2(V_+)$ as $\rho = \int_{S_2}^{\oplus} \pi_\gamma \, dm(\gamma)$. For each $\gamma \in S_2$ and for $\delta > 0$ the matrix coefficients of π_γ is a uniform limit on finite sets, as $\delta \to 0$, of coefficients of the representation $\frac{1}{\delta} \int_{U_\delta}^{\oplus} \pi_\gamma \, dm(\gamma)$ where U_δ is a neighborhhod of γ of radius δ. Therefore π_γ is weakly contained in ρ.

Part (*iii*) follows from Theorem 8.1 (see also (8.9)).

Part (*iv*) is proved as in the setting of homogeneous trees [11, 12]; see also [10, 13].

Finally, part (*v*) is clear.

References

1. P. Cartier, Harmonic analysis on trees, in *Harmonic Analysis on Homogeneous Spaces (Proc. Sympos. Pure Math., Vol. XXVI, Williams Coll., Williamstown, Mass., 1972)* (Amer. Math. Soc., Providence, 1973), pp. 419–424
2. P. Cartier, Géométrie et analyse sur les arbres, in *Séminaire Bourbaki, 24ème année*, Exp. No. 407. Lecture Notes in Math., vol. 317 (Springer, Berlin, 1973), pp. 123–140
3. E. Casadio Tarabusi, J.M. Cohen, A. Korányi, M.A. Picardello, Converse mean value theorems on trees and symmetric spaces. J. Lie Theory **8**(2), 229–254 (1998)
4. E. Casadio Tarabusi, S.G. Gindikin, M.A. Picardello, Integral geometry and harmonic analysis on homogeneous trees. arXiv 2002.09099 (monograph in preparation)

5. E. Casadio Tarabusi, M.A. Picardello, The algebras generated by the Laplace operators in a semi-homogeneous tree, in *Trends in Harmonic Analysis*. Springer INDAM Series, vol. 3 (Springer-Verlag Italia, Milan Heidelberg New York Dordrecht London, 2013), pp. 77–90. https://doi.org/10.1007/978-88-470-285-1_5
6. E. Casadio Tarabusi, M.A. Picardello, Radon transform on hyperbolic spaces and their discrete counterparts. Complex Anal. Oper. Theory **15**(1), 2021, paper no. 13. https://doi.org/10.1007/s11785-020-01055-6
7. E. Casadio Tarabusi, M.A. Picardello, Spherical functions and spectrum of the Laplacian on semi-homogeneous trees. arXiv 2208.00910 (submitted for publication)
8. J. Dixmier, *Les C^*-algèbres et leurs representations*. Cahiers Scientifiques, vol. XXIX (Gauthier-Villars, Paris, 1969)
9. J. Faraut, M.A. Picardello, The Plancherel measure for symmetric graphs. Ann. Mat. Pura Appl. **138**, 151–155 (1984)
10. A. Figà-Talamanca, C. Nebbia, *Harmonic Analysis and Representation Theory for Groups Acting on Homogeneous Trees*. London Mathematical Society Lecture Notes Series, vol. 61 (Cambridge University Press, Cambridge, Port Chester, Melbourne, and Sydney, 1991), pp. ix+151. ISBN 0-521-42444-5
11. A. Figà-Talamanca, M.A. Picardello, Spherical functions and harmonic analysis on free groups. J. Funct. Anal. **47**, 281–304 (1982)
12. A. Figà-Talamanca, M.A. Picardello, *Harmonic Analysis on Free Groups*. Lecture Notes in Pure and Applied Mathematics, vol. 87 (Marcel Dekker Inc., New York, 1983), pp. viii+145. ISBN 0-8247-7042-0
13. A. Figà-Talamanca, T. Steger, *Harmonic Analysis for Anisotropic Random Walks on Homogeneous Trees*. Mem. Amer. Math. Soc., vol. 110 (Amer. Math. Soc., Providence, 1994), pp. viii+68
14. U. Haagerup, An example of a non-nuclear C^*-algebra which has the metric approximation property. Inventiones Math. **50**, 279–293 (1979)
15. A. Iozzi, M.A. Picardello, Graphs and convolution operators, in *Topics in Modern Harmonic Analysis*, Turin/Milan, 1982, vol. I, II (INdAM, Rome, 1983), pp. 187–208
16. A. Iozzi, M.A. Picardello, Spherical functions on symmetric graphs, in *Harmonic Analysis*, Cortona, 1982. Lecture Notes in Math., vol. 992 (Springer, Berlin, 1983), pp. 344–386. ISBN 3-540-12299-0. https://doi.org/10.1007/BFb0069168
17. M.G. Kuhn, P.M. Soardi, The Plancherel measure for polygonal graphs. Ann. Mat. Pura Appl. **134**, 393–401 (1983)

Chapter 9
Trees of Fractions

A. Muhammed Uludağ

To Athanase, whose friendship transcends numbers and shapes.

Abstract This is an exposition on the submonoids of the monoid $\mathsf{GL}_2(\mathbb{N})$ and the related trees of Calkin–Wilf, Stern–Brocot, Bird, Drib, Yu–Ting–Andreev, Hanna–Czyz–Self, Kepler and Lebesgue.

9.1 The Monoids $\mathsf{SL}_2(\mathbb{N})$ and $\mathsf{GL}_2(\mathbb{N})$

By definition, $\mathsf{SL}_2(\mathbb{N})$ is the monoid of unimodular (i.e. of determinant 1) 2×2 matrices with non-negative integral entries. It is a submonoid of the famed group $\mathsf{SL}_2(\mathbb{Z})$ of all unimodular 2×2 matrices with integral entries.

Proposition 9.1 $\mathsf{SL}_2(\mathbb{N})$ *is the free monoid generated by the matrices*

$$\overline{T} := \begin{pmatrix} 1 & 1 \\ 0 & 1 \end{pmatrix}, \quad \overline{T}^t := \begin{pmatrix} 1 & 0 \\ 1 & 1 \end{pmatrix}$$

Proof Let $\overline{M} \in \mathsf{SL}_2(\mathbb{N})$ with

$$\overline{M} := \begin{pmatrix} p & q \\ r & s \end{pmatrix}.$$

A. M. Uludağ (✉)
Department of Mathematics, Galatasaray University, Beşiktaş, Istanbul, Türkiye
e-mail: muludag@gsu.edu.tr

A. M. Uludağ, A. Zeytin (eds.), *Essays on Geometry*,
https://doi.org/10.1007/978-3-031-76257-4_9

In case $qr = 0$, then $\overline{M}$ is either a power of $\overline{T}$ or $\overline{T}^t$. If $qr \neq 0$, suppose $q > p$. Then $s > r$. Then

$$\overline{M} = \begin{pmatrix} p & q-p \\ r & s-r \end{pmatrix} \begin{pmatrix} 1 & 1 \\ 0 & 1 \end{pmatrix}$$

Similarly if $p > q$, then $r > s$ so that

$$\overline{M} = \begin{pmatrix} p-q & q \\ r-s & s \end{pmatrix} \begin{pmatrix} 1 & 0 \\ 1 & 1 \end{pmatrix}$$

and it follows by induction that $\overline{M}$ is written as a word in $\overline{T}$ and $\overline{T}^t$. Suppose

$$\overline{M} = \overline{T}^{n_1}(\overline{T}^t)^{n_2} \dots \overline{T}^{n_k} = (\overline{T}^t)^{m_1}\overline{T}^{m_2} \dots (\overline{T}^t)^{m_l}$$

with $n_i, m_i > 0$. One can write $\overline{M} = \overline{T}^{n_1}(\overline{T}^t)^{n_2} \dots \overline{T}^{n_k} = \overline{XT}$ where

$$\overline{X} = \begin{pmatrix} p & q \\ r & s \end{pmatrix}$$

is a non-negative matrix. It follows that

$$\overline{M} = \overline{XT} = \begin{pmatrix} p & q+p \\ r & s+r \end{pmatrix}. \tag{9.1}$$

Similarly, one can write $\overline{M} = \overline{T}^{n_1}(\overline{T}^t)^{n_2} \dots \overline{T}^{n_k} = \overline{YT}^t$ where

$$\overline{Y} = \begin{pmatrix} a & b \\ c & d \end{pmatrix}$$

is a non-negative matrix. It follows that

$$\overline{M} = \overline{YT} = \begin{pmatrix} a+b & b \\ c+d & d \end{pmatrix}. \tag{9.2}$$

Comparing (9.2) with (9.1), we conclude that $q = a = 0$. Then

$$\overline{M} = \begin{pmatrix} b & b \\ c+d & d \end{pmatrix} \implies b = 1 \implies \det(\overline{M}) = -c \implies c = -1,$$

contradiction. □

Note that $(\overline{T}^{-1}\overline{T}^t)^6 = 1$ in $\mathsf{SL}_2(\mathbb{Z})$.

For the monoid $\mathsf{GL}_2(\mathbb{N})$ of invertible 2×2 matrices with non-negative entries one has

Corollary 9.1 $\mathsf{GL}_2(\mathbb{N})$ *is generated by the matrices* $\overline{T}, \overline{T}^t$ *and the matrix*

$$\overline{U} = \begin{pmatrix} 0 & 1 \\ 1 & 0 \end{pmatrix}.$$

Proof If $\overline{M} \in \mathsf{GL}_2(\mathbb{N})$ is of determinant 1, then $\overline{M} \in \mathsf{SL}_2(\mathbb{N})$ and it can be written as a word in $\overline{T}, \overline{T}^t$ by the previous proposition. If $\det(\overline{M}) = -1$, then $\det(\overline{U}\,\overline{M}) = 1$ and one applies the same argument. □

Note that $\overline{UTU} = \overline{T}^t$ in $\mathsf{GL}_2(\mathbb{N})$. In fact, $\mathsf{GL}_2(\mathbb{N})$ is freely generated by $\overline{U}$ with $\overline{U}^2 = 1$ and $\overline{T}$.

Exercise 9.1 Check if the following assertions are true or false:

- $\mathsf{SL}_2(\mathbb{Z}) = \langle \overline{T}, \overline{T}^t \rangle$.
- $\mathsf{PSL}_2(\mathbb{Z}) = \langle T, T^t \rangle$ (see below).
- $\mathsf{GL}_2(\mathbb{Z}) = \langle \overline{T}, \overline{U} \rangle$.
- $\mathsf{PGL}_2(\mathbb{Z}) = \langle T, U \rangle$ (see below).

9.2 The Calkin–Wilf Tree

The groups $\mathsf{SL}_2(\mathbb{Z})$ and $\mathsf{GL}_2(\mathbb{Z})$ have the projectivizations $\mathsf{PSL}_2(\mathbb{Z}) := \mathsf{SL}_2(\mathbb{Z})/\langle \pm I \rangle$ and $\mathsf{PGL}_2(\mathbb{Z}) := \mathsf{GL}_2(\mathbb{Z})/\langle \pm I \rangle$. Denoting the images of $\mathsf{SL}_2(\mathbb{N})$ and $\mathsf{GL}_2(\mathbb{N})$ under the projectivization by $\mathsf{PSL}_2(\mathbb{N})$ and $\mathsf{PGL}_2(\mathbb{N})$ respectively, we have, obviously, $\mathsf{SL}_2(\mathbb{N}) \simeq \mathsf{PSL}_2(\mathbb{N})$ and $\mathsf{GL}_2(\mathbb{N}) \simeq \mathsf{PGL}_2(\mathbb{N})$.

We denote the projectivizations of $\overline{T}, \overline{T}^t$ and $\overline{U}$ by T, T^t and U, respectively. The projectivization of a matrix will be denoted as a matrix in square brackets.

The free monoid $\mathsf{PSL}_2(\mathbb{N}) = \langle T, T^t \rangle$ acts on $\mathbb{Q}^+$ via

$$M = \begin{bmatrix} p & q \\ r & s \end{bmatrix} : x \in \mathbb{Q}^+ \to Mx := \frac{px+q}{rx+s} \in \mathbb{Q}^+.$$

Proposition 9.2 *The* $\mathsf{PSL}_2(\mathbb{N}) \simeq \mathsf{SL}_2(\mathbb{N})$ *action on* $\mathbb{Q}^+$ *is free.*

Proof Suppose $Mx = x$. Thus the equation

$$\frac{px+q}{rx+s} = x \implies rx^2 + (s-p)x - q = 0$$

has a rational solution. Hence $(s-p)^2 + 4qr = (s+p)^2 - 4$ is a square. Hence $p = s = 1 \implies qr = 0$. If $q = 0$, then $x = 0 \notin \mathbb{Q}^+$. If $r = 0$, then $x = \infty \notin \mathbb{Q}^+$. □

Let $x = a/b \in \mathbb{Q}^+$. Then

$$T\left(\frac{a}{b}\right) = \frac{a+b}{b} \iff T(x) = 1 + x \implies T^k(x) = k + x \quad (k \in \mathbb{N})$$

$$T^t\left(\frac{a}{b}\right) = \frac{a}{a+b} \iff T^t(x) = \frac{x}{1+x} \implies (T^t)^k(x) = \frac{1}{k+\frac{1}{x}} \quad (k \in \mathbb{N})$$

The $\mathsf{PSL}_2(\mathbb{N})$-action on $\mathbb{Q}^+$ is transitive in the sense that every rational is in the orbit of 1. Indeed, given $x = a/b \in \mathbb{Q}^+$, applying T^{-1} to x if $x > 1$, and $(T^t)^{-1}$ if $x < 1$, we always remain inside $\mathbb{Q}^+$ and eventually end up at 1. Going in the opposite direction, this means that we may express every reduced rational by a unique word in $\langle T, T^t \rangle$. Hence, we can form a planar tree, with 1 as its root vertex and such that every vertex has one left and one right child, generated by the rule in Fig. 9.1.

If a/b is a reduced rational, then $T(a/b)$ and $T^t(a/b)$ are also reduced, so every rational on this tree is reduced.

Suppose that $y \in \mathbb{Q}^+$ is a node of the Calkin–Wilf tree reached from the root 1 by applying n_k times T, n_{k-1} times $T^t, \ldots, n_1$ times T^t and n_0 times T. Then since $T^k(x) = k + x$ and $(T^t)^k(x) = 1/(k + 1/x)$ we have the continued fraction

$$y = T^{n_k}(T^t)^{n_{k-1}} \ldots T^{n_2}(T^t)^{n_1}T^{n_0}(1) \implies y = [n_0, n_1, \ldots, n_k], \tag{9.3}$$

where $n_0, n_k \geq 0$ and $n_j > 0$ if $j \neq 0, k$. In particular, if y is found at level N on the tree, then the sum of terms of the continued fraction expansion of y is $n_0 + \cdots + n_k = N$.

Observe that $\mathsf{SL}_2(\mathbb{N})$ has the automorphism exchanging its generator T with T^t (this is **not** the transposition operation $M \to M^t$ on $\mathsf{SL}_2(\mathbb{N})$, which is an anti-homomorphism!). This automorphism sends the y in (9.3) to

$$y = (T^t)^{n_k}T^{n_{k-1}} \ldots (T^t)^{n_2}T^{n_1}(T^t)^{n_0}(1) \implies y = [0, n_0, n_1, \ldots n_k],$$

i.e. this automorphism sends y to $1/y$. This translates as the central symmetry of the Calkin–Wilf tree in Fig. 9.2.

Reading the rows of the Calkin–Wilf tree first from left to right and then from top to bottom, we obtain an enumeration $(x_n)_{n\geq 0}$ of $\mathbb{Q}^+$, such that

$$(x_n)_{n\geq 0} = \left(\frac{1}{1}, \frac{1}{2}, \frac{2}{1}, \frac{1}{3}, \frac{3}{2}, \frac{2}{3}, \frac{3}{1}, \frac{1}{4}, \frac{4}{3}, \frac{3}{5}, \frac{5}{2}, \frac{5}{3}, \frac{3}{4}, \frac{4}{1}, \frac{1}{5}, \frac{5}{4}, \frac{4}{7}, \frac{7}{3}, \frac{3}{8}, \frac{8}{5}, \frac{5}{7} \cdots\right).$$

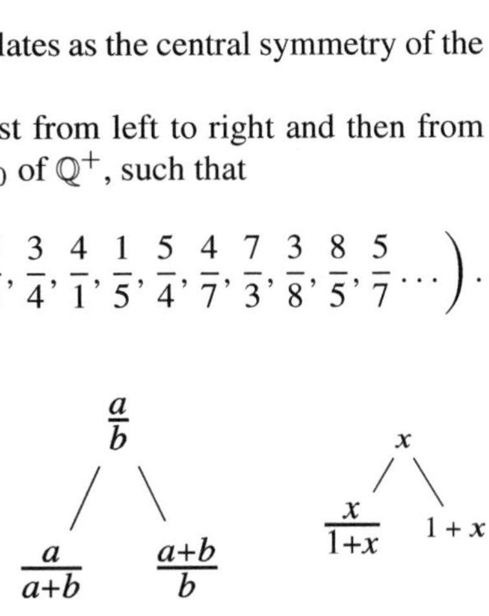

Fig. 9.1 Two equivalent versions of the generation rule of the Calkin–Wilf tree, where $x = a/b$ with $\gcd(a, b) = 1$; $a, b \in \mathbb{N}$

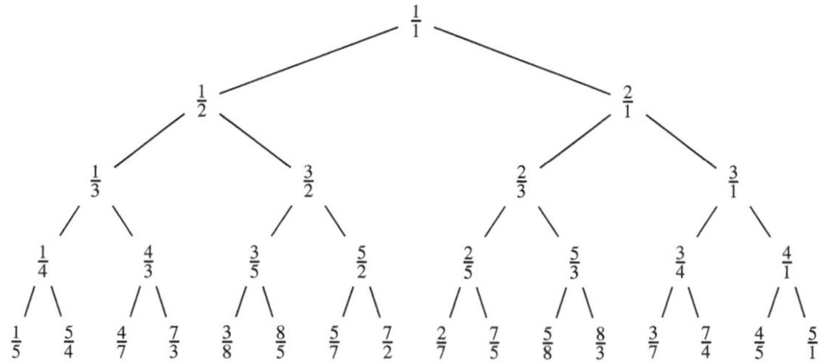

Fig. 9.2 The Calkin–Wilf tree of rationals. This tree is constructed from $1/1$ by successively applying the generation rules $a/b \to a/(a+b)$ (left) and $a/b \to (a+b)/b$ (right). It can also be obtained by evaluating at 1 the nodes of the Calkin–Wilf tree of matrices defined below

Observe that the denominator of each fraction is the numerator of the next one. To prove this, first note that if the fraction is a left child and its successor is the right child, then this is obvious by the generation rule in Fig. 9.1. If the fraction is a right child then its denominator is the same as the denominator of its parent and the numerator of its successor is the same as the numerator of the parent of its successor, hence the result follows by downward induction on the levels of the tree. Since the rightmost vertex of each row has denominator 1, as does the leftmost vertex of the next row, this proves the observation.

It follows that the nth rational number in the sequence can be written as $\text{fusc}(n)/\text{fusc}(n+1)$ $(n = 0, 1, 2, ...)$, where Dijsktra's *fusc function* $\text{fusc} : \mathbb{N} \to \mathbb{N}$ is defined as the sequence of numerators of (x_n):

$$(\text{fusc}(n))_{n\geq 0} = (1, 1, 2, 1, 3, 2, 3, 1, 4, 3, 5, 2, 5, 3, 4, 1, 5, 4, 7, ...)\,.$$

This sequence is also called *Stern's diatomic sequence*. Observe that the two children of $\text{fusc}(n)/\text{fusc}(n+1)$ are $\text{fusc}(2n+1)/\text{fusc}(2n+2)$ and $\text{fusc}(2n+2)/\text{fusc}(2n+3)$. By the generation rule, it follows that

$$\text{fusc}(2n+1) = \text{fusc}(n) \text{ and } \text{fusc}(2n+2) = \text{fusc}(n) + \text{fusc}(n+1) \quad (n = 0, 1, 2, ...).$$

These recurrences, together with $\text{fusc}(0) = 1$ completely determine fusc on all nonnegative integers. In fact $\text{fusc}(n) = b(n)$, where $b(n)$ is the number of hyperbinary representations of n, for all $n \geq 0$. Moreover,

$$x_{n+1} = \frac{1}{\lfloor x_n \rfloor + 1 - \{x_n\}}, \tag{9.4}$$

where $\{x\}$ denotes the fractional part of x (see [1]).

Exercise 9.2 Let $p, q \in \mathbb{Z}^+$ be a pair of coprime integers. Show that there exists a unique 4-tuple of non-negative integers (a, b, c, d) such that $a + b = p$, $c + d = q$ and $ad - bc = 1$.

Exercise 9.3 Prove that, in the Calkin–Wilf tree

1. The jth vertex in any given level is the reciprocal of the jth vertex from the end of that level.
2. Every vertex is the product of its children.

Exercise 9.4 Given a rational a/b in reduced form, define its *simplicity* as $s(a/b) := 1/ab$, its *complexity* as $s(a/b) := ab$, and its *trace* as $t(a/b) := a + b$. Show that the following facts hold for the Calkin–Wilf tree [11]:

1. The product of all the elements at a given level is 1.
2. The sum of simplicities of all rationals at a level is 1.
3. The product of complexities of all the elements at a level is a perfect square.
4. The sum of numerators (denominators) of all the elements at a level N is 3^{N-1}.
5. The sum of traces of all the elements at a level N is $2 \cdot 3^{N-1}$.
6. The sum of complexities at level N is equal to the sum of the squares of traces at level $N - 1$.
7. The sum of all elements at level N is $3 \cdot 2^{N-2} - \frac{1}{2}$.

Exercise 9.5 Prove recursion (9.4).

Exercise 9.6 Show that

$$\sum_{n=0}^{\infty} \frac{1}{\text{fusc}(n)^s \text{fusc}(n+1)^s} = \frac{\zeta(s)^2}{\zeta(2s)},$$

where ζ is Riemann's zeta function.

9.3 The Stern–Brocot Tree

Recall that U is the projective matrix $\begin{bmatrix} 0 & 1 \\ 1 & 0 \end{bmatrix}$, representing the transformation $x \mapsto 1/x$. By replacing each node M of the matrix version of the Calkin–Wilf tree (Fig. 9.3) by UM^tU we get another tree[1] called the *Stern–Brocot matrix tree* of Fig. 9.5.

[1] Conjugation by U serves to put the rows of the Stern–Brocot tree in increasing order from left to right; rather than decreasing order.

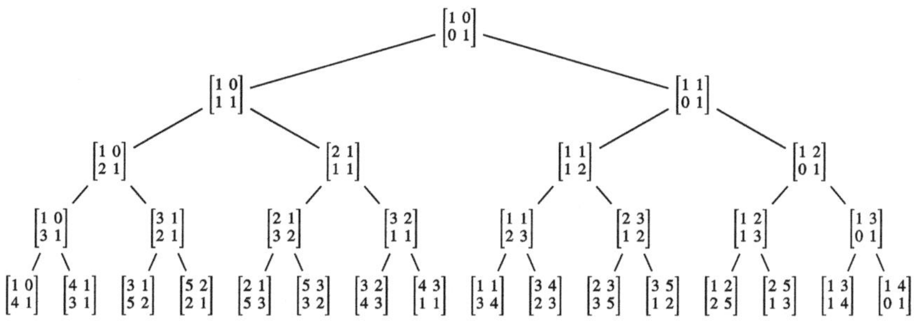

Fig. 9.3 The Calkin–Wilf tree, matrix version. The nodes of this tree are in one-to-one correspondence with the elements of $\mathsf{PSL}_2(\mathbb{N})$. The left child of the node M is T^tM, and the right child of the node M is TM. To get the Calkin–Wilf tree of rationals, evaluate each node of this tree at 1

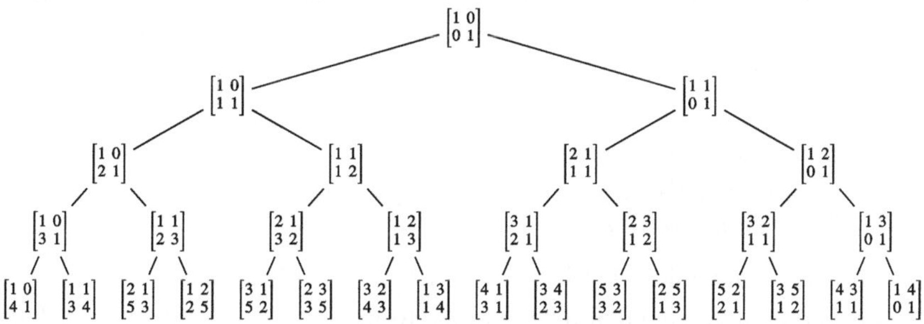

Fig. 9.4 Stern–Brocot tree, matrix version. This tree is obtained by transposing the nodes of the matrix version of the Calkin–Wilf tree. The nodes of this tree are in one-to-one correspondence with the elements of $\mathsf{PSL}_2(\mathbb{N})$. To get the Stern–Brocot tree of rationals, evaluate $M(1)$ for each node M

Let $M = T^{n_k}(T^t)^{n_{k-1}} \dots T^{n_2}(T^t)^{n_1}T^{n_0}$ be a matrix on the Calkin–Wilf matrix tree (Fig. 9.3), where $n_0, n_k \geq 0$ and $n_j > 0$ for $j \neq 0, k$. Then since $UTU = T^t$ or equivalently $UT^tU = T$,

$$UM^tU = U(T^t)^{n_0}T^{n_1} \dots (T^t)^{n_{k-1}}T^{n_k}U \tag{9.5}$$

$$= T^{n_0}(T^t)^{n_1} \dots T^{n_{k-1}}(T^t)^{n_k}. \tag{9.6}$$

On the Stern–Brocot matrix tree, the left child of the node M is MT^t and its right child is MT. For every node M, there is a planar region beneath M bounded by the edges of the tree. Label that region by M and denote it by R_M (Fig. 9.4).

By replacing each node M of the Stern–Brocot matrix tree by $M(1)$ we get another tree of rationals, called the *Stern–Brocot tree* of Fig. 9.5. Accordingly, we change the label of the region R_M to $R_{M(1)}$. This is the planar region beneath $M(1)$ bounded by the edges of the tree.

Lemma 9.1 *Let M be a node of the Stern–Brocot matrix tree and let $M(1)$ be the corresponding node on the Stern–Brocot tree of rationals. Then the region to the left*

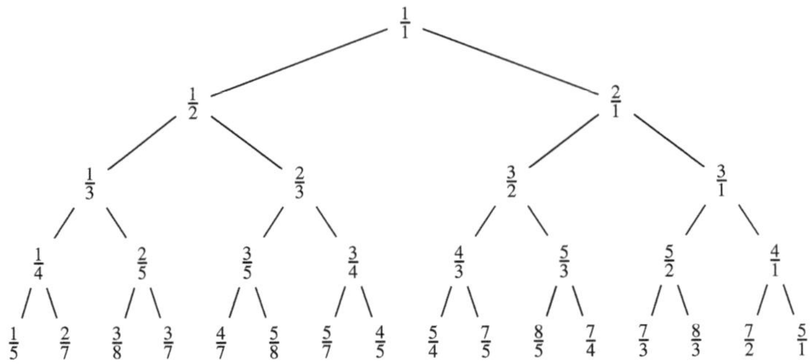

Fig. 9.5 The Stern–Brocot tree is another way of producing all positive rationals. Nodes of this tree can be obtained from the neighboring nodes via the Farey sum $\frac{p}{q} \oplus \frac{r}{s} = \frac{p+r}{q+s}$. The (invisible) leftmost and rightmost nodes are $\frac{0}{1}$ and $\frac{1}{0}$ respectively. Observe that, unlike the Calkin–Wilf tree, the rows of the Stern–Brocot tree are ordered

of x has label $M(0)$, and the region to the right of x has label $M(\infty)$. Therefore the regions $R_{M(0)}$ and $R_{M(\infty)}$ are neighbors along the edge connecting $M(1)$ to its parent, which is either $M(0)$ or $M(\infty)$.

Proof Note the fixed points $T^t(\frac{0}{1}) = \frac{0}{1}$ and $T(\frac{1}{0}) = \frac{1}{0}$. Moreover, $T(\frac{0}{1}) = T^t(\frac{1}{0}) = \frac{1}{1}$. Now observe on the tree that if M is of the form $KT^n(T^t)^m$ with $K \in \mathsf{PSL}_2(\mathbb{N})$, then

- the region to the left of the matrix node M has the matrix label $KT^{n-1} \implies M(\frac{0}{1}) = KT^n(T^t)^m(\frac{1}{0}) = KT^n(\frac{0}{1}) = KT^{n-1}(\frac{1}{1})$, which is the rational label of the node with matrix label KT^{n-1}.
- the region to the right of the matrix node M has the label $KT^n(T^t)^{m-1}$ (the parent of M) $\implies M(\frac{1}{0}) = KT^n(T^t)^m(\frac{1}{0}) = KT^n(T^t)^{m-1}(\frac{1}{1})$, which is the rational label of the node with matrix label $KT^n(T^t)^{m-1}$.

A similar argument handles the case where M is of the form $K(T^t)^nT^m$ with $K \in \mathsf{PSL}_2(\mathbb{N})$. □

Suppose that M is the matrix $\begin{bmatrix} p & q \\ r & s \end{bmatrix}$. Then by the lemma we have

$$M(0) = \frac{q}{s}, \quad M(\infty) = \frac{p}{r} \implies M(1) = M(0) \oplus M(\infty) = \frac{p+q}{r+s}$$

where by definition, the mediant of the reduced fractions p/r and q/s is

$$\frac{p}{r} \oplus \frac{q}{s} := \frac{p+q}{r+s}.$$

Note that $M(0) < M(1) < M(\infty)$ since

$$\frac{p+q}{r+s} < \frac{p}{r} \iff pr + qr < pr + ps \iff 0 < ps - qr = 1,$$

$$\frac{p+q}{r+s} > \frac{q}{s} \iff ps + qs > qr + qs \iff 1 = ps - qr > 0.$$

Define the (finite) sequence $\mathcal{S}_N$ of rationals obtained by traversing the level-N row of the Stern–Brocot tree from left to right thereby listing the regions R_x so encountered (including those we touch only at the topmost node). The leftmost region is labeled $\frac{0}{1}$ and the rightmost region is labeled $\frac{1}{0}$. There are $2^N + 1$ numbers in the sequence $\mathcal{S}_N$. For example,

$$\mathcal{S}_1 = \left(\frac{0}{1}, \quad \frac{1}{1}, \quad \frac{1}{0}\right)$$

$$\mathcal{S}_2 = \left(\frac{0}{1}, \quad \frac{1}{2}, \quad \frac{1}{1}, \quad \frac{2}{1}, \quad \frac{1}{0}\right)$$

$$\mathcal{S}_3 = \left(\frac{0}{1}, \frac{1}{3}, \frac{1}{2}, \frac{2}{3}, \frac{1}{1}, \frac{3}{2}, \frac{2}{1}, \frac{3}{1}, \frac{1}{0}\right)$$

$$\mathcal{S}_4 = \left(\frac{0}{1}, \frac{1}{4}, \frac{1}{3}, \frac{2}{5}, \frac{1}{2}, \frac{3}{5}, \frac{2}{3}, \frac{3}{4}, \frac{1}{1}, \frac{4}{3}, \frac{3}{2}, \frac{5}{3}, \frac{2}{1}, \frac{5}{2}, \frac{3}{1}, \frac{4}{1}, \frac{1}{0}\right)$$

Proposition 9.3 *The sequence $\mathcal{S}_{N+1}$ is obtained from $\mathcal{S}_N$ by inserting between each consecutive fraction x, y in $\mathcal{S}_N$ their mediant $x \oplus y$.*

For example, in $\mathcal{S}_1$ we insert between $\frac{0}{1}$ and $\frac{1}{1}$ the fraction $\frac{0}{1} \oplus \frac{1}{1} = \frac{1}{2}$ and we insert between $\frac{1}{1}$ and $\frac{1}{0}$ the fraction $\frac{1}{1} \oplus \frac{1}{0} = \frac{2}{1}$ to get $\mathcal{S}_2$.

Proof There are $2^{N+1} + 1$ fractions in $\mathcal{S}_{N+1}$. Each fraction $x = M(1)$ is written as a mediant as fractions $x = M(0) \oplus M(\infty)$, and both these fractions are above $\mathcal{S}_{N+1}$. Since there are 2^N such pairs above $\mathcal{S}_{N+1}$, each fraction in $\mathcal{S}_{N+1}$ not already occurring in $\mathcal{S}_N$ must come from a mediant. □

Hence, an alternative way to produce the Stern–Brocot tree is to start from the fractions $\frac{0}{1}$ and $\frac{1}{0}$ and recursively add mediants.

Proposition 9.4 *If $p/q < r/s$ are two consecutive elements of $\mathcal{S}_N$, then $\frac{r}{s} - \frac{p}{q} = \frac{1}{qs}$; i.e. $qr - ps = 1$. Therefore $\mathcal{S}_N$ is increasing for every N. In particular, the rows of the Stern–Brocot tree are in increasing order from left to right.*

For example, $3/5$ and $2/3$ are consecutive in $\mathcal{S}_4$, and $\frac{2}{3} - \frac{3}{5} = \frac{1}{15}$.

Proof If $p/q < r/s$ are two consecutive elements of $\mathcal{S}_N$, then either $(p/q, r/s) = (M(0), M(1))$ or $(p/q, r/s) = (M(1), M(\infty))$ for some $M \in \mathsf{PSL}_2(\mathbb{N})$. Assume the former case. Then $M = \begin{bmatrix} a\,b \\ c\,d \end{bmatrix} \Longrightarrow$

$$\frac{p}{q} = \frac{b}{d}, \quad \frac{r}{s} = \frac{a+b}{c+d} \Longrightarrow \frac{r}{s} - \frac{p}{q} = \frac{b}{d} - \frac{a+b}{c+d} = \frac{ad-bc}{d(c+d)} = \frac{1}{qs}.$$

The latter case is proved similarly. □

Definition 9.1 Let $x = M(1)$ be a node of the Stern–Brocot tree. Then the interval $I(x) := (M(0), M(\infty)) \subset \mathbb{R}$ is called the *Farey neighborhood* of x.

For the length of the Farey neighborhood of x we have

$$x = M(1) \text{ with } M = \begin{bmatrix} p & q \\ r & s \end{bmatrix} \Longrightarrow |I(x)| = M(\infty) - M(0) = \frac{p}{r} - \frac{q}{s} = \frac{1}{rs}.$$

Unless $y \in \mathbb{Q}^+$ is the root 1, it has a continued fraction representation $y = [n_0, n_1, \ldots, n_k]$ with $n_k > 2$. Then $y = [n_0, n_1, \ldots, n_k - 1]$ is the parent of y. For instance, the parent of $y = [3, 4, 5]$ is $[3, 4, 4]$, and the parent of $y = [3, 4, 2]$ is $[3, 4, 1] = [3, 5]$.

The two children of $y = [n_0, n_1, \ldots, n_k] = [n_0, n_1, \ldots, n_k - 1, 1]$ are

$$[n_0, n_1, \ldots, n_k + 1] \text{ and } [n_0, n_1, \ldots, n_k - 1, 2],$$

for instance, $y = [3, 4, 5]$ has the children $[3, 4, 6]$ and $[3, 4, 4, 2]$.

Lemma 9.2 *Let $x = [n_0, n_1, \ldots, n_k]$ be a level-N node of the Stern–Brocot tree, with $n_k > 1$ if $x \neq 1$. Let $M \in \mathsf{PSL}_2(\mathbb{N})$ be the corresponding matrix, such that $x = M(1)$. Let $M(0)$ and $M(\infty)$ be the elements of S_N preceding and following x, respectively. If k is odd, then*

$$M(0) = [n_0, n_1, \ldots, n_{k-1}] \text{ and } M(\infty) = [n_0, n_1, \ldots, n_k - 1].$$

If k is even, then

$$M(0) = [n_0, n_1, \ldots, n_k - 1] \text{ and } M(\infty) = [n_0, n_1, \ldots, n_{k-1}].$$

Proof Suppose k is odd and let $x = [n_0, n_1, \ldots, n_k]$ be a node of level N with $n_k > 1$. To reach x from the root, we turn right $n_0 \geq 0$ times, turn left $n_1 > 0$ times, $\ldots$, turn right $n_{k-1} > 0$ times, and finally turn left $n_k - 1 > 0$ times (see Fig. 9.6 and keep in mind that we start counting from 0). Hence, the region to the right of x is labelled $x^- = [n_0, n_1, \ldots, n_k - 1]$ and therefore the region to the left is labelled $x^+ = [n_0, n_1, \ldots, n_{k-1}]$. The case of even k is proved similarly. □

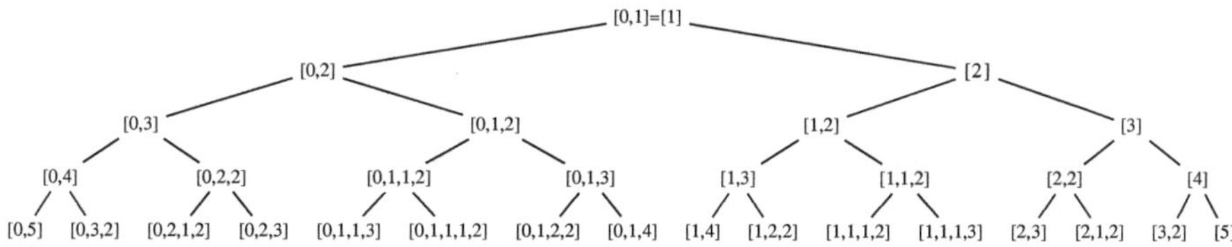

Fig. 9.6 Stern–Brocot tree, continued fraction form

Reading the rows of the Stern–Brocot tree first from left to right and then from top to bottom, we obtain another enumeration $(x_n)_{n\geq 0}$ of $\mathbb{Q}^+$:

$$(x_n)_{n\geq 0} = \left(\frac{1}{1}, \frac{1}{2}, \frac{2}{1}, \frac{1}{3}, \frac{2}{3}, \frac{3}{2}, \frac{3}{1}, \frac{1}{4}, \frac{2}{5}, \frac{3}{5}, \frac{3}{4}, \frac{4}{3}, \frac{5}{3}, \frac{5}{2}, \frac{4}{1}, \frac{1}{5}, \frac{2}{7}, \frac{3}{8}, \frac{3}{7}, \frac{4}{7}, \frac{5}{8} \cdots \right).$$

Since the level-N row of the Stern–Brocot tree is a permutation of the level-N row of the Calkin–Wilf tree, every fact about the Calkin–Wilf tree stated in Exercise 9.4 is also valid for the Stern–Brocot tree as well.

By gluing a copy of the Stern–Brocot tree to itself at its root, we obtain a tree whose nodes naturally correspond to all non-zero rationals. This tree is called the *Farey tree* and admits a natural $\mathsf{PSL}_2(\mathbb{Z})$ action. Hence, according to our convention, the Stern–Brocot tree is one half of the Farey tree which indexes only the positive rationals.

The left branch of the Stern–Brocot tree with nodes corresponding to the rationals in (0, 1) is sometimes called the *Farey tree* as well.

Remark 9.1 The *Haros–Farey sequence of order* N, denoted $\mathcal{F}_N$, is the sorted sequence of reduced fractions $p/q \in [0, 1]$ with denominator $q \leq N$. As in the case of the Stern–Brocot tree, the Farey sequences can be constructed using mediants: $\mathcal{F}_{N+1}$ is formed from $\mathcal{F}_N$ by computing the mediant of each two consecutive values in the Farey sequence of order N, keeping the subset of mediants that have denominator exactly equal to $N + 1$, and inserting these mediants between the two values from which they were computed. Thus, $\mathcal{F}_N$ is a subsequence of $\mathcal{S}_N$. The Farey sequences of orders 1 to 5 are:

$$\mathcal{F}_1 = \left(\frac{0}{1}, \frac{1}{1}\right)$$

$$\mathcal{F}_2 = \left(\frac{0}{1}, \frac{1}{2}, \frac{1}{1}\right)$$

$$\mathcal{F}_3 = \left(\frac{0}{1}, \frac{1}{3}, \frac{1}{2}, \frac{2}{3}, \frac{1}{1}\right)$$

$$\mathcal{F}_4 = \left(\frac{0}{1}, \frac{1}{4}, \frac{1}{3}, \frac{1}{2}, \frac{2}{3}, \frac{3}{4}, \frac{1}{1}\right)$$

$$\mathcal{F}_5 = \left(\frac{0}{1}, \frac{1}{5}, \frac{1}{4}, \frac{1}{3}, \frac{2}{5}, \frac{1}{2}, \frac{3}{5}, \frac{2}{3}, \frac{3}{4}, \frac{4}{5}, \frac{1}{1}\right)$$

This sequence is of importance in number theory, especially due to its connection with Riemann's hypothesis, see [12].

Beware that (1) $\mathcal{F}_n$ is not the nth row of the Stern–Brocot tree, (2) $\mathcal{F}_n$ is not the ordered list of rationals appearing in the first n rows of the Stern–Brocot tree.

Exercises

Exercise 9.7 Let $M = \begin{bmatrix} p\, q \\ r\, s \end{bmatrix} \in \mathsf{PSL}_2(\mathbb{N})$.

1. Find the matrix UM^tU.
2. Show that the map sending the node labeled $M \in \mathsf{PSL}_2(\mathbb{N})$ to the node labeled $UM^tU \in \mathsf{PSL}_2(\mathbb{N})$ induces a bijection on each row of the Calkin–Wilf tree.
3. Show that this map is not an automorphism of the Calkin–Wilf tree.

Exercise 9.8 Let x, y be two consecutive fractions in $\mathcal{S}_N$. Show that their mediant $x \oplus y$ is reduced.

Exercise 9.9 Let x, y be consecutive fractions of level N on the Stern–Borocot tree. Show that their mediant $x \oplus y$, after reduction, is the closest common ancestor of x and y. Equivalently, $x \oplus y$ is the label of the region between x and y. For example,

$$\frac{5}{8} \oplus \frac{5}{7} = \frac{3}{5} \oplus \frac{3}{4} = \frac{2}{3}, \quad \text{or} \quad \frac{8}{5} \oplus \frac{7}{4} = \frac{5}{3}.$$

Can you identify the reduction factor?

Solution Let M, N be the matrices corresponding to x and y and let K be their closest common ancestor with $K(1) = z$ the corresponding rational. Let k be their distance to the common ancestor. Then $M = K(T^t)T^{k-1}$ and $N = KT(T^t)^{k-1}$ (or vice versa).

$$K = \begin{bmatrix} a\ b \\ c\ d \end{bmatrix} \Longrightarrow$$

$$M = \begin{bmatrix} a\ b \\ c\ d \end{bmatrix} \begin{bmatrix} 1\ 0 \\ 1\ 1 \end{bmatrix} \begin{bmatrix} 1\ k-1 \\ 0\ \ 1 \end{bmatrix} = \begin{bmatrix} a+b\ b \\ c+d\ d \end{bmatrix} \begin{bmatrix} 1\ k-1 \\ 0\ \ 1 \end{bmatrix} = \begin{bmatrix} a+b\ (k-1)a+kb \\ c+d\ (k-1)c+kd \end{bmatrix}$$

$$N = \begin{bmatrix} a\ b \\ c\ d \end{bmatrix} \begin{bmatrix} 1\ 1 \\ 0\ 1 \end{bmatrix} \begin{bmatrix} 1\ \ 0 \\ k-1\ 1 \end{bmatrix} = \begin{bmatrix} a\ a+b \\ c\ c+d \end{bmatrix} \begin{bmatrix} 1\ \ 0 \\ k-1\ 1 \end{bmatrix} = \begin{bmatrix} ka+(k-1)b\ a+b \\ kc+(k-1)d\ c+d \end{bmatrix}$$

$$x = \frac{ak + b(k+1)}{ck + d(k+1)}, \quad y = \frac{a(k+1) + bk}{c(k+1) + dk}$$

$$\implies x \oplus y = \frac{(2k+1)(a+b)}{(2k+1)(c+d)} = \frac{a+b}{c+d}.$$

We see that the reduction factor is $2k + 1$, where k is the distance of x and y to their common ancestor.

Exercise 9.10 Let $x = M(1)$ be a node of the Stern–Borocot tree. Show (or rather observe on the tree) that the regions $R_{M(0)}$ and $R_{M(\infty)}$ are neighbors along the edge connecting x to its parent.

Exercise 9.11 Let $p/q, r/s$ be two siblings on the Stern–Brocot tree. Show that $qr - ps = 3$.

Solution Let

$$M = \begin{bmatrix} a & b \\ c & d \end{bmatrix}$$

be their common parent. Then these siblings are given by

$$\frac{p}{q} = MT(1) = \begin{bmatrix} a & b \\ c & d \end{bmatrix} \begin{bmatrix} 1 & 1 \\ 0 & 1 \end{bmatrix} (1) = \begin{bmatrix} a & a+b \\ c & c+d \end{bmatrix} (1) = \frac{2a+b}{2c+d}$$

and

$$\frac{r}{s} = MT^t(1) = \begin{bmatrix} a & b \\ c & d \end{bmatrix} \begin{bmatrix} 1 & 0 \\ 1 & 1 \end{bmatrix} (1) = \begin{bmatrix} a+b & b \\ c+d & d \end{bmatrix} (1) = \frac{a+2b}{c+2d}.$$

Hence

$$qr - ps = (2c+d)(a+2b) - (2a+b)(c+2d) = 3cb - 3ad = -3.$$

Exercise 9.12 Let N be a positive integer. Then any way of writing N as a sum of positive integers is called a *composition* of N (see [17]). For example, $N = 3$ has 4 compositions $3 = 1+2 = 2+1 = 1+1+1$. Let $x = [0, n_1, n_2, \ldots, n_k]$ be a node of the left-branch of the Stern–Brocot tree of level $N + 1$, where $n_k \geq 2$. To x we associate the composition $N = n_1 + n_2 + \ldots n_{k-1} + (n_k - 1)$ of N with k parts.

1. Show that the above association is a bijection between the set of compositions of N and the set of nodes of level-$N+1$ on the left-branch of the Stern–Brocot tree. Conclude that N has 2^{N-1} compositions.
2. Show that the number of compositions of N using only 1's and 2's is F_{N+1}, the $(N+1)$th Fibonacci number.

Exercise 9.13 Show that for the Farey sequence one has $|\mathcal{F}_n| = |\mathcal{F}_{n-1}| + \varphi(n)$, where $\varphi(n)$ is Euler's totient function (which counts the positive integers up to $\leq n$ that are relatively prime to n). Therefore

$$|\mathcal{F}_n| = 1 + \sum_{d=1}^{n} \varphi(m).$$

Compare $|\mathcal{F}_n|$ with $\mathcal{S}_n$.

Exercise 9.14 (Minkowski Question Mark Function) Consider a random walker on the left branch of the Stern–Brocot tree, who walks always downwards, and chooses to turn left or right with probability $1/2$ at each node. Let $x \in [0, 1]$. Show that the probability that the walker lands on the interval $[0, x]$ is given by the cumulative distribution function

$$?(x) := 2 \sum_{k=1}^{\infty} \frac{(-1)^{k+1}}{2^{n_1+n_2+\cdots+n_k}}$$

if $x = [0, n_1, n_2, \ldots]$ is irrational, and

$$?(x) := 2 \sum_{k=1}^{m} \frac{(-1)^{k+1}}{2^{n_1+n_2+\cdots+n_k}}$$

if $x = [0, n_1, n_2, \ldots n_m]$ is rational. (By replacing every node x of the Stern–Brocot with $?(x)$ we obtain what is called the *dyadic tree*.)

9.3.1 Notes

The Calkin–Wilf tree is named after Neil Calkin and Herbert Wilf's paper [10] (2000). The tree was introduced earlier by Jean Berstel and Aldo de Luca [7] as Raney tree, referring to George N. Ramey [30]. Stern's diatomic series was formulated by Moritz Abraham Stern [26], a nineteenth-century German mathematician who also invented the closely related Stern–Brocot tree. Brocot was a French clockmaker who created an array of fractions for the purpose of designing clockwork gears [9]. A similar tree appears in Kepler's Harmonices Mundi [20] (1619). There is a vast literature on these trees, one may consult [3, 4, 6, 23, 27, 32] and the papers citing them. For various generalizations, see [5, 15, 22, 24, 29].

Farey sequences are named after the British geologist John Farey [14] (1816). He conjectured the median property which was proved by Cauchy. In fact Charles Haros had published similar results in 1802. Because of this, $\mathcal{F}_n$ is re-baptised as the Farey–Haros sequence in [12].

Minkowski's question-mark function, was defined by Hermann Minkowski in 1904 [28]. It sends rational numbers to dyadics and real quadratic irrationals to rationals. It can be extended as a homeomorphism $\mathbb{P}^1(\mathbb{R}) \to \mathbb{P}^1(\mathbb{R})$ (i.e. the circle). It was conjectured by Maxim Kontsevich and proved by Michel Imbert that this extension conjugates the group $\mathsf{PPSL}_2(\mathbb{Z})$ of piecewise-$\mathsf{PSL}_2(\mathbb{Z})$ circle homeomorphisms with the group $\mathsf{PL}_2(S^1)$ of piecewise linear circle homeomorphisms with dyadic slopes [19].

9.4 The Drib Tree

Note that the automorphism α of $\mathsf{PGL}_2(\mathbb{N})$ has the following effect on the generators of $\mathsf{PSL}_2(\mathbb{N}) < \mathsf{PGL}_2(\mathbb{N})$:

$$\alpha : \begin{bmatrix} 1 & 1 \\ 0 & 1 \end{bmatrix} \to \begin{bmatrix} 1 & 1 \\ 1 & 0 \end{bmatrix}, \quad \begin{bmatrix} 1 & 0 \\ 1 & 1 \end{bmatrix} \to \begin{bmatrix} 0 & 1 \\ 1 & 1 \end{bmatrix}.$$

In other words, α sends $1 + x$ to $1 + 1/x$ and $x/(1 + x)$ to $1/(1 + x)$; i.e. α sends T to TU and T^t to T^tU. Let us momentarily call $A := TU$ and $B := T^tU$.

Applying α nodewise to the Calkin–Wilf matrix tree, we obtain what is called the *matrix Drib tree*. Note that the image $\alpha(\mathsf{PSL}_2(\mathbb{N}))$ is different from $\mathsf{PSL}_2(\mathbb{N})$ as it contains elements of determinant -1; so the matrix Calkin–Wilf tree and the matrix Drib tree are quite different. Evaluating the Drib matrix tree at 1, we get a tree of rationals called the *Drib tree*. This tree is generated by the rules in Fig. 9.7.

The nodes of the Drib tree are labeled by positive rationals. Reading the rows of the Drib tree from left to right and top to bottom, we obtain the following enumeration of rationals (Exercise 9.15) (Fig. 9.8)

$$\frac{1}{1}, \frac{1}{2}, \frac{2}{1}, \frac{2}{3}, \frac{3}{1}, \frac{1}{3}, \frac{3}{2}, \frac{3}{5}, \frac{5}{2}, \frac{1}{4}, \frac{4}{3}, \frac{3}{4}, \frac{4}{1}, \frac{2}{5}, \frac{5}{3}, \frac{5}{8}, \frac{8}{3}, \frac{2}{7}, \frac{7}{5}, \frac{4}{5}, \frac{5}{1}, \frac{3}{7}, \frac{7}{4}, \frac{4}{7}, \frac{7}{3}, \ldots$$

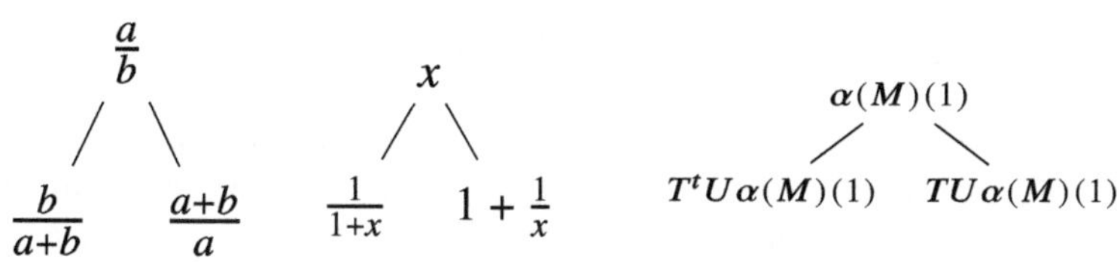

Fig. 9.7 Three equivalent versions of the generation rule of the Drib tree

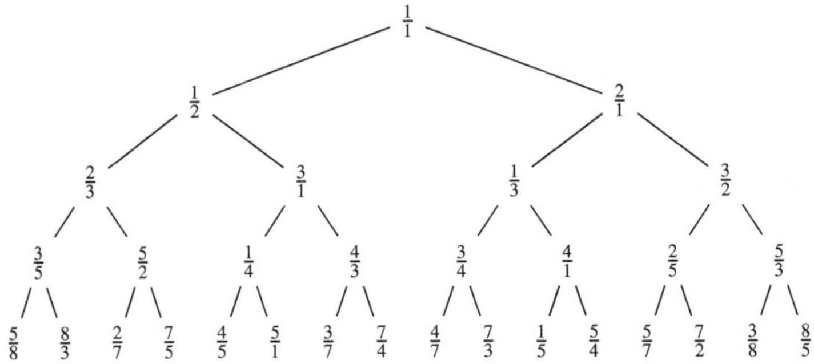

Fig. 9.8 The Drib tree of rationals. This tree is constructed from $1/1$ by successively applying the generation rules $a/b \to b/(a+b)$ (left) and $a/b \to (a+b)/a$ (right). It can also be obtained by evaluating at 1 the nodes of the Dirb tree of matrices

Exercises

Exercise 9.15 Prove that every positive rational appears exactly once on the Drib tree.

Solution Obviously, $x = 1$ appears as the root of the Drib tree. If $x > 1$, then $1/(x-1)$ is its parent on the Drib tree. If $x < 1$, then $1/x - 1$ is its parent. In each case, the continued fraction of x becomes shorter when passing to its parent. This yields a sequence of parents (including x itself) ending at 1. Hence every rational appears on the tree. Unicity is proved similarly.

Exercise 9.16 Let $x \in \mathbb{Q}^+$, let y be its parent on the Drib tree and let z be its parent on the Calkin–Wilf tree. Observe that $y \neq z$ in general.

Exercise 9.17 Prove that every depth-N row of the Drib tree is a permutation of the depth$-N$ row of the Calkin–Wilf tree.

Solution Let M be a depth-N element of the Calkin–Wilf tree. Then

$$M = T^{n_0}(T^t)^{n_1}\dots(T^t)^{n_k} \implies \alpha(M) = (TU)^{n_0}(T^tU)^{n_1}\dots(T^tU)^{n_k},$$

where $n_0 + \cdots + n_k = N$. Note that $UTU = T^t$ and $UT^tU = T$. If $\alpha(M) \in \mathsf{PSL}_2(\mathbb{N})$, then it can be written (with some work) in the form $\alpha(M) = T^{m_0}(T^t)^{m_1}\dots(T^t)^{m_k}$, with $m_0+\cdots+m_k = N$. So $\alpha(M)(1)$ is a depth-N element of the Calkin–Wilf tree as well. If $\alpha(M) \notin \mathsf{PSL}_2(\mathbb{N})$, then $\alpha(M)U \in \mathsf{PSL}_2(\mathbb{N})$ and it can be written in the above form. Because $U(1) = 1$, $\alpha(M)U(1) = \alpha(M)(1)$ is a depth-N element of the Calkin–Wilf tree as well. Since both trees have 2^{N-1} elements of depth N, this must be a bijection.

9.5 The Bird Tree

By replacing each node M of the matrix Drib tree by UM^tU we get another tree, called the *Bird matrix tree*. On the Bird matrix tree, the left child of the node N is NB and its right child is NA. Evaluating every node of this matrix tree at 1, we get a tree of rational numbers called the *Bird tree* (Fig. 9.9).

Proposition 9.5 *A node $x \in \mathbb{Q}^+$ is the parent of $y \in \mathbb{Q}^+$ in the Stern–Brocot tree if and only if $x \in \mathbb{Q}^+$ is the parent of $y \in \mathbb{Q}^+$ in the Bird tree. Therefore these two trees are isomorphic, though not as planar trees.*

Proof We will prove that $y, z \in \mathbb{Q}^+$ are the children of $x \in \mathbb{Q}^+$ in the one tree if and only if they are the children of $x \in \mathbb{Q}^+$ in the other tree. Note on the other hand that a left-child can become a right-child and vice versa; i.e. trees are not isomorphic as planar trees.

Let $M = A^{m_k}B^{m_{k-1}} \dots A^{m_2}B^{m_0}$ be a matrix on the Drib matrix tree, where $m_0, m_k \geq 0$ and $m_j > 0$ for $j \neq 0, k$. Then since $UB^tU = A \iff UA^tU = B$,

$$\begin{aligned} UM^tU &= UB^{m_0}A^{m_1} \dots B^{m_{k-1}}A^{m_k}U \\ &= A^{m_0}B^{m_1} \dots A^{m_{k-1}}B^{m_k}. \end{aligned}$$

Since $A = TU$ and $B = T^tU$ we can write

$$UM^tU = (TU)^{m_0}(T^tU)^{m_1} \dots (TU)^{m_{k-1}}(T^tU)^{m_k}.$$

Note that if the depth $m_0 + \cdots + m_k$ is even, then UM^tU is actually in $\mathsf{PSL}_2(\mathbb{N})$ and the U's in the above expression can be actually eliminated, leaving only T and T^t. If the depth is odd, then we can re-write UM^tU so that there is only one U left at the end. Suppose that the depth is odd. Then we can write

$$N := UM^tU = T^{n_0}(T^t)^{n_1} \dots T^{n_{\ell-1}}(T^t)^{n_\ell}U$$

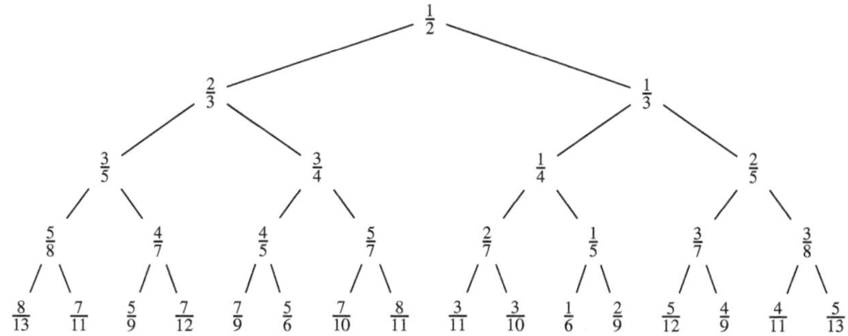

Fig. 9.9 The branch of the Bird tree corresponding to rationals in the open unit interval

(the actual computation of (n_i) in terms of (m_i) is a bit tricky; note in particular that $k \neq \ell$ in general. See Exercise 9.18). Thus the left (respectively right) child of N is NB (respectively NA), which equals

$$\begin{aligned} NB =& T^{n_0}(T^t)^{n_1} \dots T^{n_{k-1}}(T^t)^{n_k} U T^t U, \\ NA =& T^{n_0}(T^t)^{n_1} \dots T^{n_{k-1}}(T^t)^{n_k} U T U. \end{aligned}$$

Using $UTU = T^t \iff UT^tU = T$ to simplify the ends we get

$$\begin{aligned} NB =& T^{n_0}(T^t)^{n_1} \dots T^{n_{k-1}}(T^t)^{n_k} T, \\ NA =& T^{n_0}(T^t)^{n_1} \dots T^{n_{k-1}}(T^t)^{n_k+1}. \end{aligned}$$

Evaluating $NB(1)$ and $NA(1)$, we obtain the left (respectively right) child of the node $N(1)$ in the rational Bird tree. Now observe that $NB(1)$ and $NA(1)$ are respectively the right and the left children of $N(1)$ in the rational Stern–Brocot tree. Note that the right-children are swapped with the left-children here.

If the depth is even, we write $N = T^{n_0}(T^t)^{n_1} \dots T^{n_{\ell-1}}(T^t)^{n_\ell}$ and get

$$\begin{aligned} NB =& T^{n_0}(T^t)^{n_1} \dots T^{n_{k-1}}(T^t)^{n_k+1} U, \\ NA =& T^{n_0}(T^t)^{n_1} \dots T^{n_{k-1}}(T^t)^{n_k} T U. \end{aligned}$$

Evaluating $NB(1)$ and $NA(1)$, we obtain the left (respectively right) child of the node $N(1)$ in the rational Bird tree. But since $U(1) = 1$, we see that $NB(1)$ and $NA(1)$ are respectively the left and the right children of $N(1)$ in the rational Stern–Brocot tree.

This proves the "only if" part of the tree; the "if" part follows from the involutivity of α. □

Rooted tree isomorphisms preserves depth, so in the proof we have $m_0 + \cdots + m_k = n_0 + \cdots + n_\ell$.

Exercise 9.18 Recalling that $A = TU$ and $B = T^tU$, write $A^{m_0}B^{m_1} \dots B^{m_k}$ in the form $T^{n_0}(T^t)^{n_1} \dots (T^t)^{n_k}$ for the following values of (m_i): (3, 4, 7); (3, 1, 1, 1, 4, 4); (0, 1, 1, 1, 1, 1, 1, 8); (1, 3, 6, 3, 1); (2, 2, 2, 2, 2, 2, 2) and check that the depth is preserved.

Solution Use $UTU = T^t \iff UT^tU = T$. For example,

$$\begin{aligned} & (TU)^3(T^tU)^4(TU)^7 \\ =& TUTUTUT^tUT^tUT^tUT^tUTUTUTUTUTUTUTU \\ =& T \cdot UTU \cdot T \cdot UT^tU \cdot T^t \cdot UT^tU \cdot T^t \cdot UTU \cdot T \cdot UTU \cdot T \cdot UTU \cdot T \cdot UTU \\ =& T \cdot T^t \cdot T \cdot T \cdot T^t \cdot T \cdot T^t \cdot T^t \cdot T \cdot T^t \cdot T \cdot T^t \cdot T \cdot T^t \\ =& T \cdot T^t \cdot T^2 \cdot T^t \cdot T \cdot (T^t)^2 \cdot T \cdot T^t \cdot T \cdot T^t \cdot T \cdot T^t. \end{aligned}$$

Since both the Stern–Brocot and the Bird trees are planar rooted trees, they are canonically isomorphic. Identifying these two trees via this canonical isomorphism, we may view the tree isomorphism induced by α as a (non-planar) automorphism of the Stern–Brocot tree. As such, α induces an involution which sends the set of nodes (i.e. $\mathbb{Q}^+$) of the Stern–Brocot tree to itself. We call this involution $\text{ج} : \mathbb{Q}^+ \to \mathbb{Q}^+$ (Jimm). So, for example,

$$\text{ج}(1) = 1, \quad \text{ج}(2) = 2, \quad \text{ج}(1/2) = 1/2, \quad \text{ج}(1/3) = 2/3, \quad \text{ج}(23/31) = 47/107.$$

Further below we will give a formula to compute $\text{ج}(x)$ as a simple continued fraction, given the simple continued fraction expansion of x.

Lemma 9.3 *The involution* $\text{ج} : \mathbb{Q}^+ \to \mathbb{Q}^+$ *satisfies:*

$$\text{ج}(1+x) = 1 + \frac{1}{\text{ج}(x)} \quad (x \in \mathbb{Q}^+),$$
$$\text{ج}(1/x) = \frac{1}{\text{ج}(x)} \quad (x \in \mathbb{Q}^+),$$
$$\text{ج}(1-x) = 1 - \text{ج}(x) \quad (x \in \mathbb{Q}^+ \cap (0,1)).$$

Proof The node $1 + x$ is the right-child of x in the Stern–Brocot tree, whereas the node $1 + 1/\text{ج}(x)$ is the right-child of $\text{ج}(x)$ in the Bird tree. Since we've identified these two trees via their planar embeddings, this proves the first functional equation. Note that this equation can be read as $\text{ج}T = TU\text{ج}$.

Similarly, the node $x/(1 + x)$ is the left-child of x in the Stern–Brocot tree, whereas the node $1/(1 + \text{ج}(x))$ is the right-child of $\text{ج}(x)$ in the Bird tree. Since we've identified these two trees via their planar embeddings, we have the functional equation

$$\text{ج}\left(\frac{x}{1+x}\right) = \frac{1}{1+\text{ج}(x)}.$$

Using the involutivity of ج, this can be written as $\text{ج}UTU = UT\text{ج} \iff U \cdot TU\text{ج} = \text{ج}UT$. Substituting $TU\text{ج} = \text{ج}T$ from the first functional equation, we get $U\text{ج}T = \text{ج}UT$. Eliminating T from this, we get the second functional equation.

Now, note that for $y \in \mathbb{Q}^+$

$$\text{ج}\left(\frac{y}{1+y}\right) + \text{ج}\left(\frac{1}{1+y}\right) = \frac{1}{1+\text{ج}(y)} + \frac{\text{ج}(y)}{1+\text{ج}(y)} = 1.$$

Setting $x = y/(1 + y) \implies 1 - x = 1/(1 + y)$ in the above equation we get $\text{ج}(x) + \text{ج}(1 - x) = 1$, which is the third functional equation. □

Note that if we substitute $-y$ in place of x in the third functional equation, and using the first one we get $\text{ج}(1 + y) = 1 + 1/\text{ج}(y) = 1 - \text{ج}(-y)$, which implies

$\text{ج}(-y) = -1/\text{ج}(y)$. Accordingly, we extend ج to $\mathbb{Q} \setminus \{0\}$ via

$$\text{ج}(-x) := -1/\text{ج}(x) \quad (x \in \mathbb{Q}^+)$$

and call the extension ج again. Obviously, the extension is still involutive.

Theorem 9.1 $\text{ج} : \mathbb{Q} \setminus \{0\} \to \mathbb{Q} \setminus \{0\}$ *satisfies the functional equations*

$$\begin{aligned} &\text{(I)} \quad \text{ج}(1/x) = 1/\text{ج}(x) \iff \text{ج}U = U\text{ج} \\ &\text{(II)} \quad \text{ج}(1-x) = 1 - \text{ج}(x) \quad (x \neq 1) \iff \text{ج}K = K\text{ج} \\ &\text{(III)} \quad \text{ج}(-x) = -1/\text{ج}(x) \iff \text{ج}V = UV\text{ج}. \end{aligned}$$

Hence, ج is equivariant with respect to Dyer's automorphism; i.e. $\text{ج}(Mx) = \alpha(M)\text{ج}(x)$ for $M \in \mathsf{PGL}_2(\mathbb{Z})$. Consequently, the map from the group

$$\mathrm{Aut}(\mathsf{PGL}_2(\mathbb{Z})) \simeq \langle V, K, J \mid V^2{=}J^2{=}K^2{=}(KJ)^2 = (VJ)^4 = (K(VJ)^2)^3 = 1\rangle$$

to the set of maps $\mathbb{Q} \setminus \{0\} \to \mathbb{Q} \setminus \{0\}$, sending $V \mapsto -x$, $K \mapsto 1-x$ and $J \mapsto \text{ج}(x)$ is a faithful representation.

(We are abusing terminology by considering U, V, K both as elements of $\mathrm{Aut}(\mathsf{PGL}_2(\mathbb{Z}))$ and as elements of $\mathsf{PGL}_2(\mathbb{Z})$, i.e. as operators acting on $\mathbb{Q}\setminus\{0\}$.)

Proof Equation III simply follows from the definition of the extension of ج to $\mathbb{Q} \setminus \{0\}$. By using this and the involutivity of ج, it is easy to check that $U = (V\text{ج})^2$ (recall that $U = (VJ)^2$ in $\mathrm{Aut}(\mathsf{PGL}_2(\mathbb{Z}))$). Hence Equation I simply says $\text{ج}(V\text{ج})^2 = (V\text{ج})^2\text{ج}$, which simply says that $U^2 = 1$.

This leaves Equation II. If $x = -y < 0$, then this is equivalent to $\text{ج}(1 + y) = 1 - \text{ج}(-y) = 1 + 1/\text{ج}(y)$, which is known from Lemma 9.3. If $x \in (0, 1)$, this is also known from Lemma 9.3. Finally, suppose for $x > 0$ we have $\text{ج}(1 - x) = 1 - \text{ج}(x)$. Then, one has

$$\text{ج}(1 - (x + 1)) = \text{ج}(-x) = -1/\text{ج}(x)$$

which equals

$$1 - \text{ج}(x + 1) = 1 - (1 + 1/\text{ج}(x)) = -1/\text{ج}(x),$$

hence Equation III follows by induction for $x > 1$ as well.

To establish the faithfulness of the representation, it suffices to observe that ج is non-trivial, and that the restriction to the index-2 subgroup $\mathsf{PGL}_2(\mathbb{Z})$ of $\mathrm{Aut}(\mathsf{PGL}_2(\mathbb{Z}))$ generated by $U = (JV)^2$, V and K is faithful. □

Remark 9.2 We've defined ج on $\mathbb{Q} \setminus \{0\}$. Suppose we want to extend ج to $\mathbb{Q}$, so that the functional equations are still valid. Then Equation III would imply $\text{ج}(0)^2 = -1$, whereas Equation II would imply $1 = \text{ج}(1) = \text{ج}(1-0) = 1 - \text{ج}(0) \implies \text{ج}(0) = 0$.

Hence there is no consistent way to define $\zeta(0)$. Since by Equation I we would have $\zeta(\infty) = \zeta(1/0) = 1/\zeta(0)$, defining $\zeta(\infty)$ is not possible, neither.

Corollary 9.2 *Let $x = [n_0, n_1, \ldots, n_k]$ be the simple continued fraction expansion of $x \in \mathbb{Q}^+$. Denote by 1_k the sequence $1, 1, \ldots, 1$ of length k. Then*

$$\zeta(x) = [1_{n_0-1}, 2, 1_{n_1-2}, 2, 1_{n_2-2}, 2, \ldots 2, 1_{n_{k-1}-2}, 2, 1_{n_k-1}] \tag{9.7}$$

with the rules: $[\ldots, n, 1_0, m, \ldots] := [\ldots, n, m, \ldots]$ and $[\ldots, n, 1_{-1}, m, \ldots] := [\ldots, n+m-1, \ldots]$. (A 1_{-1} at the beginning of a sequence is replaced by 0.)

Proof Use the functional equations $\zeta(1 + 1/x) = 1 + \zeta(x)$ and $\zeta(1/x) = 1/\zeta(x)$ repeatedly to get

$$\begin{aligned}\zeta([n_0, n_1, \ldots, n_k]) &= \zeta(1 + [n_0 - 1, n_1, \ldots, n_k]) \\ = [1, \zeta([n_0 - 1, n_1, \ldots, n_k])] &= [1_2, \zeta(1 + [n_0 - 3, n_1, \ldots, n_k])] \\ \cdots = [1_{n_0-1}, \zeta(1 + [0, n_1, \ldots, n_k])] &= [1_{n_0-1}, \zeta(1 + 1/[n_1, \ldots, n_k])] \\ = [1_{n_0-1}, 1 + \zeta([n_1, \ldots, n_k])] &= [1_{n_0-1}, 2, \zeta([n_1 - 1, n_2, \ldots, n_k])] \ldots\end{aligned}$$

□

To compute the value of ζ at a number $x < 0$, use the corollary with the functional equation $\zeta(x) = -1/\zeta(-x)$.

Exercises

Exercise 9.19 Compute $\zeta(x)$, where x is one of the following: $[3, 3, 3, 3, 3, 3]$; $[2, 2, 2, 2, 2, 2]$; $[0, 1, 1, 1, 1, 1, 1]$ and check that $\zeta(\zeta(x)) = x$.

Solution

$$\begin{aligned}\zeta([3, 3, 3, 3, 3, 3]) =& [1_2, 2, 1_1, 2, 1_1, 2, 1_1, 2, 1_1, 2, 1_2] \\ &= [1, 1, 2, 1, 2, 1, 2, 1, 2, 1, 2, 1, 1]\end{aligned}$$

and likewise

$$\begin{aligned}&\zeta([1, 1, 2, 1, 2, 1, 2, 1, 2, 1, 2, 1, 1]) = \\ &[1_0, 2, 1_{-1}, 2, 1_0, 2, 1_{-1}, 2, 1_0, 2, 1_{-1}, 2, 1_0, 2, 1_{-1}, 2, 1_0, 2, 1_{-1}, 2, \\ &1_0, 2, 1_{-1}, 2, 1_0] = \\ &[2, 1_{-1}, 2, 1_0, 2, 1_{-1}, 2, 1_0, 2, 1_{-1}, 2, 1_0, 2, 1_{-1}, 2, 1_0, 2, 1_{-1}, 2, \\ &1_0, 2, 1_{-1}, 2, 1_0] =\end{aligned}$$

$$[3, 1_0, 2, 1_{-1}, 2, 1_0, 2, 1_{-1}, 2, 1_0, 2, 1_{-1}, 2, 1_0, 2, 1_{-1}, 2, 1_0, 2, 1_{-1}, 2, 1_0] =$$
$$[3, 2, 1_{-1}, 2, 1_0, 2, 1_{-1}, 2, 1_0, 2, 1_{-1}, 2, 1_0, 2, 1_{-1}, 2, 1_0, 2, 1_{-1}, 2, 1_0] =$$
$$[3, 3, 1_0, 2, 1_{-1}, 2, 1_0, 2, 1_{-1}, 2, 1_0, 2, 1_{-1}, 2, 1_0, 2, 1_{-1}, 2, 1_0] =$$
$$\dots$$
$$[3, 3, 3, 3, 3, 3]$$

Exercise 9.20 Show that the following two-variable forms of functional equations are valid, where $x, y \in \mathbb{Q} \setminus \{0\}$:

$$x + y = 1 \iff \mathfrak{C}(x) + \mathfrak{C}(y) = 1, \tag{9.8}$$

$$xy = 1 \iff \mathfrak{C}(x)\mathfrak{C}(y) = 1, \tag{9.9}$$

$$x + y = 0 \iff \mathfrak{C}(x)\mathfrak{C}(y) = -1, \tag{9.10}$$

$$\frac{1}{x} + \frac{1}{y} = 1 \iff \frac{1}{\mathfrak{C}(x)} + \frac{1}{\mathfrak{C}(y)} = 1. \tag{9.11}$$

Exercise 9.21 Let $x \in \mathbb{Q}^+$ be a node of the Stern–Brocot tree and consider the path γ_x on the tree connecting the root to x. Let $y = \mathfrak{C}(x)$. Then $\mathfrak{C}$ sends the path γ_x to γ_y. Show (or rather observe) that $\mathfrak{C}$ sends the straight segments of γ_x (i.e. those segments of the path where it always turns in the same direction) to zig-zag segments (i.e. those segments where the path changes direction at each node it encounters) and vice versa.

Exercise 9.22 Let den : $\mathbb{Q}^+ \to \mathbb{Z}$ be the denominator function.

1. Show that den satisfies the system

$$\mathrm{den}(1 + x) = \mathrm{den}(x),$$
$$\mathrm{den}\left(\frac{1}{1+x}\right) = \mathrm{den}(x) + \mathrm{den}(1/x).$$

   Moreover, den is uniquely defined by this system, if we set $\mathrm{den}(1) = 1$.
2. Define the *codenominator* function $\mathsf{F} : \mathbb{Q}^+ \to \mathbb{Z}^+$ by $\mathsf{F}(x) := \mathrm{den}(\mathfrak{C}(x))$. Show that F satisfies the system

$$\mathsf{F}(1 + 1/x) = \mathsf{F}(x),$$
$$\mathsf{F}\left(\frac{1}{1+x}\right) = \mathsf{F}(x) + \mathsf{F}(1/x).$$

   Moreover, F is uniquely defined by this system, if we set $\mathsf{F}(1) := 1$.
3. Prove that $\mathfrak{C}(x) = \mathsf{F}(1/x)/\mathsf{F}(x)$ for $x \in \mathbb{Q}^+$.

4. Show that one has $\mathsf{F}(x+2) = \mathsf{F}(x+1) + \mathsf{F}(x)$ for $x \in \mathbb{Q}^+$; i.e. F extends the Fibonacci sequence to $\mathbb{Q}^+$ (the Fibonacci sequence (F_n) is defined for $n \in \mathbb{Z}$ by $F_{n+2} = F_{n+1} + F_n$, with $F_1 = F_2 = 1$).
5. Let $x \in (0, 1) \cap \mathbb{Q}$. Show: $\mathsf{F}(x) = \mathsf{F}(1 - x)$.
6. Let $n \in \mathbb{Z}$ and $x \in \mathbb{Q}^+$ such that $n + x > 0$. Prove:

$$\mathsf{F}(n + x) = F_n \mathsf{F}(1 + x) + F_{n-1} \mathsf{F}(x).$$

7. Let $n \in \mathbb{Z}^+$ and $x \in \mathbb{Q}^+$. Prove:

$$\mathsf{F}\left(\frac{F_n + F_{n+1}x}{F_{n-1} + F_n x}\right) = \mathsf{F}(x).$$

For some other properties of the codenominator function, see [36].

Exercise 9.23 (Project) Study the limit distribution of $\daleth(\mathcal{F}_n)$, where $\mathcal{F}_n$ is the Haros–Farey sequence (see Remark 9.1)

9.5.1 Extension of Jimm to the Real Line

Let x be an irrational number. Then it has a non-terminating continued fraction expansion $x = [n_0, n_1, \dots]$. For $x > 0$ define

$$\daleth(x) := [1_{n_0-1}, 2, 1_{n_1-2}, 2, 1_{n_2-2}, 2, \dots 2, 1_{n_{k-1}-2}, 2, 1_{n_k}, \dots], \tag{9.12}$$

with the usual rules (cf. Corollary 9.2)

$$[\dots, n, 1_0, m, \dots] \to [\dots, n, m, \dots]$$
$$[\dots, n, 1_{-1}, m, \dots] \to [\dots, n + m - 1, \dots]$$

(a 1_{-1} at the beginning of a sequence is replaced by 0). Set $\daleth(x) := -1/\daleth(x)$ for $x < 0$. This defines an extension $\daleth : \mathbb{R} \setminus \{0\} \to \mathbb{R} \setminus \{0\}$.

Note that if $\varphi := [1_\infty]$ is the golden ratio, then a repeated application of the second rule gives

$$\begin{aligned}
\daleth(\varphi) =& [1_0, 2, 1_{-1}, 2, 1_{-1}, 2, 1_{-1}, 2, \dots] \\
=& [2, 1_{-1}, 2, 1_{-1}, 2, 1_{-1}, 2, \dots] \\
=& [3, 1_{-1}, 2, 1_{-1}, 2, 1_{-1}, 2, \dots] \\
=& [4, 1_{-1}, 2, 1_{-1}, 2, 1_{-1}, 2, \dots] = \dots = \infty,
\end{aligned}$$

from which we deduce $\daleth(-1/\varphi) = -\daleth(\varphi) = \infty$, $\daleth(1/\varphi) = 1/\daleth(\varphi) = 0$ and $\daleth(-\varphi) = -1/\daleth(\varphi) = 0$.

Recall that a *noble (or golden) number* is a number x whose continued fraction expansion terminates with 1_∞ (equivalently, x is in the $\mathsf{PGL}_2(\mathbb{Z})$-orbit of φ). Denote by $\mathcal{N}$ the set of noble numbers.

Lemma 9.4 *$\mathfrak{Z}$ sends noble numbers onto rational numbers. Moreover, $\mathfrak{Z}$ is 2-1 on $\mathcal{N}$.*

Proof Any rational x has two equivalent continued fraction expansions

$$x = [n_0, \dots, n_k, \infty] = [n_0, \dots, n_k - 1, 1, \infty]$$

with $n_k > 1$ unless $x = 1$ (in which case we have $1 = [1] = [0, 1]$).

Applying the definition (9.12) of $\mathfrak{Z}$ to these two expansions, we obtain two different noble numbers:

$$\begin{aligned} \mathfrak{Z}([n_0, \dots, n_k, \infty]) &= [\dots, 1_{n_k-2}, 2, 1_\infty], \\ \mathfrak{Z}([n_0, \dots, n_k - 1, 1, \infty]) &= [\dots, 1_{n_k-3}, 2, 1_{-1}, 2, 1_\infty] \\ &= [\dots, 1_{n_k-3}, 3, 1_\infty]. \end{aligned}$$

Conversely, we have

$$\begin{aligned} \mathfrak{Z}([\dots, 1_{n_k-2}, 2, 1_\infty]) =& [\dots, \underbrace{2, 1_{-1}, 2, \dots 1_{-1}, 2}_{n_k-2 \text{ times } 1_{-1}}, 1_0, 2, 1_{-1}, 2, 1_{-1}, 2, \dots] \\ =& [\dots, n_k, \infty] = x \end{aligned}$$

and similarly

$$\begin{aligned} \mathfrak{Z}([\dots, 1_{n_k-3}, 3, 1_\infty]) =& [\dots, \underbrace{2, 1_{-1}, 2, \dots 1_{-1}, 2}_{n_k-3 \text{ times } 1_{-1}}, 1, 2, 1_{-1}, 2, 1_{-1}, 2, \dots] \\ =& [\dots, n_k - 1, 1, \infty] = x. \end{aligned}$$

□

We refer the reader to [33] for proofs of the theorems below as well as further properties of $\mathfrak{Z}$.

Theorem 9.2

1. *$\mathfrak{Z}$ is involutive on $\mathbb{R} \setminus (\mathcal{N} \cup \{0\})$.*
2. *$\mathfrak{Z}$ is continuous on $\mathbb{R} \setminus \mathbb{Q}$ with jumps at rationals. Moreover, it is almost everywhere differentiable with derivative vanishing almost everywhere on $\mathbb{R}$.*
3. *$\mathfrak{Z}$ satisfies the functional equations on $\mathbb{R} \setminus \mathbb{Q}$*

$$\mathfrak{Z}(1/x) = 1/\mathfrak{Z}(x), \quad \mathfrak{Z}(-x) = -1/\mathfrak{Z}(x), \quad \mathfrak{Z}(1 - x) = 1 - \mathfrak{Z}(x).$$

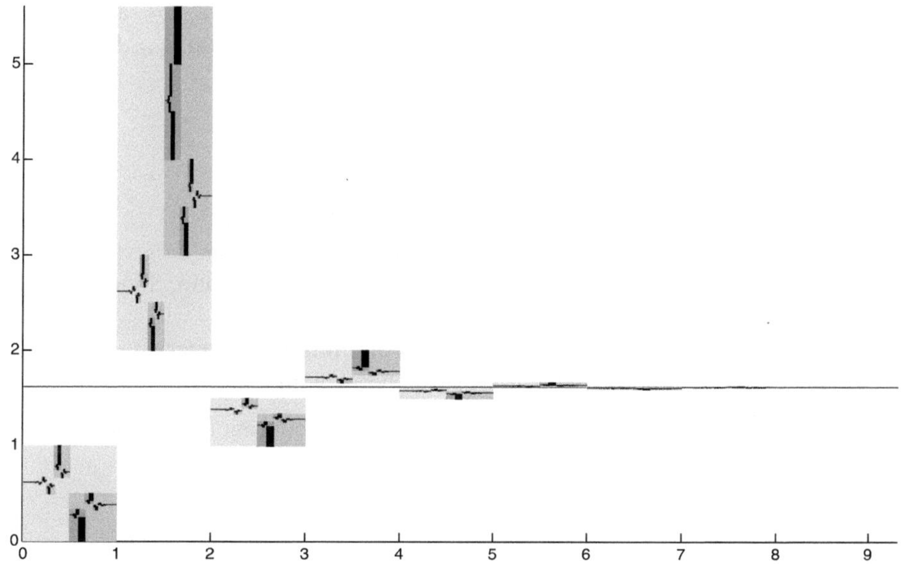

Fig. 9.10 Plot of Jimm on $\mathbb{R}^+$. The asymptote is the line $y = \varphi$

4. *Ꝣ is equivariant with respect to the* $\mathsf{PGL}_2(\mathbb{Z})$*-action on* $\mathbb{R} \setminus \mathbb{Q}$*; i.e.*

$$\text{Ꝣ}(Mx) = \alpha(M)\text{Ꝣ}(x), \quad (M \in \mathsf{PGL}_2(\mathbb{Z})),$$

where α is Dyer's outer automorphism of $\mathsf{PGL}_2(\mathbb{Z})$.

Theorem 9.3 *Ꝣ sends non-noble quadratic irrationals onto non-noble real quadratic irrationals. Moreover, Ꝣ commutes with the Galois action on real quadratic irrationals; i.e. with $a, b, c, d \in \mathbb{Q}$ and $b, d > 0$ non-square one has (Fig. 9.10)*

$$\text{Ꝣ}(a + \sqrt{b}) = c + \sqrt{d} \implies \text{Ꝣ}(a - \sqrt{b}) = c - \sqrt{d}.$$

Exercises

Exercise 9.24 (Binary Representation of Continued Fractions) One can represent the finite or infinite continued fraction $x = [n_0, n_1, \dots]$ by the infinite binary sequence

$$(0_{n_0}, 1_{n_1}, 0_{n_2}, \dots),$$

which we again denote by x for simplicity. Note that any $x \in \mathbb{Q}^+$ admits two representations; one ending with 1_∞ and the other ending with 0_∞. Define the operation $\oplus$ on the binary representations via

$$(x \oplus y)_i := x_i + y_i \mod 2.$$

(Equivalently, $\oplus$ is the 'bitwise exclusive or' operation).

1. For y fixed, let $f_y(x) := x \oplus y$. Show that f_y is an involution.
2. Let $\varphi := (01)_\infty$ be the binary representation of the golden ratio. Show that (the binary representation of) $\mathfrak{C}(x)$ is precisely the function f_φ.
3. Show that the two different representations of a rational number are sent under f_φ to two different binary sequences, each ending with $(01)_\infty$.

Exercise 9.25 Show that the two preimages of a non-zero rational number x under $\mathfrak{C}$ are Galois conjugates.

9.6 More Trees

9.6.1 Yu–Ting–Andreev Tree

Consider the submonoid of $\mathsf{PGL}_2(\mathbb{Z})$ defined by

$$X := \left\langle \begin{bmatrix} 1 & 1 \\ 0 & 1 \end{bmatrix}, \begin{bmatrix} 0 & 1 \\ 1 & 1 \end{bmatrix} \right\rangle = \langle T, UT \rangle.$$

We may form the corresponding tree of matrices by starting from the root which is the identity matrix; such that the left-child of the node labeled M is TM and the right-child of the node labeled M is UTM. This tree is called the *Yu–Ting–Andreev* tree. Hence, the nodes of this tree are in one-to-one correspondence with the elements of the submonoid $\langle T, UT \rangle$ of $\mathsf{PGL}_2(\mathbb{Z})$. Note that this submonoid is not related to $\mathsf{PSL}_2(\mathbb{N})$ via an automorphism of $\mathsf{PGL}_2(\mathbb{N})$ as in the case of the Drib tree (Fig. 9.11).

Evaluating each node at 1, we get the Yu–Ting–Andreev tree of rationals (Fig. 9.12).

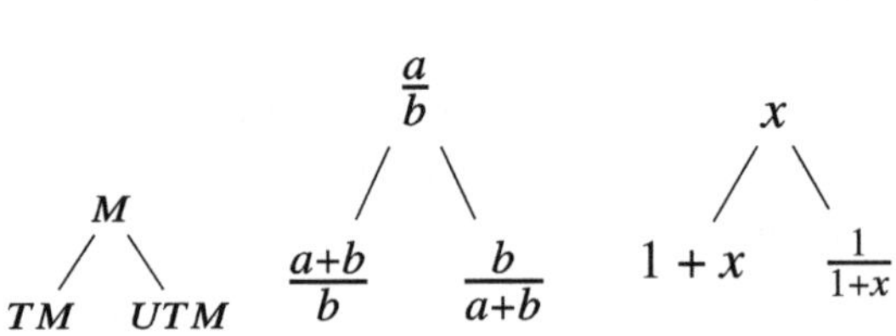

Fig. 9.11 Generation rules for the Yu–Ting–Andreev matrix tree and its rational version

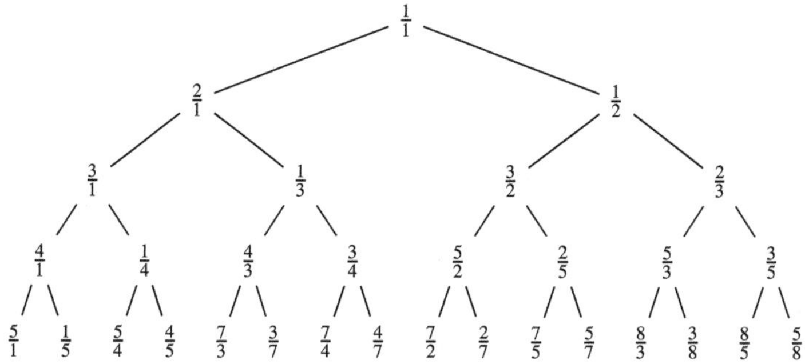

Fig. 9.12 Yu–Ting–Andreev tree

Fig. 9.13 The Hanna–Czyz–Self tree

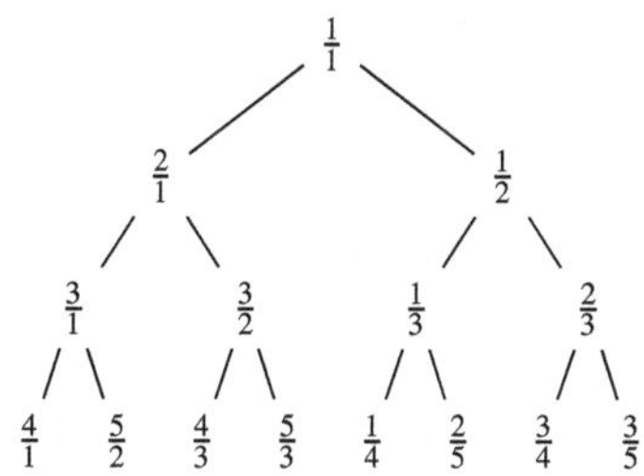

9.6.2 *Hanna–Czyz–Self Tree*

By replacing each node M of the Yu–Ting–Andreev matrix tree by UM^tU we get another tree, called the *Hanna–Czyz–Self* tree (Fig. 9.13).

9.6.3 *Kepler Tree*

Consider the submonoid $\langle UTU, UT\rangle < \mathsf{PGL}_2(\mathbb{N})$. We may form a tree of matrices by starting from the root which is the identity matrix, such that the left-child of the node labeled M is $UTUM$ and the right-child of the node labeled M is UTM. This tree is called the *matrix Kepler tree*. The tree obtained by evaluating its nodes at 1 is called the *Kepler tree*. This latter tree enumerates rationals in the unit interval (0, 1) (Figs. 9.14, 9.15, and 9.16).

HARMONICIS LIB. III. 27

prà pro numeratore, & unitas infrà pro denominatore: deinde uterq; numerus seorsim ponitur suprà, & summa utriusq; utrinq; infrà; ut ex una qualibet fractione duo oriantur rami, eousq; dum ex summa fiat numerus index figuræ indemonstrabilis.

Has sectiones chordæ septem primùm auditu indice inveni, totidem nempe, quot sunt Harmoniæ non majores unâ Diapason: postea caussas & sectionum singularum, & universarum Numeri, non sine labore ex penitissimis Geometriæ fontibus erui. Legat curiosus lector, quæ de his sectionibus ante annos 22. scripsi in Mysterio Cosmographico, Capite XII. & perpendat, quomodo fuerim illo loco hallucinatus super caussis sectionum & Harmoniarum; perperam nisus

CAP. II.

Caussæ septenarij hujus successivè inventæ.

Fig. 9.14 Kepler's tree in his book Harmonices Mundi

Fig. 9.15 The generation rule of the Kepler tree

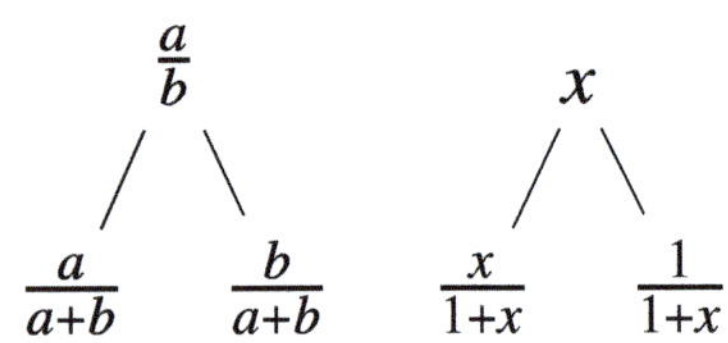

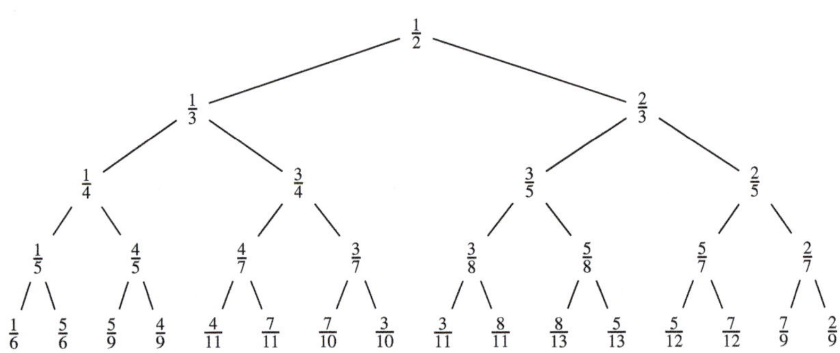

Fig. 9.16 The Kepler tree of rationals in (0, 1). The two children of p/q are $p/(p+q)$ and $q/(p+q)$

9.6.4 Lebesgue Tree

Replace each node x of the Kepler tree by $1 - x$, take two copies of the tree so obtained, and connect the roots of these two trees to a new root labeled 1. The resulting tree is called the *Lebesgue tree*.

The Lebesgue tree is not obtained from a submonoid of $\mathsf{PGL}_2(\mathbb{N})$ by evaluating the nodes of the corresponding tree at 1.

Via their planar embeddings, we can identify the Lebesgue tree with the left-branch of the Stern–Brocot tree indexing the rationals in $(0, 1)$. Let x be a node of the Stern–Brocot tree and let y_x be corresponding node on the Lebesgue tree. Recall the definition of the Farey neighborhood $I(x)$ of x (Definition 9.1).

Theorem 9.4 *The length (i.e. the Lebesgue measure) of $I(x)$ is the product of all ancestors of y_x on the Lebesgue tree.*

For a proof, see [35] (Fig. 9.17).

9.6.5 Notes

The Bird and Drib trees have been named in 2009 by Ralf Hinze [18] after Richard S. Bird [8] ("Drib" is the inverse of "Bird"). The Hanna–Czyz–Self and Yu–Ting–Andreev trees have been named by Kevin Ryde [31] (2014) after the works [13], [2], [16], and [37]. The Kepler tree already appears in Kepler's Harmonice Mundi [20]. To our knowledge, the Lebesgue tree first appeared in [35]. For more details about these trees see [21] and [25].

The involution Jimm was first introduced in [34]. For more details, see [36], [35], and [33].

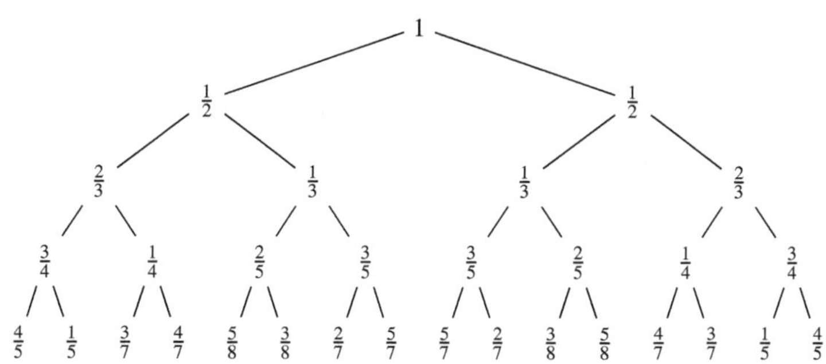

Fig. 9.17 The "Lebesgue tree" $\mathcal{L}$. The two nephews (offsprings of a sibling) of p/q are $p/(p+q)$ and $q/(p+q)$ in $\mathcal{L}$. Multiplying the numbers in a lineage yields always a reciprocal integer [35]

Exercises

Exercise 9.26 Prove the following observations about the Kepler tree:

1. The sum of siblings is always 1.
2. Siblings have the same denominator.
3. Let x, y be two consecutive nodes in a row, which are not siblings. Then x and y have the same numerator. Moreover, $1/x + 1/y$ is an integer. Can you identify this number?

Exercise 9.27 Consider the Yu–Ting–Andreev monoid $\langle T, UT \rangle$.

1. Since $U\langle T, UT\rangle U = \langle UTU, TU\rangle$, we can form a tree of rationals, starting from the root 1, such that the left child of the node x is $UTU(x) = x/(1+x)$ and the right child is $TU(x) = 1 + 1/x$. Show that this tree is obtained from the Yu–Ting–Andreev tree by replacing each node by its reciprocal $1/x$. (Therefore this tree enumerates all positive rationals.)
2. Since $\alpha\langle T, UT\rangle = \langle TU, UTU\rangle$, we can form a tree of rationals, starting from the root 1, such that the left child of the node x is $TU(x) = 1 + 1/x$ and the right child is $UTU(x) = x/(1+x)$. Show that this tree is obtained from the Yu–Ting–Andreev tree by replacing each node by ج(x). (Therefore this tree enumerates all positive rationals.)

Exercise 9.28 Prove that the only submonoids of $\mathsf{PGL}_2(\mathbb{Z})$ such that, when evaluated at 1, produce all rationals are

1. $\langle UTU, T\rangle$ (Calkin–Wilf monoid),
2. $\langle UT, TU\rangle = \alpha\langle UTU, T\rangle$ (Drib monoid),
3. $\langle T, UT\rangle$ (Yi–Ting–Andreev monoid)
4. $\langle UTU, TU\rangle = U\langle T, UT\rangle U = \alpha\langle UT, T\rangle$ (reciprocal Yi–Ting–Andreev monoid)

(Note that the tree is determined not just by the monoid but also by the choice of generators defining the descent to left and right).

Exercise 9.29 Identify the Hanna–Czyz–Self tree with the Stern–Brocot tree via their planar embeddings. Let x be a node of the Stern–Brocot tree and let $f(x)$ be the node with the same address in the Hanna–Czyz–Self tree as the address of x in the Stern–Brocot tree. Give a formula expressing $f(x)$ as a map of continued fractions, similar to the formula we gave for ج(x).

Acknowledgments Author was supported by TÜBİTAK grants 119F405 and 221N171 as well as a GSU research grant FIR-2022-1089 during the preparation of this manuscript.

References

1. M. Aigner, G.M. Ziegler, *Proofs from the Book* (Springer, Berlin, 1999)
2. D.N. Andreev, On a wonderful numbering of the positive rational numbers (in Russian). Math. Prosveshchenie Ser. **3**(1), 126–134 (1997)
3. R. Backhouse, J.F. Ferreira, Recounting the rationals: twice!, in *International Conference on Mathematics of Program Construction* (Springer, 2008), pp. 79–91
4. R. Backhouse, J.F. Ferreira, On Euclid's algorithm and elementary number theory. Sci. Comput. Program. **76**(3), 160–180 (2011)
5. B. Bates, T. Mansour, The q-Calkin–Wilf tree. J. Combin. Theory Ser. A **118**(3), 1143–1151 (2011)
6. B. Bates, M. Bunder, K. Tognetti, Linking the Calkin–Wilf and Stern–Brocot trees. Eur. J. Combin. **31**(7), 1637–1661 (2010)
7. J. Berstel, A. De Luca, Sturmian words, Lyndon words and trees. Theor. Comput. Sci. **178**(1–2), 171–203 (1997)
8. R.S. Bird, Loopless functional algorithms, in *International Conference on Mathematics of Program Construction* (Springer, 2006), pp. 90–114
9. A. Brocot, *Calcul des rouages par approximation: nouvelle méthode* (A. Brocot, 1862)
10. N. Calkin, H.S. Wilf, Recounting the rationals. Am. Math. Mon. **107**(4), 360–363 (2000)
11. K.S. Choudary, A.S. Reddy, Calkin-Wilf tree. Resonance **25**(7), 1001–1013 (2020)
12. C. Cobeli, A. Zaharescu, The Haros-Farey sequence at two hundred years. Acta Univ. Apulensis Math. Inform **5**, 1–38 (2003)
13. J. Czyz, W. Self, The rationals are countable—Euclid's proof. College Math. J. **34**(5), 367–369 (2003)
14. J. Farey, LXXIX. On a curious property of vulgar fractions. Philos. Mag. **47**(217), 385–386 (1816)
15. S. Han, A.M. Masuda, S. Singh, J. Thiel, The (u, v)-Calkin–Wilf forest. Int. J. Number Theory **12**(05), 1311–1328 (2016)
16. P.D. Hanna, entries a071585 and a071766. *Online Encyclopedia of Integer Sequences* (2002)
17. S. Heubach, T. Mansour, *Combinatorics of Compositions and Words* (Chapman and Hall/CRC, 2009)
18. R. Hinze, The bird tree. J. Funct. Program. **19**(5), 491–508 (2009)
19. M. Imbert, *Sur l'isomorphisme du groupe de richard thompson avec le groupe de ptolemee*. London Mathematical Society Lecture Note Series (1997), pp. 313–324
20. J. Kepler, M. Caspar, W. von Dyck, F. Hammer, *Harmonice mundi* (CH Beck'sche Verlagsbuchhandlung, 1940)
21. G.D. Knott, *The Stern-Brocot Tree*. Civilized Software (2019), p. 12109
22. R.A. Kucharczyk, Enumerating trees. Preprint. arXiv:1201.1851 (2012)
23. J.C. Lagarias, C.P. Tresser, A walk along the branches of the extended Farey tree. IBM J. Res. Dev. **39**(3), 283–294 (1995)
24. T. Mansour, M. Shattuck, Two further generalizations of the Calkin–Wilf tree. J. Combin. **2**(4), 507–524 (2011)
25. Y.Y. Mendizabal, Enumeration systems of positive rationals (q+) based on stern's sequence-stern-en segidan oinarritutako q+ zenbaki arrazional positiboen zenbait zenbakitze sistema. EKAIA EHUko Zientzia eta Teknologia aldizkaria (2022)
26. F. Mertens, Über eine zahlentheoretische funktion. Sitzungsberichte Akad. Wiss. Wien IIa **106**, 761–830 (1897)
27. H. Mestanoğlu, Stern-brocot sayı dizisi ve modüler grup. Master's thesis, Balıkesir Üniversitesi Fen Bilimleri Enstitüsü, 2021
28. H. Minkowski, *Zur geometrie der zahlen. verhandlungen des III. internationalen mathematiker-kongresses*. Heidelberg, Berlin (1904), pp. 171–172
29. M.B. Nathanson, Forests of complex numbers. Int. J. Number Theory **13**(01), 15–25 (2017)
30. G.N. Raney, On continued fractions and finite automata. Math. Ann. **206**(4), 265–283 (1973)

31. K. Ryde, *Trees of rationals*. (2014)
32. K.E. Stange, An arborist's guide to the rationals. Preprint. arXiv:1403.2928 (2014)
33. A.M. Uludağ, On the involution Jimm, in *Topology and Geometry.: A Collection of Essays Dedicated to Vladimir G. Turaev*, ed. by A. Papadopoulos. IRMA Lectures in Mathematics and Theoretical Physics (EMS Press, 2021), pp. 561–578
34. A.M. Uludağ, H. Ayral, An involution of reals, discontinuous on rationals, and whose derivative vanishes ae. Turk. J. Math. **43**(3), 1770–1775 (2019)
35. A.M. Uludağ, H. Ayral, A subtle symmetry of Lebesgue's measure. J. Theor. Probab. **32**(1), 527–540 (2019)
36. A.M. Uludağ, B.E. Gökmen, The conumerator and the codenominator. Bull. Sci. Math. **180**, 103192 (2022)
37. S. Yu-Ting, A "natural" enumeration of non-negative rational numbers—an informal discussion. Am. Math. Mon. **87**(1), 25–29 (1980)

Chapter 10
Binary Quadratic Forms: Modern Developments

Ayberk Zeytin

To Athanase, whose brilliance, encouragement, energy, and partnership are deeply appreciated.

Abstract In this chapter, we offer a historical overview through the vast topic of binary quadratic forms. We begin with a quick review of their history and then an overview of contemporary algebraic developments on the subject.

10.1 Introduction

It seems that the first encounter of humanity with binary quadratic forms[1] is through the famous Pythagorean triples, as we will see below. In fact, the Babylonian tablet Plimpton 322, dating back to around 1800 BCE, represents the earliest documented interaction between humans and binary quadratic forms. As far as the author is aware, there is a gap until Euclid, who lived around 300 BCE. Euclid studied quadratic irrationalities, which may be viewed as an indirect study of the topic. Diophantus of Alexandria (circa third century CE) made some early investigations into quadratic forms in his work "Arithmetica". Namely, Diophantus dealt with special cases of quadratic equations and their solutions. The Indian mathematician Brahmagupta (circa 598 CE–668 CE) made significant contributions to the study of indeterminate equations, including quadratic ones. For instance, he used the famous formula for the roots of quadratic equations.

Then, until Fermat (1607–1665), no significant contributions to the topic exists to our knowledge. He worked not only on Pythagorean triples but also on some equa-

[1] A binary quadratic form is a homogeneous element of the ring $\mathbf{Z}[x, y]$ of degree two.

A. Zeytin (✉)
Department of Mathematics, Galatasaray University, Beşiktaş, Istanbul, Türkiye
e-mail: azeytin@gsu.edu.tr

A. M. Uludağ, A. Zeytin (eds.), *Essays on Geometry*,
https://doi.org/10.1007/978-3-031-76257-4_10

tions of Pell type. It seems that binary quadratic forms were studied heavily during and after the eighteenth century, especially with contributions from many prominent mathematicians, including Euler (1707–1783), Lagrange (1736–1813), Legendre (1752–1833), Gauss (1777–1855), and Dirichlet (1805–1859). One addition that we propose to this list is Poincaré,[2,3] whose contributions are mostly overlooked.

The theory of binary quadratic forms can be viewed, at least historically, as the birth of algebraic number theory. Indeed, it was the observation that $X^2 + Y^2$ is enough for representing primes that are congruent to 1 modulo 4, whereas a similar statement has to include not only X^2+5Y^2 but also $2X^2+2XY+3Y^2$, that gave rise to many questions. Such problems led to the definition of classes of binary quadratic forms. Gauss carried this idea further and defined a binary operation, called Gauss product, on the set of classes of primitive binary quadratic forms of discriminant Δ, denoted by $\mathcal{F}(\Delta)$. It turned out that $\mathcal{F}(\Delta)$ endowed with the Gauss product is an abelian group which is isomorphic to the narrow class group of a corresponding quadratic number field (whenever Δ is a fundamental discriminant) in our modern terminology. Contemporary algebraic number theory is so well developed that its one of the many raison d'être, namely the representation problem of binary quadratic forms, is not mentioned quite often and is not visible to an inexperienced eye. In this work, we focus more on the modern algebraic developments of the binary quadratic forms, leaving many topics untouched, for instance, the analytic aspects and, in particular, the theta functions of definite and indefinite binary quadratic forms. Reader interested in such topics is invited to consult [49] and [55] as a starting point.

The organization of the paper is as follows: The next section is a quick review of the history. This part is neither complete nor detailed. We take refuge in the existence of many wonderful historical accounts on number theory (e.g. [14–16, 50]). However, we give a little bit detailed account about Poincaré's contributions to the field, which seem to have been overlooked. The last section is devoted to some central and contemporary developments concerning the algebraic aspects of indefinite binary quadratic forms.

[2] The mainstream LLMs do not recognize Poincaré in a list of mathematicians who made contributions to the topic of binary quadratic forms. In fact, here is an output of ChatGPT, V3.5 when asked about Poincaré's contributions: "Henri Poincaré is not primarily known for his work on binary quadratic forms. In fact, the study of binary quadratic forms is a field of mathematics that was extensively developed by other mathematicians, such as Gauss, Lagrange, and Jacobi."

[3] This text was finalized before the appearance of [21], to which the reader is encouraged to consult for a dedicated and complete treatise.

10.2 A Very Brief History of Binary Quadratic Forms

Babylonian tablet Plimpton 322, dated nearly as 1800 BCE, is the first known documented encounter between mankind and binary quadratic forms.[4] It contains the *first* list of 15 Pythagorean triples, that is integral solutions of the equation $X^2 + Y^2 = Z^2$ with a couple mistakes. The list contains the triple (12709, 13500, 18541) which appears to be the largest. It turns out that Babylonians knew the *rule* in producing such triplets but interested merely in some particular cases. They have also made certain attempts in the direction of proof, [24], yet whether the Babylonians knew the proof or not is still in debate. The book [30] seems to be the first text where the tablet in question is first analyzed. Reader interested in Babylonian mathematics is invited to consult [29] and references therein.

Book X of Euclid's *Elements* (circa 300 BCE) is reserved for the study of quadratic irrationalities. This is the book where Euclid describes his famous algorithm (Proposition 3). The same proposition appears in Book VII of Elements, as well. Though it is closely related, Euclid's viewpoint did not pay much attention to binary quadratic forms. We refer to [25] for a treatment of Book X of Elements in modern terms.

The Cattle problem, attributed to Archimedes (approximately around 250 BCE), is another instance where we encounter binary quadratic forms. For the exact statement of the problem, the interpretation of the problem given by Wurm (sometimes called *Wurm's problem*) and its reduction to the equation:

$$1 + 4729494u^2 = t^2;$$

(where $2|u$ and $4657|u$) can be found in [48]. There was a small error in the solution of Amthor, [26], which came as late as 1880. This mistake was then corrected in [22] more than 100 years later!

Some place for binary quadratic forms has also been reserved in "Arithmetica" of Diophanthus of Alexandria[5] (third century CE). This is a book comprised of 130 problems. There, once again, we meet Pythagorean triples, with a somewhat more systematic treatment. In fact, starting from two Pythagorean triples, Diophantus

[4] The tablet was found in Iraq and is now at Columbia University, see http://www.math.ubc.ca/~cass/courses/m446-03/pl322/pl322.html.

[5] One should mention that the exact dates of birth and death of Diophantus are still in dispute, although we know, from an old riddle composed by Metrodorus, that he died when he was 84. The exact riddle reads: "This tomb holds Diophantus. Ah, how great a marvel! the tomb tells scientifically the measure of his life. God granted him to be a boy for the sixth part of his life; and adding a twelfth part to this he clothed his cheeks with down; He lit him the light of wedlock after a seventh part, and five years after his marriage He granted him a son. Alas! late-born wretched child, after attaining the measure of half his father's life, chill Fate took him. After consoling his grief by this science of numbers for four years, he ended his life." [32, p. 93]. It is also arguable, though probably/apparently unnecessary, that the name Diophantus is in fact an ancient analogue of Bourbaki, i.e. a collective author, [44].

knew how to derive a third one using the identity

$$(X^2 + Y^2)(X'^2 + Y'^2) = (XX' \pm YY')^2 + (XY' \mp X'Y)^2. \tag{10.1}$$

This is related, in particular, to problems 8[6] and 9[7] of Book II of Arithmetica. Indeed, Diophanthus refers to his book "Porismata" (auxiliary theorems) quite frequently in "Arithmetica" and according to Weil it is perhaps not too far-fetched to imagine that (10.1) may have been one of his porismata', [50, p. 11]. More on the history of sums of two squares can be found in [15, Chapter 6], as well.

As far as the author is aware, there is a gap in the development of binary quadratic forms until the time of Fermat except for Leonardo of Pisa (1170–1250), who is rather known by the name Fibonacci. His *Liber Quadratorum* from 1225 (see [19] for an English translation) concerns mostly with square numbers and miscellaneous problems around squares, starting with the famous problem of the representability of integers by sums of two squares. It was Euler who proved relevant statements for the forms $x^2 + 2y^2$ and $x^2 + 3y^2$. For a trained eye, the generalization of (10.1) to

$$(X^2 + NY^2)(X'^2 + NY'^2) = (XX' \pm NYY')^2 + N(XY' \mp X'Y)^2$$

is quite clear, but we had to wait Euler and his book algebra for this and its variant

$$(X^2 - NY^2)(X'^2 - NY'^2) = (XX' \pm NYY')^2 - N(XY' \pm X'Y)^2$$

to appear. Interested reader is invited to consult [27] and [51] for further details on the history of the Pell equation. Pell's equation has other applications in solving other Diophantine equations, see for instance, [28, Chapter 8].

Similar ideas did not lead to a solution for the case $x^2 + 5y^2$. This was known to Fermat. Euler had also failed to explain this phenomenon. It was Lagrange who came up with the idea to separate the $\mathrm{PSL}_2(\mathbf{Z})$ classes of binary quadratic forms of discriminant D according to the elements they represent in $(\mathbf{Z}/|D|\mathbf{Z})^\times$. Legendre (1752–1833) took the studies of Lagrange further and generalized what Lagrange achieved. Indeed, the identity

$$(2x^2 + 2xy + 3y^2)(2x'^2 + 2x'y' + 3y'^2) = X^2 + 5Y^2;$$

where $X = 2xx' + xy' + yx' + 3yy'$ and $Y = xy' - yx'$ is used by Lagrange to prove quadratic reciprocity concerning the binary quadratic form $X^2 + 5Y^2$. This identity is in fact one of the first examples of composition of two binary quadratic forms. Legendre recognized that similar identities can be reproduced for any two binary quadratic forms having the same discriminant laying the foundations of composition. The missing pieces of this composition that prevented the set of binary

[6] To split a given square in two squares.

[7] To split a given number which is the sum of two squares into two other squares.

quadratic forms of fixed discriminant being a group (including the composition as defined by Legendre being multi-valued) has been completed by Gauss in his *Disquisitiones Arithmeticae*, [20]. This corner stone contribution—the author believes—is the first work where binary quadratic forms are treated not as tools to solve other problems but rather as objects of interest. Gauss had also treated many other concepts such as *ambiguous forms*[8] and reduced forms. Gauss had computed class groups of certain imaginary quadratic number fields and put forward his famous class number problems. One should also mention Dirichlet, especially [17, 18], where the famous class number formula for imaginary quadratic number fields is given.

To the above list of prominent mathematicians, we would like to propose an addition: Poincaré (1854–1912). Poincaré's works are not only on binary but also on ternary and other quadratic as well as cubic forms of several variables. In what follows, we offer a look at his contributions to only binary quadratic forms in chronological order.

The first work of Poincaré concerning quadratic forms is "Sur quelques propriétès des formes quadratiques", [38] where he announces that to each form he can associate a corresponding complex number defined as a series. Four properties of these complex numbers are listed. The first one claims that they can be computed by evaluating definite integrals. The remaining three properties are announced so as to attack the classical problems pertaining to quadratic forms; namely, the solution of the corresponding Pell equation, equivalence problem: determining whether two forms are equivalent, and minimum problem: determining the smallest positive integer represented by the quadratic form. These problems were already addressed by Gauss in [20]. More precisely, in § 198 the Pell equation[9] and its solution is discussed, in § 158 the equivalence problem[10] is discussed and in § 172 the definition of reduced forms is given which is closely related to the minimum problem.

In the continuation of this work, [37], his work on the equivalence problem is announced. Here, Poincaré mentions the lattice corresponding to a given definite quadratic form. Given two equivalent forms, he mentions that the associated lattices differ by a rotation of angle θ, which he knows how to compute. He then introduces the series

$$\sum_{(x,y)\in \mathbf{Z}\times\mathbf{Z}\setminus\{(0,0)\}} \frac{1}{(ax^2+bxy+cy^2)^k} \tag{10.2}$$

[8] A binary quadratic form $f = (a, b, c)$ is called ambiguous if there is some $\gamma \in \mathrm{PSL}_2(\mathbf{Z})$ so that $\gamma \cdot f = (a, -b, c)$.

[9] Given a form (M, N, P) whose determinant is D, and m the greatest common divisor of the numbers $M, 2N, P$: to find the smallest numbers t, u satisfying the indeterminate equation $t^2 - Du^2 = m^2$.

[10] Given any two forms having the same determinant, we want to know whether or not they are equivalent, whether or not they are properly or improperly equivalent, or both (for this can happen too).

as an arithmetic invariant[11] of a given definite binary quadratic form $aX^2 + bXY + cY^2$. The fact that such an invariant can only be a necessary condition is immediately noted. This gap is filled by defining what is called a covariant function and two such examples are given. It must be noted that the series given in (10.2) appear also in the above mentioned works of Dirichlet, e.g. [17, § 7], as well. In addition, Poincaré announces that using these methods he can write a prime number which is congruent to 1 modulo 4 as a sum of two squares.

The article [39] is where Poincaré uses his correspondence between binary quadratic forms (both definite and indefinite) and lattices and interprets known facts in terms of the geometry (mostly fundamental triangles) of the lattices. The first two sections are devoted to developing algebraic operations (addition, multiplication) on lattices in **C**. In the following section, given a binary quadratic form $f = aX^2+bXY+cY^2$, the lattice generated by $(b/\sqrt{a}, \sqrt{|b^2 - 4ac|/Da})$ and $(\sqrt{a}, 0)$ is associated to the form; where D is the square-free part of the discriminant of f. Various facts concerning this correspondence is discussed. In addition, the reduction algorithm (both for definite and indefinite forms) is also interpreted—using fundamental triangles—in terms of the theory developed in first two sections. The next section is devoted to defining a group operation which is a geometric version of the group operation defined by Gauss, see the note following this article in [31] by A. Châtelet. The last section of this work is related to Kummer's (and Dedekind's) theory of ideal numbers. Here, Poincaré explains how one deals with factorization of ideals in prime ideals in terms of lattices.

In his work "Sur les invariants arithmetiques",[12] [40], Poincaré describes a series associated to a linear form. Then, he explicitly describes the series (10.2) that he announced in [37] in addition to another series denoted by θ by writing the form as a product of two (conjugate) linear forms. Poincaré argues (without proof) that these series admit an integral representation, as well. Poincaré ends this paper with the remark that both for definite and indefinite binary quadratic forms, the series he introduced can be used to distinguish the equivalence of two binary forms.

Poincaré's works on quadratic forms focuses on generalization of various known results to forms of more variables and/or of higher degree, for instance, [35, 36, 42, 43] between 1881 and 1905. This does not mean that Poincaré did not think about such problems, but rather that he thinks of problems related to arithmetic from a mostly function theoretical viewpoint.[13] His 1905 article entitled "Sur les invariants arithmetiques", [43], is a very good example of this viewpoint. Indeed, after discussing arithmetic invariants of linear forms, Poincaré uses this theory to develop similar invariants for quadratic forms. The definite and the indefinite cases are treated separately. Poincaré interprets and also extends Dirichlet's results using

[11] According to Poincaré an arithmetic invariant of a form is a quantity invariant under the action of $SL_2(\mathbf{Z})$.

[12] Poincaré published another article with the same title in 1905, [43] which will be discussed this below.

[13] This was remarked by Poincaré, as well.

the theory developed. We refer to [2] for a detailed account of this article using modern terms.

10.3 Modern Era

In this section, we will concentrate on the more modern developments—by modern, we mean the main contribution in the last 50 years. More precisely, we will talk about the work of Bhargava where the Gauss product of two forms is interpreted in terms of what is called Bhargava cubes inspired from the group law on elliptic curves, Zagier's alternative reduction theory, Penner's geometric interpretation of Gauss product for definite binary quadratic forms and the contribution of the author where the class of an indefinite binary quadratic form is interpreted as a bipartite ribbon graph with specific features.

10.3.1 Notation and Terminology

For an element $f(x, y) = ax^2 + bxy + cy^2 \in \mathbf{Z}[x, y]$ we will write $f = (a, b, c)$, for short. The coefficients a, b and c will be viewed—by abuse of notation—as functions on forms, that is, for $f = (a, b, c)$ we have $a(f) = a$, $b(f) = b$ and $c(f) = c$. We define the *discriminant* of f to be $\Delta(f) := b^2 - 4ac$, which is always assumed to be non-zero and non-square, hence f cannot be written as a product of two linear forms with integral coefficients. f is called *positive* (resp. *negative*) *definite* if $\Delta(f) < 0$ and $a > 0$ (resp. $a < 0$). If $\Delta(f) > 0$; f is called *indefinite*. $f = (a, b, c)$ is called *primitive* if 1 is the largest integer dividing a, b and c at the same time. For any binary quadratic form f and any given integer N, the *set of representations* of N by f is defined as:

$$R_f(N) := \{(X, Y) \in \mathbf{Z}^2 \colon f(X, Y) = N\}.$$

Whenever $R_f(N) \neq \emptyset$ we say that N is *represented* by f. The set of all integers that are represented by f becomes

$$R_f = \{N \in \mathbf{Z} \colon R_f(N) \neq \emptyset\}.$$

The *modular group* $\mathrm{PSL}_2(\mathbf{Z})$ acts on this set by change of variables; namely, for any $f = (a, b, c)$ and $\gamma = \begin{pmatrix} p & q \\ r & s \end{pmatrix}$ we define

$$\gamma \cdot f = (f(p, r), 2apq + b(ps + qr) + 2crs, f(q, s)).$$

The equivalence class of $f = (a, b, c)$ is denoted by $[f] = [a, b, c]$. The above concepts, the discriminant (hence f being positive/negative definite or indefinite), primitivity, the set R_f are invariant under this action. In the search for a unique representative of each equivalence class, we define a definite binary quadratic form $f = (a, b, c)$ to be *reduced* if the following inequalities are satisfied:

$$|b| \leq a \leq c.$$

If further one of the inequalities is an equality then we require $b \geq 0$. We say that an indefinite binary quadratic form f is *G-reduced* if we have:

$$|\sqrt{|\Delta(f)|} - 2|a|| < b < \sqrt{|\Delta(f)|}.$$

We also say that an indefinite binary quadratic form is *semi-reduced* if $ac < 0$. In the case of positive definite forms, it turns out that every equivalence class has a unique representative. However, we will see in Sect. 10.3.5 that whenever f is indefinite, the equivalence class $[f]$ has at least two different reduced forms.

Let us denote the set of forms of discriminant Δ as $\mathcal{F}(\Delta)$. Gauss has defined a binary operation which will be referred to as *Gauss product*. In modern terms, $\mathcal{F}(\Delta)$ together with Gauss product turns out to be an abelian group. To the experts (see [1, 10, 12, 13]), the definition of Gauss seemed unmotivated. In Sect. 10.3.4 we will mention Bhargava's new approach to the topic.

10.3.2 Minus Continued Fractions: Zagier Reduction

In this section, we discuss Zagier's contribution to the theory. Let us begin with:

Definition 10.1 ([52, p. 122]) An indefinite binary quadratic form $f = (a, b, c)$ is called *Z-reduced* if

$$a, c > 0 \text{ and } b > a + c.$$

This definition is slightly different from the definition given by Gauss. For instance, if $f = (a, b, c)$ is G-reduced, then $ac < 0$ is satisfied; whereas for Z-reduced forms, $ac > 0$. Given any indefinite binary quadratic form f, the transformation

$$S(L^2S)^n = \begin{pmatrix} n & 1 \\ -1 & 0 \end{pmatrix}$$

is applied successively; where $n = \lceil \frac{b+\sqrt{\Delta(f)}}{2a} \rceil$. This results in an eventually periodic sequence of forms. Recall that given an indefinite binary quadratic form $f = (a, b, c)$, one can recover the periodic part of the standard continued fraction of the root of the equation $ax^2+bx+c = 0$ from the period of the form f that appear in the reduction process. Zagier reduced forms recover the mentioned correspondence in the case of minus continued fractions.[14,15]

The main application of this theory is an interesting result—known as Hirzebruch theorem—concerning class numbers of imaginary quadratic number fields:

Theorem 10.1 ([52, §14, Satz 3]) *Let $p > 3$ be a prime number which is congruent to -1 modulo 4. Write $\sqrt{p} = [[a_0; \overline{a_1, a_2, \ldots, a_r}]]$ with the part $\overline{a_1, a_2, \ldots, a_r}$ representing the minimal period of the minus continued fraction expansion. If $\mathbf{Q}(\sqrt{p})$ has narrow class number 1, then the class number of $\mathbf{Q}(\sqrt{-p})$ is*

$$\frac{1}{3}(a_1 + a_2 + \ldots + a_r) - r.$$

One should consult [45] for further discussion regarding Z-reduced forms, which completes the analogy concerning the reduction algorithm associated to this viewpoint.

10.3.3 Geometry of Gauss Product: Penner's Work

In [33] Penner discusses a geometric interpretation of the Gauss law. In order to describe such an interpretation one needs to revisit product law on binary quadratic forms as defined by Dirichlet. Let us recall that given two binary quadratic forms $f_1 = (a_1, b_1, c_1)$ and $f_2 = (a_2, b_2, c_2)$, we say that f_1 and f_2 are *unitable* if $\Delta(f_1) = t_1^2\Delta$ and $\Delta(f_2) = t_2^2\Delta$—that is when the square-free part of the discriminants agree. In the case when $\Delta(f_1) = \Delta(f_2)$, we further say that f_1 and f_2 are *concordant* if $b_1 = b_2$ and if $a_2|c_1$ and $a_1|c_2$. That is, if $f_1 = (a_1, b, a_2c)$ and

14 The minus continued fraction is the process of expressing a given real number as a sum of the form

$$[[a_0; a_1, a_2, a_3]] := a_0 - \cfrac{1}{a_1 - \cfrac{1}{a_2 - \cfrac{1}{a_3 - \cfrac{1}{\ddots}}}};$$

where $a_1, a_2, \ldots$ are all elements of $\mathbf{Z}_{\geq 2}$.

15 Minus continued fractions enjoy properties both similar to (such as the characterization of a real number x being a rational number being having a finite continued fraction expansion) and different from (such as roots of quadratic equations admitting eventually periodic expansions in the standard case whereas the terms in expansion being equal to 2 eventually in the minus continued fractions case) standard continued fractions.

$f_2 = (a_2, b, a_1 c)$. In such a case, the product $f_1 \cdot f_2$ is defined simply as $(a_1 a_2, b, c)$. It is proven by Dirichlet that given two forms, say Q_1, Q_2, of discriminant Δ one can always find forms $f_1 \in [Q_1]$ and $f_2 \in [Q_2]$ so that f_1 and f_2 are concordant, see for instance [9, p. 335]. Moreover, one can already see that unitable forms are concordant but the converse is not true.

Given a positive definite binary quadratic form $f = (a, b, c)$, the equation $aX^2 + bX + c = 0$ has exactly two conjugate roots: $\omega_f = \frac{-b+\sqrt{\Delta(f)}}{2a} \in \mathcal{H} = \{z \in \mathbf{C} \mid \mathrm{Im}(z) > 0\}$ and $\overline{\omega_f} \in \overline{\mathcal{H}}$. As $a, b, c \in \mathbf{Z}$, $\omega_f \in \mathcal{H} \cap \mathbf{Q}(\sqrt{-1})$. Conversely, given an element $\omega = \frac{\alpha}{\beta} + \frac{\gamma}{\delta}\sqrt{-1} \in \mathcal{H} \cap \mathbf{Q}(\sqrt{-1})$, the integral positive definite binary quadratic form $f_\omega = (\beta^2\delta, -2\alpha\beta\delta, \alpha^2\delta + \beta^2\gamma)$ admits ω as its corresponding element. When $\gcd(\alpha, \beta)$ and $\gcd(\gamma, \delta)$ are 1, f_ω is primitive. One computes that $\Delta(f_\omega) = -4\beta^4\gamma\delta$.

In the view of the above correspondence, given two different points $\omega_1, \omega_2 \in \mathcal{H} \cap \mathbf{Q}(\sqrt{-1})$ the forms f_{ω_1} and f_{ω_2} are unitable if and only if there are constants $\lambda_1, \lambda_2 \in \mathbf{Z}$ so that $\Delta(f_{\omega_1})/\lambda_1^2 = \Delta(f_{\omega_2})/\lambda_2^2$. In particular, if ω_1 and ω_2 are solutions of the equation

$$\frac{\Delta(f_\omega)}{a(f_\omega)^2} = -\alpha^2 \quad \text{or} \quad \frac{\Delta(f_\omega)}{b(f_\omega)^2} = -\beta^2 \quad \text{or} \quad \frac{\Delta(f_\omega)}{c(f_\omega)^2} = -\gamma^2 \tag{10.3}$$

for some $\alpha, \beta, \gamma \in \mathbf{Z}_+$, then the forms f_{ω_1} and f_{ω_2} are unitable. The functions mentioned in (10.3) induce the following equations in $\mathcal{H}$

$$v = \frac{\alpha}{2} \quad \text{or} \quad v = \pm\beta u \quad \text{or} \quad u^2 + \left(v - \frac{1}{\gamma}\right)^2 = \frac{1}{\gamma^2}, \tag{10.4}$$

respectively. Considered with its hyperbolic structure, these loci become horocycles centered at $\infty \in \partial\mathcal{H}$, hypercycle (to the imaginary axis) through $0 \in \partial\mathcal{H}$ and horocycle centered at $0 \in \partial\mathcal{H}$, respectively. We conclude

Proposition 10.1 ([33, Corollary 8]) *Given $f_1 = (a_1, b_1, c_1)$ and $f_2 = (a_2, b_2, c_2)$, f_1 and f_2 are unitable if and only if they lie on one of the loci mentioned in (10.4).*

We will refer to these loci as hypercycles. Suppose now that two forms f_1 and f_2 and concordant, hence unitable. Then by Proposition 10.1 we know that there is a hypercycle, say h, containing ω_{f_1} and ω_{f_2}. The following result which gives us an insight about the geometry of the product $[f_1] \cdot [f_2]$:

Theorem 10.2 ([34, Corollary 3.11]) *If f_1 and f_2 are two positive definite, primitive concordant binary quadratic forms lying on a common hypercycle h, then the point $\omega_{[f_1]\cdot[f_2]} \in h$ is the point closest to the origin.*

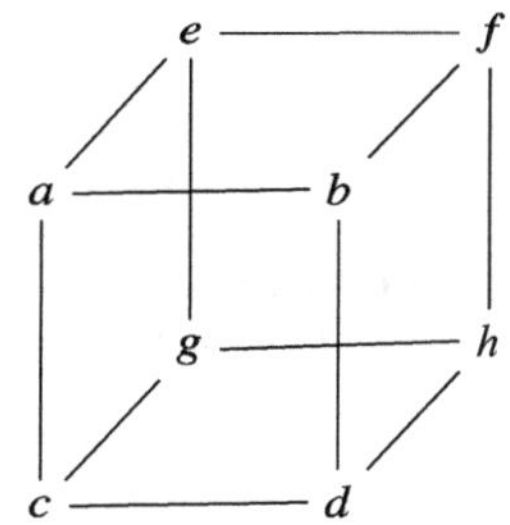

Fig. 10.1 An octuple of integers arranged as a cube

10.3.4 Bhargava's Cubes: Composition Laws

In his thesis, [3], which is accompanied by four articles, [4–7], Bhargava introduced generalization of Gauss product on the set of binary quadratic forms of fixed discriminant. The main tool was elements of the 8 dimensional $\mathbf{Z}$-module $C := \mathbf{Z}^2 \otimes \mathbf{Z}^2 \otimes \mathbf{Z}^2$ whose elements are identified with cubes of integers or equivalently with octuples $(a, b, \ldots, h)$ of integers (Fig. 10.1):

One may partition such a cube, called a *Bhargava cube*, into three pairs of 2×2 matrices using the three linearly independent directions up-down, left-right and front-back:

$$\begin{array}{lll} \text{up-down} & U = \begin{pmatrix} a & e \\ b & f \end{pmatrix} & D = \begin{pmatrix} c & g \\ d & h \end{pmatrix} \\ \text{left-right} & L = \begin{pmatrix} a & c \\ e & g \end{pmatrix} & R = \begin{pmatrix} b & d \\ f & h \end{pmatrix} \\ \text{front-back} & F = \begin{pmatrix} a & b \\ c & d \end{pmatrix} & B = \begin{pmatrix} e & f \\ g & h \end{pmatrix} \end{array} \tag{10.5}$$

From such a partition, we may obtain three binary quadratic forms by declaring

$$f_{UD} = -\det(Ux + Dy), \quad f_{LR} = -\det(Lx + Ry), \quad f_{FB} = -\det(Fx + By)$$

For the pairs of matrices (U, D), (L, R) and (F, B) given in (10.5) we have an action of $\mathrm{PSL}_2(\mathbf{Z})$:

$$\begin{pmatrix} p & q \\ r & s \end{pmatrix} \cdot (M, N) \mapsto (pM + qN, rM + sN)$$

inducing an action of $(\mathrm{PSL}_2(\mathbf{Z}))^3$ on C. Using this action, one can check that $\Delta(f_{UD}) = \Delta(f_{LR}) = \Delta(f_{FB})$, which will be called the discriminant of the cube C giving rise to these forms. We will denote this discriminant by $\Delta(C)$. The discriminant of a cube is an invariant of the action of $(\mathrm{PSL}_2(\mathbf{Z}))^3$. Primitivity[16] of a cube is another invariant of this action. One has the following result which can be verified with a messy but straightforward computation:

Theorem 10.3 *Let $C \in \mathcal{C}$ be a cube giving rise to the forms f_{UD}, f_{LR} and f_{FB}. The product of the three forms f_{UD}, f_{LR} and f_{FB} is the identity, that is*

$$f_{UD} \cdot f_{LR} \cdot f_{FB} = f_I;$$

where f_I is the identity form—$f_I = (1, 0, -\Delta(C)/4)$ if $\Delta(C) \equiv 0 \pmod 4$, and $(1, 1, -(\Delta(C) - 1)/4)$ if $\Delta(C) \equiv 1 \pmod 4$.

Following the footprints of the geometric description of the group law on elliptic curves, such an identity can be used to define a group operation, which agrees with the Gauss product by its construction.

The converse is also possible. That is, given any pair f_1, f_2 of primitive binary quadratic forms of discriminant Δ there exists a primitive cube $C \in \mathcal{C}$ of discriminant Δ with the property that $f_{UD} = f_1$ and $f_{LR} = f_2$.

10.3.5 Çarks: A Combinatorial Viewpoint

In recent years, the author and his collaborators, [47, 53], have given an interpretation of the theory of primitive indefinite binary quadratic forms as a part of the theory of dessins d'enfants. In order to describe the theory, let first fix the elements:

$$S = \begin{pmatrix} 0 & -1 \\ 1 & 0 \end{pmatrix} \text{ and } L = \begin{pmatrix} 1 & -1 \\ 1 & 0 \end{pmatrix}$$

which generate the group $\mathrm{PSL}_2(\mathbf{Z})$ freely. Consider the Farey tree, denoted by $\mathcal{F}$, whose topological realization can be described as follows: let $\bullet$ denote the fixed point of L (that is $\exp(2\pi\sqrt{-1}/6)$), $\circ$ denote the fixed point of S (that is, $\sqrt{-1}$) and let g denote the geodesic in $\mathcal{H}$ joining $\bullet$ to $\circ$. The set of *black* (resp. white) vertices of $\mathcal{F}$ is the orbit of $\bullet$ (resp. $\circ$) and the set of edges is the orbit of g under $\mathrm{PSL}_2(\mathbf{Z})$, see Fig. 10.2. The result of such a construction is a bipartite plane tree where each edge $W \cdot g$ is labeled with the element W.

The tree $\mathcal{F}$ is endowed with the natural translation action of $\mathrm{PSL}_2(\mathbf{Z})$ defined as left multiplication, which is also compatible with the usual action of $\mathrm{PSL}_2(\mathbf{Z})$ on $\mathcal{H}$. The automorphism group of any binary quadratic form $f = (a, b, c)$, denoted by $\mathrm{Aut}(f)$, is defined as those elements of $\mathrm{PSL}_2(\mathbf{Z})$ which act trivially on f, i.e. $\mathrm{Stab}_{\mathrm{PSL}_2(\mathbf{Z})}(f)$.

[16] One may define a cube $C \in \mathcal{C}$ to be primitive if the forms f_{UD}, f_{LR} and f_{FB} are primitive.

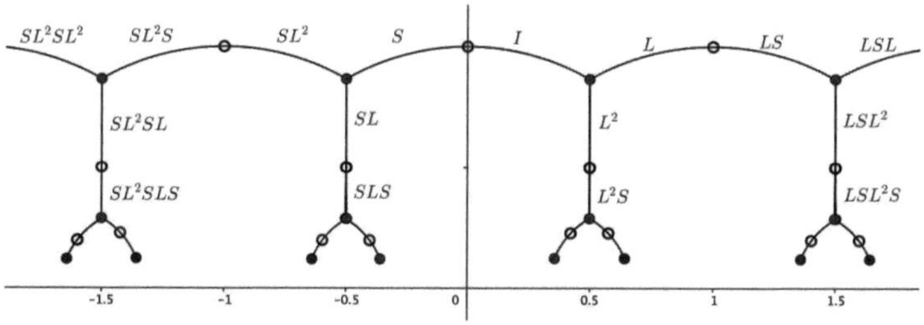

Fig. 10.2 The labeled bipartite Farey tree, $\mathcal{F}$

Theorem 10.4 ([8, Theorem 6.12.4]) *Given any indefinite binary quadratic form $f = (a, b, c)$ the group* $\mathrm{Aut}(f)$ *is generated by one hyperbolic element in* $\mathrm{PSL}_2(\mathbf{Z})$*, that is* $\mathrm{Aut}(f) \cong \langle W_f \rangle$.

Definition 10.2 Given an indefinite binary quadratic form f, the infinite bipartite planar graph $\mathcal{F}/\mathrm{Aut}(f) = \mathcal{F}/\langle W_f \rangle$ is defined as the çark[17] associated to f and is denoted by $\mathcal{G}_f$.

As a result of its construction, the çark $\mathcal{G}_f$ associated to f enjoys many important features, all of which are explained in Fig. 10.3 for the form $f = (25, 111, -33)$ whose automorphism group is generated by

$$W_f = (L^2S)^3(LS)(L^2S)(LS)^2(L^2S)(LS)^4 = \begin{pmatrix} 7 & 33 \\ 25 & 118 \end{pmatrix}$$

- ▶ each edge of $\mathcal{G}_f$ can be labeled with a coset of $\mathrm{Aut}(f)$. The edge labeled with $\mathrm{Aut}(f)$ may be regarded as a distinguished (or base) edge of $\mathcal{G}_f$. This is a one to one correspondence between the conjugacy class of the generator W_f of $\mathrm{Aut}(f)$ and the set of edges of $\mathcal{G}_f$.
- ▶ the conjugacy class of W_f admits a one to one correspondence with the equivalence class of f. As a result, each edge of $\mathcal{G}_f$ can be labeled with a unique element of $[f]$.
- ▶ the vertices of type • (resp. ∘) on graph $\mathcal{G}_f$ are of degree 3 (resp. 2).
- ▶ being a bipartite graph, there is a well defined notion of left and right on $\mathcal{G}_f$. A path is called a left-turn path if $(n+1)$st edge of the path is to the left of the nth. A bi-infinite[18] path is called a face of the çark.

[17] The word çark is Turkish and means wheel. It is pronounced as chark.

[18] A *path* on a graph is defined as an ordered sequence of adjacent non-backtracking edges. The first edge is called *initial* and the last edge is called *terminal*. A path is called *bi-infinite* if it does not have neither an initial, nor a terminal edge.

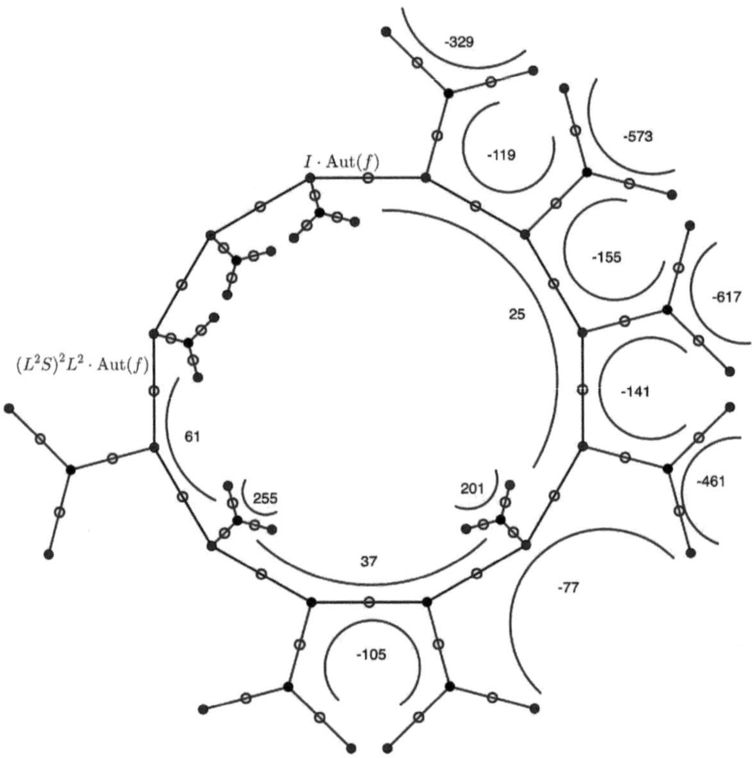

Fig. 10.3 The çark corresponding to $f = (25, 111, -33)$ which is of discriminant $15621 = 3 \cdot 41 \cdot 127$

- the graph $\mathcal{G}_f$ has a unique cycle[19] called the *spine* of $\mathcal{G}_f$. If a vertex, an edge, or a face is on the spine, then it will be referred to as a spinal vertex, edge or face, respectively.
- the spine is the boundary between positive (inner) labeled faces and negative (outer) labeled faces. Spine of $\mathcal{G}_f$ is the quotient of the river in Conway's topograph by the automorphism of f, [11]. This viewpoint is treated in a more geometric way in [23].
- there are finitely many Farey branches[20] attached to the spine at each spinal vertex of type •. Consecutive Farey branches that point in the same direction is called a Farey bunch. For instance, the çark in Fig. 10.3 contains 6 Farey bunches containing, 3,1,1,2,1,4 Farey branches, respectively. Notice that one may use the

[19] A path is called closed if initial and terminal edges agree. A cycle on a graph is defined as a finite closed path.

[20] A Farey branch is an infinite bipartite rooted tree which is isomorphic as a bipartite ribbon graph to the quotient $\mathcal{F}/\langle L\rangle$.

Farey branches in the Farey bunches together with the natural cyclic order of the spine to specify the çark. This is merely the periodic part of the continued fraction expansion of the root of $aX^2 + bX + c = 0$. Farey bunches allows to depict the çarks in a more compact way, see Fig. 10.4, called the Farey bunch representation.

The theory around indefinite binary quadratic forms can be interpreted in terms of çarks. For instance, a form is semi-reduced if and only if it is the label of a spinal edge, a form is Gauss-reduced if it is the label of a spinal edge which is adjacent to the root of an inner Farey bunch. As there are at least two such edges in any given çark, this implies that there are at least two reduced forms in any equivalence class. The reduction procedure of Gauss is merely the process of moving the base edge to a base edge, [47]. The graph $\mathcal{G}_f$ is symmetric with respect to its spine if and only if the form f is ambiguous.

The representation problem admits an algorithmic solution, [53, 54], as well. To explain this, let us define distance between two vertices to be the number of edges in between, and the root of a face to be the unique vertex (of type $\bullet$) which is closest to the spine. Then, via an *arithmetic progression* on the labels of faces, one may show that these labels increase in absolute value as the distance between the face and the spine increases. Therefore, given any integer N, there are only finitely many possibilities for the faces of $\mathcal{G}_f$ to be labeled N. Therefore, one may solve the representation problem by a search on the faces of $\mathcal{G}_f$.

Actions on Çarks

The combinatorial/geometric interpretation provided by çarks leads to interesting actions. For instance, given a primitive indefinite binary quadratic form f we let $[a_1, b_1, \ldots, a_k, b_k]$ be the encoding of the Farey bunches attached to the spine of $\mathcal{G}_f$. Then, the çark corresponding to f^{-1} is encoded by $[b_k, a_k, \ldots, b_1, a_1]$.

Another map that acts on the set of çarks is the map $\mathbf{J}$ studied as a map from $\mathbb{R}\cup\{\infty\}$ to itself in [46]. Indeed, Dyer's outer automorphism of the $\mathrm{PGL}_2(\mathbf{Z})$ induces such an involution. Despite being discontinuous at the rationals, this involution satisfies a surprising number of functional equations. Indeed, given any continued fraction $[a_0; a_1, a_2, \ldots]$, with $a_0 \geq 0$ and $a_i \geq 1$, is mapped to the continued fraction $[1_{a_0-1}, 2, 1_{a_1-2}, 2, 1_{a_2-2}, \ldots]$; where 1_n denotes the n-tuple $(1, 1, \ldots, 1)$. By convention 1_k is eliminated if it is in the beginning of the expansion and the expansion $[\ldots, a, 1_{-1}, b, \ldots]$ is replaced by $[\ldots, a + b - 1, \ldots]$. In addition, 1_0 is eliminated from the continued fraction expansion, i.e. $[\ldots, a, 1_0, b, \ldots]$ is $[\ldots, a, b, \ldots]$. Away from finite continued fractions (i.e. rational numbers) and noble numbers (i.e. those real numbers whose continued fraction expansion eventually becomes $[1, 1, 1, \ldots]$) the map $\mathbf{J}$ satisfies the functional equation:

$$\mathbf{J}(-x) = -\frac{1}{x};$$

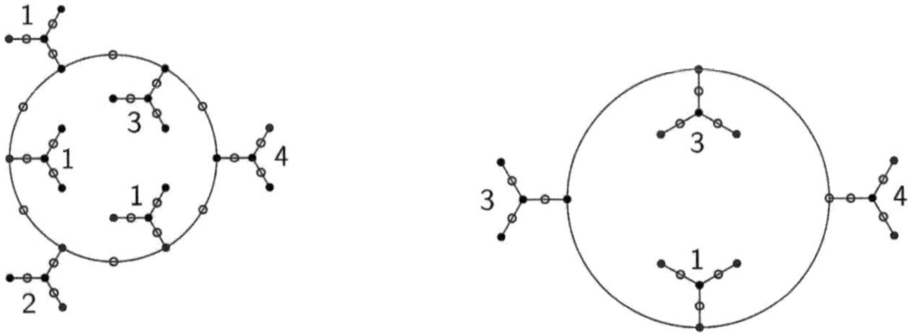

Fig. 10.4 The çark corresponding to $f = (25, 111, -33)$ and $\mathbf{J}(f)$, respectively

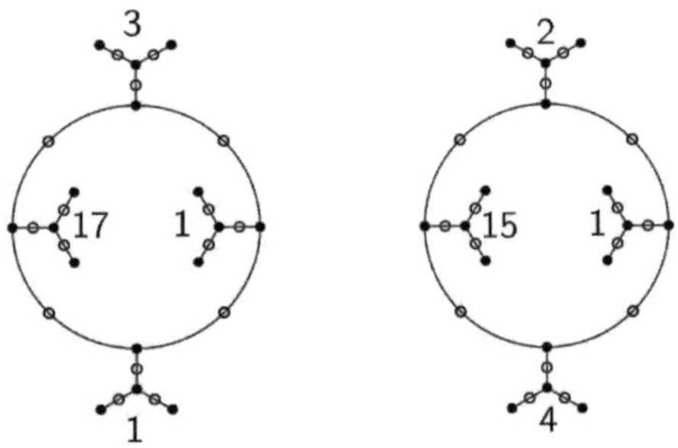

Fig. 10.5 The çark corresponding to $g = (-55, 89, 35)$ and $\mathbf{J}(g) = (-63, 48, 38)$, respectively

using which we may define $\mathbf{J}$ for those real numbers whose continued fraction begins with a negative integer.

The map $\mathbf{J}$ induces a map on the equivalence classes of binary quadratic forms as it preserves those real numbers having eventually periodic continued fraction expansion. For instance, the form $f = (25, 111, -33)$ is mapped onto the form $\mathbf{J}(f) = (-43, 46, 13)$ which is of discriminant $4352 = 2^8 \cdot 17$. Figure 10.4 shows the Farey bunch representation of çarks f and $\mathbf{J}(f)$.

The form $g = (-55, 89, 35)$ is also of discriminant 15621 and $g \notin [f]$ as seen in Fig. 10.5. One can quickly compute that $\mathbf{J}(g) = (-63, 48, 38)$ whose discriminant is $11880 = 2^3 \cdot 3^3 \cdot 5 \cdot 11$.

From the above observations, we may see that the action of $\mathbf{J}$ on the class groups can be defined when we consider class groups of all the quadratic orders at once.

Acknowledgments Author is supported by TUBITAK 1001 Research grant 119F405 during the preparation of this manuscript.

References

1. F. Arndt, Zur Theorie der binären kubischen Formen. J. Reine Angew. Math. **53**, 309–321 (1857)
2. N. Bergeron, Les "invariants arithmétiques" de Poincaré. Grad. J. Math. **3**(1), 1–14 (2018)

3. M. Bhargava, *Higher composition laws*. ProQuest LLC, Ann Arbor, MI, Thesis (Ph.D.)–Princeton University, 2001
4. M. Bhargava, Higher composition laws. I: A new view on Gauss composition, and quadratic generalizations. Ann. Math. (2) **159**(1), 217–250 (2004)
5. M. Bhargava, Higher composition laws. II: On cubic analogues of Gauss composition. Ann. Math. (2) **159**(2), 865–886 (2004)
6. M. Bhargava, Higher composition laws. III: The parametrization of quartic rings. Ann. Math. (2) **159**(3), 1329–1360 (2004)
7. M. Bhargava, Higher composition laws. IV: The parametrization of quintic rings. Ann. Math. (2) **167**(1), 53–94 (2008)
8. J. Buchmann, U. Vollmer, *Binary Quadratic Forms: An Algorithmic Approach*, volume 20 of Algorithms and Computation in Mathematics (Springer, Berlin, 2007)
9. J.W.S. Cassels, *Rational Quadratic Forms*, volume 13 of London Mathematical Society Monographs (Academic Press [Harcourt Brace Jovanovich, Publishers]/London, 1978)
10. M.A. Cayley, Mémoire sur les Hyperdéterminants. J. Reine Angew. Math. **30**, 1–37 (1846)
11. J.H. Conway, *The Sensual (quadratic) Form*, volume 26 of Carus Mathematical Monographs (Mathematical Association of America, Washington, DC, 1997). With the assistance of Francis Y. C. Fung
12. R. Dedekind, Über binäre trilineare Formen und die Komposition der binären quadratischen Formen. J. Reine Angew. Math. **129**, 1–34 (1905)
13. R. Dedekind, Über Komposition der quadratischen Formen. Gött. Nachr. **1907**, 86–100 (1907)
14. L.E. Dickson, *History of the Theory of Numbers. Vol. I: Divisibility and Primality* (Chelsea Publishing, New York, 1966)
15. L.E. Dickson, *History of the Theory of Numbers. Vol. II: Diophantine Analysis* (Chelsea Publishing, New York, 1966)
16. L.E. Dickson, *History of the Theory of Numbers. Vol. III: Quadratic and Higher Forms*. With a chapter on the class number by G. H. Cresse (Chelsea Publishing, New York, 1966)
17. P.G.L. Dirichlet, Sur l'usage des séries infinies dans la théorie des nombres. J. Reine Angew. Math. **18**, 259–274 (1838)
18. P.G.L. Dirichlet, Recherches sur diverses applications de l'Analyse infinitésimale à la Théorie des Nombres. J. Reine Angew. Math. **19**, 324–369 (1839)
19. L.P. Fibonacci, *The Book of Squares. An Annotated Translation into Modern English by L. E. Sigler*, vol. XX (Academic Press/Harcourt Brace Jovanovich Publishers, Boston, 1987), 124 p.
20. C.F. Gauss, *Disquisitiones arithmeticae* (Springer, New York, 1986). Translated and with a preface by Arthur A. Clarke, Revised by William C. Waterhouse, Cornelius Greither and A. W. Grootendorst and with a preface by Waterhouse
21. C. Goldstein, Poincaré and arithmetic revisited, in *Working Paper* or preprint, March (2023)
22. C.C. Grosjean, H.E. De Meyer, A new contribution to the mathematical study of the cattle-problem of Archimedes, in *Constantin Carathéodory: An International Tribute, Vol. I, II* (World Sci. Publ., Teaneck, NJ, 1991), pp. 404–453
23. A. Hatcher, *Topology of Numbers* (American Mathematical Society, Providence, 2022)
24. J. Høyrup, Pythagorean "rule" and "theorem": mirror of the relation between Babylonian and Greek mathematics. Math. Semesterber. **45**(2), 141–156 (1998)
25. W. Knorr, "La croix des mathématiciens": the Euclidean theory of irrational lines. Bull. Amer. Math. Soc. (N.S.) **9**(1), 41–69 (1983)
26. B. Krummbiegel, A. Amthor, Das Problema bovinum des Archimedes. Schlömilch Z., Hl. A. **25**, 121–136 (1880)
27. Literatur-Berichte: Geschichte der Gleichung t^2—$Du^2 = 1$. Monatsh. Math. Phys. **13**(1), A32–A33 (1902). Von Dr. H. Konen. Mit 2 Figuren im Text. 132 S. 8°. Hirzel, Leipzig (1902) Ladenpreis 4 M
28. L.J. Mordell, *Diophantine Equations*. Pure and Applied Mathematics, vol. 30 (Academic Press, London, 1969)

29. O. Neugebauer, Zur geometrischen Algebra. (Studien zur Geschichte der antiken Algebra. III.). Quell. Stud. Geschichte Math. Astron. Phys. B **3**, 244–259 (1936)
30. O. Neugebauer, A.J. Sachs, *Mathematical Cuneiform Texts* (American Oriental Society and the American Schools of Oriental Research, New Haven, 1945). With a chapter by A. Goetze
31. *Œuvres de Henri Poincaré. Tome V* (Gauthier-Villars, Paris, 1950). Publié avec la collaboration de Albert Chalêlet
32. W.R. Paton, *Greek Anthology, Volume V: Book 13: Epigrams in Various Metres* (Harvard UP, 1918)
33. R.C. Penner, *The Geometry of the Gauss Product*, vol. 81 (1996), pp. 2700–2718. Algebraic geometry, 4
34. R.C. Penner, *Decorated Teichmüller Theory* (European Mathematical Society (EMS), Zürich, 2012)
35. H. Poincaré, Sur les formes cubiques ternaires et quaternaires. J. l'École Polytech. **50**, 190–253 (1881).
36. H. Poincaré, Sur les formes cubiques ternaires et quaternaires (2nde partie). J. l'École Polytech. **51**, 45–91 (1882).
37. H. Poincaré, Sur les formes quadratiques. C. R. Acad. Sci. Paris **89**, 897–899 (1879)
38. H. Poincaré, Sur quelques propriétès des formes quadratiques. C. R. Acad. Sci. Paris **89**, 344–346 (1879)
39. H. Poincaré, Sur une mode nouveau de représentation géométrique des formes quadratiques définies ou indéfinies. J. l'École Polytech. **47**, 177–245 (1880). Oeuvres V, 117–184
40. H. Poincaré, Sur les invariants arithmétiques, in *Comptes rendus de la 10e session - Alger 1881* (Association française pour l'avancement des sciences, Paris, 1882), pp. 109–117
41. H. Poincaré, Sur la reproduction des formes. C. R. Acad. Sci. Paris **97**, 949–951 (1884)
42. H. Poincaré, Les fonctions Fuchsiennes et l'Arithmétique. J. Math. (4) **3**, 405–464 (1887)
43. H. Poincaré, Sur les Invariants Arithmétiques. J. Reine Angew. Math. **129**, 89–150 (1905)
44. N. Schappacher, Wer war Diophant? Math. Semesterber. **45**(2), 141–156 (1998)
45. B.R. Smith, Constructing minimal periods of quadratic irrationalities in Zagier's reduction theory. J. Number Theory **187**, 1–26 (2018)
46. A.M. Uludağ, H. Ayral, An involution of reals, discontinuous on rationals, and whose derivative vanishes a.e. Turk. J. Math. **43**(3), 1770–1775 (2019)
47. A.M. Uludağ, A. Zeytin, M. Durmuş, Binary quadratic forms as dessins. J. Théor. Nombres Bordx. **29**(2), 445–469 (2017)
48. I. Vardi, Archimedes' cattle problem. Amer. Math. Mon. **105**(4), 305–319 (1998)
49. M.-F. Vignéras, Séries thêta des formes quadratiques indéfinies, in *Séminaire Delange-Pisot-Poitou, 17e année: 1975/76, Théorie des nombres: Fasc. 1* (Secrétariat Math., Paris, 1977), pages Exp. No. 20, 3
50. A. Weil, *Number Theory*. Modern Birkhäuser Classics (Birkhäuser Boston, Boston, 2007). An approach through history from Hammurapi to Legendre, Reprint of the 1984 edition
51. E.E. Whitford, *The Pell Equation*, ProQuest LLC, Ann Arbor, MI. Thesis (Ph.D.)–Columbia University, 1912
52. D.B. Zagier, *Zetafunktionen und quadratische Körper. Eine Einführung in die höhere Zahlentheorie* (1981)
53. A. Zeytin, On reduction theory of binary quadratic forms. Publ. Math. Debrecen **89**, 203–221 (2016)
54. A. Zeytin, InfoMod: a visual and computational approach to Gauss' binary quadratic forms. Appl. Algebra Engrg. Comm. Comput. **33**(4), 437–455 (2022)
55. S. Zwegers, *Mock theta functions*. PhD thesis, Universiteit Utrecht, 2002

Chapter 11
A Note on Reversibility of Unipotent Matrices

Krishnendu Gongopadhyay and Chandan Maity

Abstract Let $\mathbb{D} = \mathbb{R}$, $\mathbb{C}$ or $\mathbb{H}$. Let $\mathrm{U}_n(\mathbb{D})$ be the group of unipotent upper triangular matrices over $\mathbb{D}$. Let $\mathfrak{u}_n(\mathbb{D})$ be the Lie algebra of $\mathrm{U}_n(\mathbb{D})$ that consists of $n \times n$ upper triangular matrices with 0 in all the diagonal entries. In this chapter, we consider the adjoint action of the extended group $\mathrm{U}_n^{\pm 1}(\mathbb{D})$ that consists of all upper triangular matrices over $\mathbb{D}$ having diagonal elements 1 or -1, and construct a large class of strongly $\mathrm{Ad}_{\mathrm{U}_n^{\pm 1}(\mathbb{D})}$-real elements in $\mathfrak{u}_n(\mathbb{D})$.

11.1 Introduction

Let G be a Linear group with Lie algebra $\mathfrak{g}$. Consider the natural adjoint action of G on $\mathfrak{g}$ given by $\mathrm{Ad}(g)(X) = gXg^{-1}$. First we recall the notion of *adjoint reality* in the Lie algebra set-up.

Definition 11.1 ([2, Definition 1.1]) Let G be a Lie group with Lie algebra $\mathfrak{g}$. An element $X \in \mathfrak{g}$ is called Ad_G*-real* if $-X = \mathrm{Ad}(g)X$ for some $g \in G$. An Ad_G-real element X is called *strongly* Ad_G*-real* if $-X = \mathrm{Ad}(\tau)X$ for some involution $\tau \in G$.

The above definition provides an infinitesimal version of the notion of 'real' elements in a group that are conjugate to their own inverses. The real elements are also known as 'reversible' or 'reciprocal elements' in the literature, cf. [4–6]. The real elements are closely related to the 'strongly real' elements which are conjugate to their inverses by involutions. In [2], the authors classified the 'adjoint real' and 'strongly adjoint real' nilpotent orbits in the classical Lie algebras. This gives the classification of the real and strongly real unipotent elements in the classical Lie groups via the exponential maps.

K. Gongopadhyay (✉)
Indian Institute of Science Education and Research (IISER) Mohali, S.A.S. Nagar, India
e-mail: krishnendu@iisermohali.ac.in

C. Maity
Indian Institute of Science Education and Research (IISER) Berhampur, Berhampur, India
e-mail: cmaity@iiserbpr.ac.in

A. M. Uludağ, A. Zeytin (eds.), *Essays on Geometry*,
https://doi.org/10.1007/978-3-031-76257-4_11

Let $\mathbb{D} = \mathbb{R}$, $\mathbb{C}$ or $\mathbb{H}$. Let $\mathrm{T}_n(\mathbb{D})$ be the subgroup of $\mathrm{GL}_n(\mathbb{D})$ consisting of $n \times n$ upper triangular matrices over $\mathbb{D}$. Let $\mathrm{U}_n(\mathbb{D})$ be the subgroup of $\mathrm{T}_n(\mathbb{D})$ consisting of matrices with 1 in the diagonal entries. Let $\mathfrak{u}_n(\mathbb{D})$ be the Lie algebra of $\mathrm{U}_n(\mathbb{D})$ that consists of $n \times n$ upper triangular matrices with 0 in all the diagonal entries. By a unipotent element $u \in \mathrm{GL}_n(\mathbb{D})$ we mean a non-identity matrix with 1 as only eigenvalue.

Considering the adjoint reality of the Lie algebra $\mathfrak{u}_n(\mathbb{D})$, induced by the conjugacy action of the linear Lie group $\mathrm{U}_n(\mathbb{D})$, we observe the following.

Proposition 11.1 *There is no non-trivial* $\mathrm{Ad}_{\mathrm{U}_n(\mathbb{D})}$*-real element in* $\mathfrak{u}_n(\mathbb{D})$.

The proof follows from the following observation: Let $g \in \mathrm{GL}_n(\mathbb{D})$ and $X \in \mathrm{M}_n(\mathbb{D})$ be such that $-X = gXg^{-1}$. If $\lambda \in \overline{\mathbb{D}}$ is an eigenvalue of g, then $-\lambda$ is also an eigenvalue of g. In particular, g cannot be a unipotent matrix.

When $\mathbb{F} = \mathbb{R}$ or $\mathbb{C}$, this result for $\mathrm{U}_n(\mathbb{F})$ was proved in [7, Theorem 1.3] using a different method that involves matrix computations. However, with the above notion of adjoint reality, it becomes significantly simpler and shorter.

Since any nilpotent Lie group N can be embedded in $\mathrm{U}_n(\mathbb{D})$ for some n, as an immediate consequence we have the following result which was also observed in [1, Lemma 4.6].

Corollary 11.1 *There is no non-trivial real element in a nilpotent Lie group* N.

Although $\mathrm{U}_n(\mathbb{D})$ contains no real unipotent element, some of the elements in $\mathrm{U}_n(\mathbb{D})$ may become real in an enlarged group containing $\mathrm{U}_n(\mathbb{D})$. Let $\mathrm{U}_n^{\pm 1}(\mathbb{D})$ be the group of all upper triangular matrices over $\mathbb{D}$ having diagonal elements 1 or -1. Note that if an upper triangular matrix g in $\mathrm{T}_n(\mathbb{D})$ is strongly real, then g has to be in $\mathrm{U}_n^{\pm 1}(\mathbb{D})$. This motivates us to consider the conjugation action of $\mathrm{U}_n^{\pm 1}(\mathbb{D})$ on the Lie algebra $\mathfrak{u}_n(\mathbb{D})$.

Let $\mathfrak{u}_n^{\star}(\mathbb{D})$ be the subset of $\mathfrak{u}_n(\mathbb{D})$ consisting of the elements with non-zero entries at $(i, i+1)$th place for $i = 1, \dots, n-1$ (entries above the main diagonals are non-zero). We prove the following.

Theorem 11.1 *Every element in* $\mathfrak{u}_n^{\star}(\mathbb{D})$ *is strongly* $\mathrm{Ad}_{\mathrm{U}_n^{\pm 1}(\mathbb{D})}$*-real.*

In [7] some class of real unipotent elements in $\mathrm{U}_n^{\pm 1}(\mathbb{F})$ has been constructed for $\mathbb{F} = \mathbb{R}$ or $\mathbb{C}$. Applying the above theorem, we get the following corollary that gives certain types of strongly real elements in $\mathrm{U}_n^{\pm 1}(\mathbb{D})$. This extends the class of real elements in [7] over the quaternions.

Let $\mathrm{U}_n^{\star}(\mathbb{D})$ be the set of elements in $\mathrm{U}_n(\mathbb{D})$ consisting of elements with non-zero entries at $(i, i+1)$th place for $i = 1, \dots, n-1$ (entries above the main diagonals are non-zero).

Corollary 11.2 *Every element in* $\mathrm{U}_n^{\star}(\mathbb{D}) \sqcup \big(- \mathrm{U}_n^{\star}(\mathbb{D})\big)$ *is strongly real in* $\mathrm{U}_n^{\pm 1}(\mathbb{D})$.

11.2 Proof of Theorem 11.1

In this section, we will consider adjoint action of $\mathrm{U}_n^{\pm 1}(\mathbb{D})$ on $\mathfrak{u}_n(\mathbb{D})$. Let $E_{i,j}^{(n)}$ be the $n \times n$ matrix with the sole non-zero entry 1 in the (i, j)th place. Let $X_n := (x_{i,j})_{n \times n} \in \mathfrak{u}_n^{\star}(\mathbb{D})$ be any element. Then $x_{i,i+1} \neq 0$ for $i = 1, \ldots, n-1$. Set

$$Y_n := \sum_{i=1}^{n-1} x_{i,i+1} E_{i,i+1}^{(n)} . \tag{11.1}$$

The following lemma reduces the problem into a simpler form.

Lemma 11.1 *For $X_n \in \mathfrak{u}_n^{\star}(\mathbb{D})$, let Y_n be as defined in* (11.1). *Then there exists $g_n \in \mathrm{U}_n(\mathbb{D})$ such that $g_n X_n g_n^{-1} = Y_n$.*

Proof We will use induction on n, the order of the matrix. For $n = 2$, we may take $g_2 = \mathrm{Id}$ as $X_2 = Y_2 \in \mathfrak{u}_2(\mathbb{D})$. Now we will consider $n = 3$. Let $X_3 = \begin{pmatrix} 0 & x_{1,2} & x_{1,3} \\ & 0 & x_{2,3} \\ & & 0 \end{pmatrix}$. Then $Y_3 := \begin{pmatrix} 0 & x_{1,2} & 0 \\ & 0 & x_{2,3} \\ & & 0 \end{pmatrix}$; see (11.1). By taking $g_3 := \begin{pmatrix} 1 & 0 & 0 \\ & 1 & x_{1,2}^{-1} x_{1,3} \\ & & 1 \end{pmatrix}$, $g_3 X_3 g_3^{-1} = Y_3$. Now we will assume that the statement of this lemma is true for every element in $\mathfrak{u}_{n-1}^{\star}(\mathbb{D})$. We will prove it for $\mathfrak{u}_n^{\star}(\mathbb{D})$. For this write $X_n = \begin{pmatrix} X_{n-1} & \mathbf{x} \\ & 0 \end{pmatrix} \in \mathfrak{u}_n^{\star}(\mathbb{D})$, where $X_{n-1} \in \mathfrak{u}_{n-1}^{\star}(\mathbb{D})$, and $\mathbf{x} \in \mathbb{D}^{n-1}$. Write $\mathbf{x} = (x_1, \ldots, x_{n-1})^T \in \mathbb{D}^{n-1}$. Then $x_{n-1} \neq 0$. Since $X_{n-1} \in \mathfrak{u}_{n-1}^{\star}(\mathbb{D})$, by induction hypothesis there exists Y_{n-1}, as in (11.1), and $g_{n-1} \in \mathrm{U}_n(\mathbb{D})$ such that $g_{n-1} X_{n-1} g_{n-1}^{-1} = Y_{n-1}$.

Set $\hat{\mathbf{x}} := (0, \ldots, 0, x_{n-1})^T \in \mathbb{D}^{n-1}$, $Y_n := \begin{pmatrix} Y_{n-1} & \hat{\mathbf{x}} \\ & 0 \end{pmatrix}$, and $g_n := \begin{pmatrix} g_{n-1} & \mathbf{a} \\ & 1 \end{pmatrix} \in \mathrm{U}_n(\mathbb{D})$, for some $\mathbf{a} \in \mathbb{D}^{n-1}$. We want to choose $\mathbf{a}$ such that

$$g_n X_n g_n^{-1} = Y_n .$$

This relation is equivalent to the following two equations:

$$g_{n-1} X_{n-1} = Y_{n-1} g_{n-1} \tag{11.2}$$

$$Y_{n-1} \mathbf{a} = g_{n-1} \mathbf{x} - \hat{\mathbf{x}} . \tag{11.3}$$

Note that (11.2) is already true from the induction hypothesis. Thus we want to choose $\mathbf{a}$ such that (11.3) holds. The $(1, n-1)$th entry from both sides of (11.3) is equal to 0, and $Y_{n-1} \mathbf{a} = (x_{1,2} a_2, x_{1,3} a_3, \ldots, x_{1,n-1} a_{n-1}, 0)^T$ where $\mathbf{a} = (a_1, \ldots, a_{n-1})^T \in \mathbb{D}^{n-1}$. The system (11.3) always has a solution as $x_{i,i+1} \neq 0$. This completes the proof. □

Proof of Theorem 11.1 Let $X_n \in \mathfrak{u}_n^\star(\mathbb{D})$ be any arbitrary element, and Y_n be as in (11.1). Define $\sigma := \mathrm{diag}(1, -1, \ldots, (-1)^{n-1})_{n\times n}$. Then $\mathrm{Ad}(\sigma)Y_n = -Y_n$, and $\sigma^2 = \mathrm{Id}$. By Lemma 11.1, $gX_ng^{-1} = Y_n$ for some $g \in \mathrm{U}_n(\mathbb{D})$. Now $g^{-1}\sigma g \in \mathrm{U}_n^{\pm 1}(\mathbb{D})$ is an involution and $\mathrm{Ad}(g^{-1}\sigma g)X_n = -X_n$. This completes the proof. □

The next lemma is a basic result. We include proof for the sake of completeness.

Lemma 11.2 *The exponential map* $\exp\colon \mathfrak{u}_n^\star(\mathbb{D}) \longrightarrow \mathrm{U}_n^\star(\mathbb{D})$ *is bijective.*

Proof The injectivity follows from the fact that the exponential map

$$\exp\colon \mathfrak{u}_n(\mathbb{D}) \longrightarrow \mathrm{U}_n(\mathbb{D})$$

is injective. For surjectivity, let $u \in \mathrm{U}_n^\star(\mathbb{D})$. Then $(\mathrm{Id} - u)^n = 0$. Define

$$X := -(\mathrm{Id} - u) - \frac{(\mathrm{Id} - u)^2}{2} - \cdots - \frac{(\mathrm{Id} - u)^{n-1}}{n-1}.$$

Then $X \in \mathfrak{u}_n^\star(\mathbb{D})$ and $\exp X = u$. □

Proof of Corollary 11.2 Let $g \in \mathrm{U}_n^\star(\mathbb{D})$. Then $g = \exp X$ for some $X \in \mathfrak{u}_n^\star(\mathbb{D})$; see Lemma 11.2. By Theorem 11.1, $-X = \tau X\tau^{-1}$ for some involution $\tau \in \mathrm{U}_n^{\pm 1}(\mathbb{D})$. Thus $g^{-1} = \tau g\tau^{-1}$. Next if $h = -g \in -\mathrm{U}_n^\star(\mathbb{D})$, then $h^{-1} = \tau h\tau^{-1}$. This completes the proof. □

We will conclude with another class of strongly real elements in $\mathrm{U}_n^{\pm 1}(\mathbb{D})$.

Example 11.1 Let $X \in \mathfrak{u}_n(\mathbb{D})$ be an element in the following form

$$X = \sum_{i<j,\, i+j \text{ is odd}} x_{i,j} E_{i,j}^{(n)}.$$

Then $\mathrm{Ad}(\sigma)X = -X$, for $\sigma = \mathrm{diag}(1, -1, \ldots, (-1)^{n-1})_{n\times n}$. Thus X is strongly $\mathrm{Ad}_{\mathrm{U}_n^{\pm 1}(\mathbb{D})}$-real. Thus the unipotent element, $\exp X$ is a strongly real element in $\mathrm{U}_n^{\pm 1}(\mathbb{D})$.

Acknowledgments We have been fascinated by the versatility of the work and the dynamism of Athanase Papadopulos. Athanase's suggestion for early career mathematicians to think about interesting problems but, at the same time, unconventional ones has been very inspiring to us. We note here that the concept of reversibility also appeared in the work of Papadopoulos. In [3], McCarthy and Papadopoulos proved that every element in the mapping class group of a compact, connected surface is a product of involutions. They also investigated the geometric properties of elements which are products of two involutions, i.e. the strongly real elements.

The first named author thanks Muhammed Uludağ for inviting him to the conference in honor of Athanase Papadopulos and for providing warm hospitality.

Gongopadhyay acknowledges partial support by the SERB core research grant CRG/2022/003680. Maity is supported by an NBHM PDF during the course of this work.

References

1. P. Chatterjee, On abstract homomorphism of algebraic groups, Preprint, https://arxiv.org/abs/1511.06180
2. K. Gongopadhyay, C. Maity, Reality of unipotent elements in simple Lie groups. Bull. Sci. Math. **185**, 103261 (2023)
3. J. McCarthy, A. Papadopoulos, Involutions in surface mapping class groups. Enseign. Math. (2) **33**(3–4), 275–290 (1987)
4. A.G. O'Farrell, I. Short, *Reversibility in Dynamics and Group Theory*, London Mathematical Society Lecture Note Series, vol. 416 (Cambridge University Press, Cambridge, 2015)
5. P. Sarnak, Reciprocal geodesics, in *Analytic Number Theory*, Clay Math. Proc., 7 (Amer. Math. Soc., Providence, RI, 2007), pp. 217–237
6. M.B. Sevryuk, *Reversible Systems*, Lecture Notes in Mathematics, 1211 (Springer, Berlin, 1986), vi+319 pp.
7. R. Slowik, Real elements in $\mathrm{T}_n(K)$. Linear Multilinear Algebra **61**(5), 667–677 (2013)

Chapter 12
"Le complément supérieur": On the Poetics of Mathematics

Arkady Plotnitsky

Abstract This chapter considers mathematics, with a special emphasis on geometry, as combining mathematical and artistic thinking. By artistic thinking I do not refer to aesthetic aspects of mathematics or mathematical aspects of art or aesthetics, which subjects are only marginally addressed in the article. Instead, I focus on creative aspects of mathematics as the invention of new concepts, theories, or fields. As my subtitle indicates, this argument, beginning with the term "poetics," follows that of Aristotle's Poetics [*Peri poietikẽs*], a treatise on ancient Greek poetry, the title of which is derived from the ancient Greek word *poeien* meaning "making," putting something together. Aristotle's *Poetics* is about how literature is made or composed. Aristotle did not apply the term poetics to mathematics and did not consider mathematics in this way, focusing instead, in his other works, on logical aspects of mathematics. By contrast, I argue, under the heading of the composition principle, that, as a creative endeavor, mathematics is primarily defined by its compositional nature, rather than by its logical or calculational aspects, essential as the latter are. Of course, while compositional, mathematics is not the same as literature and art. In particular, the poetics of mathematics is the poetics of concepts, which play a more limited role in literature and art, and ally mathematics or mathematical sciences, such as physics, more with philosophy. Two other principles define mathematics, as well as science, in the present view: the continuity principle (found in literature and art as well) and the unambiguity principle (not necessary in literature and art). The chapter introduces yet another principle in considering the nature of reality, the "reality without realism" (RWR) principle, which in the case of mathematics, where the primary reality considered in mental, becomes the "ideality without idealism" (IWI) principle.

A. Plotnitsky (✉)
Literature, Theory, and Cultural Studies Program, Philosophy and Literature Program, Purdue University, W. Lafayette, IN, USA
e-mail: plotnits@purdue.edu

A. M. Uludağ, A. Zeytin (eds.), *Essays on Geometry*,
https://doi.org/10.1007/978-3-031-76257-4_12

One of my personal beliefs is that fascination with symmetries and groups is one way of coping with frustrations of life's limitations: we like to recognize symmetries which allow us to recognize more than what we can see. In this sense the study of geometric group theory is a part of culture, and reminds me of several things that Georges de Rham practiced on many occasions, such as teaching mathematics, reciting Mallarmé, or greeting a friend.
–Pierre de la Harpe, *Topics in Geometric Group Theory* [14, p. 3]

[We] shall not consider pure mathematics as a separate branch of knowledge, but rather a refinement of general language.
–Niels Bohr, *The Philosophical Writings on Niels Bohr* [8, p. 68]

Lui [le vers], philosophiquement rémunére le défaut des langues, complément supérieur [Philosophically, it is verse that makes up for the insufficiency of languages, as a superior complement].
–Stéphane Mallarmé, *Crise de vers* [47, p. 847]

12.1 Introduction

Mathematics is assuredly part of culture, as Pierre de la Harpe says in the statement used as my first epigraph. Admittedly, de la Harpe refers more specifically to geometric groups and how they relate to our experience by "allow[ing]" us "to recognize more than what we can see," [14, p. 3] on which this chapter will reflect as well, by giving this recognition a more radical meaning. His statement, nevertheless, implies the same claim for mathematics as a whole, as suggested by his invocation of Georges de Rham's teaching of mathematics, along with de Rham's reciting Stéphane Mallarmé and greeting a friend, representing high and everyday culture, respectively. By the same token, however, and just as assuredly, culture is part of mathematics, including pure and abstract mathematics, such as geometric group theory.

The term "abstract mathematics" will refer here, more broadly, to mathematics' separation or abstraction from dealing with anything other than itself, and thus also from applied mathematics or any field, such as physics, that uses mathematics, in which case a more common term is pure mathematics. (I shall sometimes use the term pure mathematics if the occasion may lead to a confusion.) Abstract mathematics more commonly refers to still more "abstract" fields within pure mathematics, such as abstract algebra. What I designate, again, more generally (thus including most of pure mathematics), abstract mathematics is sometimes also designated "modern mathematics," a term explained below. In any event, as I shall argue, so defined, abstract mathematics is akin to poetry or literature, viewed as an abstraction from general language, for example, as considered and practiced by Mallarmé, who invokes a parallel between literature and mathematics in this connection. So, it is not surprising that de Rham would have recited Mallarmé's poetry, which is not commonly recited because of its esoteric and *abstract* character.

Aiming at this abstraction need not imply, as the word "pure" might suggest, an unconditional separation or superiority of pure mathematics over the rest of mathematics or mathematical sciences, such as physics, or literature or art over the rest of culture, manifested in the difference between literature and general or everyday language. This aim is sometimes aspired to, although rarely by deeper thinkers and practitioners in either domain, including initiators of major projects, defined by their abstract nature, such as Évariste Galois in mathematics, Werner Heisenberg in physics (where this type of project has been pursued as well), Wassily Kandinsky in painting, or Mallarmé in literature. This is why I prefer the terms abstraction or, as explained later, refinement to purity. Aiming at this kind of abstraction defines a particular project, a specific form of thought and culture, also enabling new ways of connecting abstract mathematics or art to other forms of, respectively, mathematics and science or of art. This is certainly true about geometric group theory, as de la Harpe suggests or Galois theory, a highly abstract algebraic theory, which nevertheless found its applications beyond algebra in mathematics, pure and applied, or in physics, as in quantum field theory. As will be seen, beginning with Heisenberg, quantum mechanics and then quantum field theory (which deals with high-energy quantum phenomena) are remarkable examples of reconnecting to physics abstract mathematics born from its break from physics.

There are not so many of those whose thinking exemplify this mutual reciprocity of mathematics and culture, and of mathematics and art as distinctive parts of culture, better than Athanase Papadopoulos, to whom I am honored to dedicate this essay. I shall not discuss his specific reflections on culture or, which would be especially inviting, art—literature, music, or painting—leaving these reflections to the readers of his own writings, including those dealing with the history of mathematics, where this reciprocity is manifested as well. Instead, as what appears to me a fitting tribute, I would like to consider mathematics, with a special emphasis on geometry (Athanase's main mathematical field), as combining mathematical and artistic thinking. Athanase's work, however, will be cited and its significance felt throughout in this chapter.

By *artistic* thinking I do not refer to *aesthetic* aspects of mathematics or mathematical aspects of art or aesthetics, which subjects will only be marginally addressed here. Instead, as indicated by my choice of the term "artistic," rather than "aesthetic," which has more to do with our *feelings* concerning the beautiful or the sublime, I shall consider a creative nature of mathematics as the invention of new concepts, theories, or fields, which has primarily to do with *thinking*. As my subtitle indicates, my argument follows that of Aristotle's *Poetics* [*Peri poietikẽs*] concerning literature, [1]. This is our Latin term, now used more broadly, but to be reserved for literary arts, such as poetry, drama, or fiction, sometimes using it interchangeably with poetry, given the latter a similarly broader sense. (In Aristotle's times, virtually all literature was written in verse.) Aristotle's book, in its surviving part, focuses on tragedy, but also comments on epic poetry, while seeing tragedy as a superior genre. Aristotle's title derived from the ancient Greek word *poeien*,"making." Aristotle's *Poetics* is about how literature is made or (my preferred term, again, Latin in origin) *composed*. The term may refer, as it will be here, both to the practice of composition in the corresponding domain and to an investigation of this practice, such as one undertaken in this chapter. (The difference between these two meanings will be clear from the context, but will be noted whenever necessary.)

Aristotle did not apply this term to mathematics and did not consider mathematics in this way, focusing instead, in his other works, on logical aspects of mathematics. By contrast, I argue, under the heading of the composition principle (defined below), that as a creative endeavor, mathematics is primarily defined by its compositional nature, rather than by its logical or calculational aspects, essential as the latter are for mathematics. Aristotle's mentor and sometimes antagonist (especially in considering poetry) Plato did see mathematics as compositional and more creative than poetry, which Plato ultimately came to see as merely imitating already existing objects or compositions. For Plato, it was philosophy that was the most creative, ultimately the only truly creative, form of thought, capable of reaching or at least touching the ultimate nature of reality, assumed to be mental, and divine, rather than human. Of course, while equally compositional, mathematics is not the same as literature and art. In particular, as I shall argue, the poetics of mathematics is the poetics of concepts, of *composing* concepts, which play a more limited role and sometimes no role at all in literature and art, and ally mathematics or mathematical sciences, such as physics, more closely with philosophy. Mathematical theories are compositions of concepts, and mathematical fields, such as geometry, are compositions of theories. Two other principles define mathematics, as well as science and philosophy, in the present view, the continuity principle, found in literature and art as well, and the unambiguity principle, not necessary in literature and art, which, conversely, often favor ambiguity.

A new composition entails a degree of *discontinuity* with the preceding situation in the corresponding area of mathematics, thus, defining a new situation there, from that of one's personal work (in which case a new composition may also be something learned rather than created) to a given theory or field or even mathematics itself, either more immediately or over time. Thus, the discovery by the Pythagoreans of the incommensurable magnitudes changed ancient Greek mathematics, Niels Henrik Abel's and Galois's work changed algebra, and Bernhard Riemann's work changed geometry and helped to usher in topology. I shall, nevertheless, consider such transformative events as primarily defined by "the composition principle," rather than "the discontinuity principle," operative as the latter may be in transformations of mathematics or other fields. Composition is a Latin term, from *componere*, meaning compose of elements, which is close to the ancient Greek *poeien* as used by Aristotle, who considered a tragedy as composed of such elements as plot, characters, language, or thought, with sub-elements within each, and each different in literature vs. their daily use. While it is not identity, this proximity between *poeien* and composition is sufficient to use these terms interchangeably for my purposes.

The term composition, as now used in art, originates in music, which is sometimes compared to mathematics because of its abstract nature. Another reference here is abstract painting, originating with Kandinsky's works, often entitled "compositions," borrowing the designation from music, although they are also related, sometimes expressly, to mathematics. While Kandinsky commented on these relations in his writings on art, they are more significantly manifested in his paintings, the *abstract* spaces of which may, for example, be associated with Riemannian geometry or Einstein's general relativity, based on Riemannian geometry. This association is also found in Claude Monet's representational works, such as the Nymphéas [Water Lillies] in the Musée de l'Orangerie, which have a certain abstract quality to them, and perhaps, painted late in Monet's life (1918), were influenced by abstract art (e.g., [61]).

The composition principle is well suited for considering the *discontinuous* aspects of mathematics and science (or philosophy and art) in combination with their *continuity* in time or, at any given time, in space with the preceding history of each field, equally necessary for understanding the history and functioning of geometry or mathematics in general. This continuity allows us, as individuals (even those with a limited knowledge of mathematics) or communities, to share mathematics. Otherwise, mathematics could not function any more than could science, or as science, if one uses the term "science" in the broader sense, akin to French "*science*" or German "*Wissenschaft*." (English tends to separate mathematics and science.) I define the assumption of this continuity as "the continuity principle." This continuity or mathematics and science, in the first place, are grounded, arguably most essentially, even if not exclusively, by "the unambiguity principle." The principle states the necessity of unambiguous definitions and communication, at least as unambiguous as possible, of the compositions created in mathematics and science, such as concepts. This principle is not generally applicable in literature or art, the meaning of which is, while shared sometimes, may be ambiguous. An individual human experience is ultimately unique, even in mathematics or science. Mathematics or science, however, need to reduce this uniqueness in their functioning, although not, or less so, in an actual experience of them by mathematicians and scientists. On the other hand, literature and art embody this uniqueness of human experience.

In any event, the composition principle allows for the interplay of continuities and discontinuities in mathematics and science. Thus, while components of a composition—a concept, a theory, or a discipline, such as geometry—may be unique, they need not be. Some concepts, such as group by Galois or manifold by Riemann, are more radically innovative and, as such, are more disconnected from earlier concepts, but never entirely so. As discussed in detail in Sect. 12.5, a composition can borrow elements from earlier compositions, and in this sense be a partial *mimesis* of them, while the relationships between parts can be new and must be new for the composition to be new.

My argument, beginning with the composition principle (as partially juxtaposed to the discontinuity principle), is also a response to Manin's view challenging, in the case of mathematics, Thomas Kuhn's famous distinction between revolutionary and normal practice in science, [45]. Manin argued that there is more continuity in mathematical practice than scientific ones, for example, in physics, [49–51]. As his last illustration of this thesis (advocated by him over decades), Manin offered an elegant analysis of the history of the mathematical theory of time and periodicity, from Ptolemy's epicycles to Schrödinger's quantum amplitude interference and Feynman path integrals, with several key junctures in between, such as Fourier sums and integrals, [51]. One cannot deny this overarching continuity and local continuities along the way, traced by Manin. It is also true that such continuities deserve more consideration than they are commonly given, vs. revolutionary breaks within this history, which have attracted most attention. It appears to me, however, that it is difficult to see this history apart from such breaks, sometimes radical, even in mathematics itself, such as those of Fourier's analysis or the calculus of variations, used by Erwin Schrödinger as something more continuous with the preceding mathematics or classical physics. These mathematical concepts have also shaped major changes in theoretical physics during the history that Manin traced, most significantly the shift from classical to quantum theory. Manin acknowledged such developments in physics (including in their use of mathematics) as "more sensitive to respective 'revolutions' and 'paradigm shifts'," while contending that "those parts that are closer to 'pure mathematics' show rather a kind of continuous development, as [he] argue[s] in [his] paper", [51].

I would, by contrast, argue that either history shows the interplay between continuities and breaks, an interplay arguably more complex than that between normal and revolutionary practices in Kuhn's sense [45]. Some of them in mathematics exhibit changes, defined by new concepts, that are as revolutionary as anything found in physics. Such changes do have their history and, hence, continuities with preceding concepts or ways of thinking, which is why I adopt the continuity principle. In physics, too, one can see quantum mechanics and quantum field theory, two theories that define quantum theory now, as, mathematically, a quantum-theoretical transformation (a "reinterpretation," as it was called initially) of Hamiltonian and Lagrangian formulations of classical mechanics. Nevertheless, quantum mechanics and then quantum field theory (which especially deals with high-energy quantum phenomena) represent a radical change in fundamental physics, including as concerns the mathematics used, especially, and centrally for this chapter, from finite-dimensional spaces over $\mathbb{R}$ in classical physics, or relativity, to infinite-dimensional Hilbert spaces over $\mathbb{C}$ in quantum mechanics. In fact, while a quantum theory, quantum field theory also breaks, in some respects radically, both physically and mathematically with quantum mechanics [e.g., [64, pp. 283–306]].

It is also true that major accomplishments are possible by primarily, even if not entirely (which, I would argue, is not possible), by means of continuities with preceding developments, for example, with new ones being compositions of previous concepts or components of concepts and thus by more continuous applications of the composition principle. This can, however, happen in physics as well: there are cases when continuity succeeded, at least more so than expected, while attempts at revolutions failed, which also may (but need not) mean that the failure was in not finding the right revolutionary path.[1] In general, as Kuhn realized and argued in considering Max Planck's discovery of quantum theory, [44], it is not always clear and may be a matter of interpretation where, or whether, a normal practice ends and a revolutionary practice begins. I would, accordingly, see the balance between them as contingent, shaped by the circumstances of a given case.

Thus, while set theory and its concepts are part of the history of category theory and its concepts, as Manin contends [49] in considering Grothendieck's works in algebraic geometry, there is, in my view, just as much discontinuity between them. Grothendieck's use of these theories, while by then established parts of algebraic topology and thus continuous with then, was radically new even vis-à-vis the previous history of algebraic geometry, as based on set theory, as in the case of Weil, who even disparaged category theory [81, p. 358]. There is still a continuity involved because Jean-Pierre Serre already started to bridge both fields by using sheaf theory. That, however, does not diminish the revolutionary nature of Grothendieck's work.

The philosophy adopted by this chapter is Heraclitean, a philosophy of becoming: we cannot step in the river of geometry or mathematics twice, because neither is the same and we are not the same. The question is the nature of this becoming, which is not restricted to history in its conventional sense of recording actual events. Thus, the history of concepts proceeds along entirely different lines and rhythms than the history of mathematics or science as technical, including institutional practice. The invention of each new concept is still an event in the history of mathematics, but not necessarily a properly recorded, archived, or even reachable event. The becoming, historical (in its multiple senses) or still other, of geometry or mathematics has a complex and multiple nature, in part by making both shaped by culture and being part of culture, multiple in turn, extending beyond mathematics. Thus, modern mathematics and science are the product of the culture of modernity, defined by several major transformations, some exterior to mathematics and science (in art, philosophy, economics, and politics) extending from, roughly, the sixteenth century on. Modernity is, however, inconceivable apart from mathematics and science.

To emphasize this reciprocity, I shall, while following the common meaning of the designation "modern physics," historically extend the common meaning of the designation "modern mathematics" to the same period, from the rise of modernity to our own time. Modern physics, defined by its mathematical-experimental nature, is seen as having emerged along with modernity itself, although, as such, modern physics has a long prehistory, on which I shall comment below. On other hand, "modern mathematics" commonly refers to the mathematics that began to emerge roughly around 1800 and consolidated by roughly 1900. It is defined by a movement

[1] The history of renormalization in quantum electrodynamics (QED) offers an example. It was believed by many, including Paul Dirac, the founder of the theory, that its problems, having to do with the appearance of infinities (divergent integrals) there, could only be resolved by means of an alternative theory and hence by a revolutionary break from QED. As it happened, by 1950 the renormalization program was able to handle these difficulties within the standard QED. While this did not satisfy Dirac and has continued to remain a concern for some, renormalization has remained an essential part of all high-energy (relativistic) quantum field theories.

towards making mathematics independent from a relation to nature and thus from physics, and, in part correlatively, from our daily phenomenal intuition and concepts, making modern mathematics abstract in this double sense.

Most of my analysis here is in accord with this periodization. I extend this designation to the rise of modernity because some of this separation is already found at the rise of modernity, in algebra, analytic geometry, and Gottfried Leibniz's version of calculus, or his project of "universal characteristic." The latter was a universal language envisioned by Leibniz as able to express mathematical, scientific, and philosophical concepts, a language usable by a universal logical calculator, *calculus rationcinator*. The idea is a major precursor of the present-day computer languages and the view of their universality, which would avoid the irreducible heterogeneity and ambiguity of general language and its multiple incarnation. This and related areas of inquiry became central to the emergence of modern mathematics and led to the idea of an independent mathematical reality, a sense transpiring, via the algebraization of geometry, in René Descartes's *La géométrie* [21] or Pierre Fermat's work in analytic geometry and algebra. Both geometry and analysis had kept closer relationships with representing nature and thus with physics. As such they also kept closer connections with our daily thinking and its representation of nature, a representation that classical physics refines and mathematically idealizes, including by using geometry and calculus, or indeed algebra. Still, both links were weakened before 1800, which makes the standard use of modern mathematics applicable, with due qualifications, to the mathematics at the rise of modernity. This joint separation of mathematics from physics and from our general phenomenal thinking became, however, more pronounced and more dominant institutionally in the nineteenth century.

Major founding figures at the rise of the mathematics and science of modernity, Johannes Kepler, Descartes, Fermat, and Galileo, and later Sir Isaac Newton and Leibniz, all exemplify this mutual interfolding of modernity with mathematics and physics. Descartes and Leibniz were, along with Baruch Spinoza, also major philosophers, who defined what we call modern philosophy, vs. ancient Greek or Medieval philosophy. While the boundaries between ancient Greek and modern philosophy are manifestly and deliberately crossed, those between Medieval and modern philosophy are not easily defined either. The difficulties to do so is exemplified, for example, by Kepler's thought, in view of a more strongly theological orientation of his philosophical and sometimes even scientific thought, modern as the latter was in its mathematical nature.

Galileo was a major philosophical thinker as well. His dialogues hold their own (it is difficult to achieve more) even against those of Plato in their philosophical and literary qualities. He was also, even if implicitly or in practice, a philosopher of physics, by making physics mathematical, geometrical, and a philosopher of geometry, by extending the domain of geometry—*geo*-metry—to cosmology. In this, it is true, he followed his modern predecessors, such as Copernicus or Kepler, or even the ancient Greeks, who already expanded geometry from the earth to the solar system, in Ptolemy as still a geocentric cosmos, but still a cosmos. (The ancient Greeks also considered a heliocentric cosmos.) In a more radical, essentially modern, gesture, however, Galileo introduced the idea of a mathematical, geometrical representation of nature, at all scales, famously appealing to the book of nature written in the language of mathematics [26, pp. 183–184]. Indeed, as Edmund Husserl argued, in doing so, and thus separating geometry, as such and in its use in physics, from philosophy, Galileo became the primary figure in initiating, in Husserl's famous title, "the crisis of European science" [36] (hereafter the *Crisis*). I shall discuss this subject in Sects. 12.2 and 12.3.

To complete this introduction, I give the definitions of some of the key terms of this chapter. Geometry or, especially, mathematics cannot be given a single or strictly precise definition. Accordingly, I shall only state my understanding of geometry and other major mathematical fields, assuming their existence and practice conform to the major principles assumed here—the unambiguity, continuity, and composition principle. This is, however, sufficient for my purposes and is in accord with how these fields are generally viewed. Indeed, this understanding already involves more general concepts, such as spatiality, only given properly analytical definitions in mathematics and physics.

I understand geometry as the mathematical formalization of spatiality, especially in terms of measurement, and topology as the mathematization of the structure of spatial-like objects apart from measurements, through their continuity or discontinuity. I understand algebra as the mathematical formalization of the relationships between symbols, with arithmetic as dealing specifically with numbers. Numbers are ultimately symbols as well, less abstract than algebraic ones, although it may depend on how one defines numbers. The corresponding mathematical fields are geometry, topology, algebra, and number theory. Analysis deals with questions of limit and related concepts, such as continuity, differentiation, and so forth. There are multiple intersections between these fields, and fields that branch off them, such as algebraic geometry. Defining algebra as the mathematical formalization of the relationships between symbols makes it part of all modern mathematics. Geometrical and topological mathematical objects always have algebraic components, while algebraic objects may, but need not, have geometrical or topological components. The term algebra has other meanings, such as that (standing at the origins of algebra as a mathematical discipline) of the study of algebraic equations, or that of algebraic structures, such as groups. In defining algebra as the formalization of the relationships between symbols, however, it is not only a matter of having an algebraic component as part of the mathematical structure of geometrical or topological objects, but also of defining these objects themselves algebraically.

A significant portion of the article deals with physics, especially with modern, mathematical-experimental, physics, which emerged in the work of Descartes, Kepler, Galileo, and others, again, along with modernity itself. Physics becomes a mathematical-experimental science that deals with the material constitution of nature, in fundamental physics, such as classical mechanics, relativity, and quantum theory, with the ultimate constitution of nature. Although this is not always sufficiently recognized, mathematics comes first in this conjunction, initially, as geometry, which Newton was still compelled to use in presenting his new mechanics, discovered by means of calculus, in *Principia*, [53].

In reflecting on the rise of modern physics with Descartes, Galileo, and others, Martin Heidegger observed: "Modern science is experimental because of its mathematical project" [31, p. 93]. This observation follows Husserl's argument in the *Crisis*, discussed later in this chapter. Heidegger's view is, however, significant independently of this connection or Heidegger's own (generally critical) assessment of modern science. Modern science, specifically modern physics, is experimental not only or primarily because of the role of measurement vs. Aristotle's physics as a physics of qualitative observations, as is often argued, not entirely accurately. Ancient Greek geometry, geo-*metry*, was a science [*episteme*] of physical space, such as that of surfaces of the ground, and measurements in space. Modern (and even some ancient Greek) scientists extended geometry to spaces beyond earth, ultimately to the cosmological scale, still defined, in post-Big-Bang extending universe, geometrically. This geometrical view was restored to cosmology by Einstein's general relativity (1915), which made the physics of gravity a geometry, by using modern mathematics, Riemannian geometry, quickly (around 1917) used by Einstein on the cosmological scale.

By that time, modern geometry was separated from physics, a separation amplified by the role of topology, a modern discipline, although it has an earlier genealogy, in particular, in the work Leonhard Euler. Euler was one of the key figures, along with Jean-Baptiste le Rond d'Alembert and Pierre Simon de Laplace, in the history of the relationships between mathematics and physics in the eighteenth century, before the emergence of modern mathematics defined by its separation from physics (e.g., [54]). As Euler's work on topology indicates, the situation is more complex, in both directions, those of separating mathematics from physics and that of connecting, or reconnecting it, to physics. This complexity is also found in the work of such major figures, more commonly seen as mathematicians, as Karl Friedrich Gauss, Riemann, and Henri Poincaré in the nineteenth century, and David Hilbert, Hermann Weyl, and John von Neumann, in the twentieth century, with Poincaré also crossing into the twentieth century, a symbolic figure in this respect as well.

The most fundamental reason for Heidegger's claim that modern physics is experimental because of its mathematical project is the following. From its emergence on, modern physics—classical physics, relativity, special and general, and quantum theory, the main types of fundamental theories (theories dealing with the ultimate constitution of matter) now, with several theories comprising each—uses mathematics to relate to and especially to predict the quantitative data found in experimentally observed phenomena. René Thom even argued that this is the case already in Aristotle, insofar as one sees his physics as in effect (qualitative) topology [55, 56, 76]. This argument would imply that, contrary to the view expressed by Heidegger, all physics, at least from Aristotle on, is experimental because of its mathematical project. The main difference would, then, be that, unlike Aristotle's qualitative topology of physics, the mathematics of modern physics is quantitative, as is required in physics if it needs to relate to measurements of the quantitative data observed in quantum phenomena, rather than only phenomena themselves. That, however, does not invalidate a stronger claim that, from Aristotle on, all physics is ultimately defined by mathematics, initially especially geometry or even proto-topology, except that the Pythagoreans defined their physics or at least their cosmology by both geometry and arithmetic. The Pythagorean harmony of the spheres represented by proportions, or in our terms, rational numbers, the view shattered by the Pythagoreans' discovery of the incommensurable magnitudes, such as those of the diagonal and the size of a square.

Modern mathematics, beginning with the mathematics emerging at the rise of modernity, restored the role of both geometry and (by then) algebra to physics and made physics modern. This is how we still see the universe now, by extending the representation of its geometry and topology, from the Riemannian spaces, accompanied by the tensor calculus on them, of general relativity, to such, topologically, geometrically, and algebraically, stratospheric objects as Calabi–Yau manifolds. While the expanding macroscopic space of the universe appears, as things stand now to be on average flat, its origin in the Big Bang or what happened before it is a separate matter, especially given that this early history may be quantum in nature. If so, it may, by the RWR principle, not be possible to speak or even conceive of its ultimate constitution, its ultimate *reality*, including as either continuous or discrete.

Philosophically, one of my concerns here is the nature of reality, material, as in physics, or mental, as in mathematics, and our capacity to deal with this reality, by representing, knowing, or conceiving of it, or the impossibly of achieving either. "Reality" and (a necessary counterpart) "existence" are assumed here to be primitive concepts and not given any analytical definition, partially given in the case of other (relatively) primitive concepts, such as space or time, at least as the latter appear in mathematics and physics. By "reality" I refer to that which is assumed to exist, without making any claims concerning the *character* of this existence, claims that, as explained below, define what is called realism, or in the case of mathematics, Platonism (although "mathematical realism" is used as well). On the other hand, the absence of such claims allows one to place this character beyond representation or, which is the view assumed in this article, even beyond conception.

This placement is considered here under the heading of reality without realism (RWR) or, in the case, of mathematics ideality without idealism (IWI), following [64, 66]. As such, this placement defines the RWR/IWI principle, the designation combining its material and mental incarnations, in physics and mathematics, respectively. I understand existence as a capacity to have effects on the world. The assumption that something is real, including of the RWR/IWI-type, is made, by inference, on the basis of such effects, as experiential phenomenal effects, rather than something as merely imagined. By the same token, the RWR/IWI view allows for and even requires a representation of these effects but not a representation or even a conception of how they come about. As RWR/IWI, the ultimate reality that makes them possible cannot be experienced as such, but only reveals, or more accurately, makes possible the assumption of, its existence through these effects.

The RWI/IWI view originated in the foundations of mathematics, where it was suggested by Henri Lebesgue, one of the founders of modern integration and measure theory, in the wake of the paradoxes of Georg Cantor's set theory, although the idea has remained marginal and rarely, if ever, adopted. Indeed, its origin there has been barely noticed as such, if at all, at least in what it signals philosophically, as considered here. It has been better known, if debated, in fact always remaining a minority view, in physics, following quantum mechanics, especially Niels Bohr's RWR-type interpretation of it, the type of interpretation assumed here as well, technically in a more radical form than in Bohr, following [64, 66–68]. It is also somewhat easier to express the concept of RWR in this case of physics, although doing so poses difficulties there as well, as was noted by Bohr, [8, v2, p. 59].

RWR interpretations place *the ultimate reality responsible* for quantum phenomena beyond conception. At the same time, hence, my emphasis, these interpretations assume that such a conception or even representation and knowledge is possible in considering quantum phenomena as observed phenomena, which are effects of this ultimate reality. The existence of this reality is inferred from these effects, predicted by the mathematics of quantum mechanics. Importantly for this chapter, while it was fundamentally defined by new (noncommutative) algebra, quantum mechanics also introduced a new form of geometry, that of Hilbert spaces over $\mathbb{C}$, into physics. Just as the algebra of quantum mechanics, this geometry had already existed in mathematics. It was one of Hilbert's many contributions to geometry, although usually seen as that to functional analysis, which is of course true as well. Initially, quantum mechanics did not use this concept, and it was only formalized, in fact axiomatized, in these terms by von Neumann, [78], who also introduced the term Hilbert space in mathematics. The concept was invented, in a less abstract form, by Hilbert. It is worth noting that quantum mechanics is a key part of the genealogy of noncommutative geometry, invented by Alain Connes, [10]. In the case of classical physics and relativity, a mathematical, in fact (while involving algebra and analysis, in this case commutative) geometrical representation is possible, at any level of reality considered.

The assumption of the independent existence of nature or matter essentially amounts to the assumption that it existed before we existed and will continue to exist when we will no longer exist. This assumption has been challenged, even to the point of denying that there is any material reality vs. mental reality, with Plato as the most famous ancient case and Bishop Berkeley as the most famous modern case. Such views are useful in suggesting that any conception of how anything exists, or even that it exists, including as independent of human thought, belongs to thought. It need not follow, however, that something which such concepts represent, or to which they relate otherwise than by representing it, possibly placing it beyond representation or even conception, does not exist. Berkeley assumed that God does exist as this type of independent reality. Plato also saw the ultimate nature of ideal reality theologically, as divine, even if connected to human thought. The nature of this connection has also been debated since Plato's own time and still is.

In any event, that any conception of how anything exists, or even that it exists (including as beyond thought), still belongs to thought, need not imply that something beyond thought does not exist. This was the point made, thus in effect implying the RWR/IWI view, by Lebesgue in 1905, in considering the context of set theory. This was two decades before quantum mechanics, which, introduced in 1925–1926, by Heisenberg and Schrödinger, brought the RWR view into the debate concerning fundamental physics. Quantum theory, discovered by Max Planck in 1900, was barely introduced by the time of Lebesgue's comment, and whether Lebesgue knew about it or not, it could not have been his source, because it had not posed this type of possibility then. Bohr's 1913 atomic theory used the RWR view in dealing with the so-called quantum jumps, still a decade after Lebesgue's comment, the reason for which was purely mathematical. Lebesgue observed (in commenting on the paradoxes of set theory, shaking the foundations of mathematics then) that the fact that we cannot imagine or mathematically define objects, such as "sets," that are neither finite nor infinite, does not mean that such objects do not exist [29, pp. 261–273], [13, p. 258].

Lebesgue did not specify in what type of domain, material or mental, such entities might exist. His observation is, however, a profound reflection of the limit of our thought concerning the nature of reality, either material or mental. Similarly, the fact that we cannot conceive of entities that are neither continuous nor discontinuous does not mean that such entities do not exist, including in nature, a possibility brought about by quantum physics. It does not follow either that such entities do exist. The possibility of their existence, however, reflects the limits of our thought, but also our thought's capacity to conceive of this possibility. According to this conception, there is a stratum of mental reality that is at least in part and perhaps ultimately responsible for thought, for the reality of thought, including mathematical thought, but that no thought can reach.

Lebesgue's insight was a response to the problem of the continuum and the debates concerning it shaped by Cantor's work and especially his continuum hypothesis that was given a deeper and more radical understanding, defining our view now, by Gödel's incompleteness theorems of 1931 and then Cohen's proof of the undecidability of the continuum hypothesis in the 1960s. The undecidability of Cantor's continuum hypothesis suggests that the concept of continuum may not be realizable mathematically and does not exist in nature, for example, as space, time, or motion. It need not follow, however, that the ultimate constitution of physical reality, spatial or temporal, is discrete, although, as discussed below, this view has been proposed. Accordingly, while in RWR interpretations this reality is beyond human thought, it is still assumed to exist in nature, in accord with Lebesgue's observation that something that cannot be thought by us can, nevertheless, exist as real in nature or in thought, possibly even mathematical thought, as Gödel's incompleteness theorems confirmed.

As, however, Lebesgue appears to have understood as well, it is equally impossible to be certain that any such reality does exist. Assuming this reality, even if only with a practical justification (no other justification is claimed in this chapter or is possible in the RWR/IWI view) is a philosophical wager. It was, however, this wager that led Heisenberg to QM, defined by a new mathematics combining both continuity and discreteness, and correlatively geometry and algebra, the mathematics enabling quantum mechanics to predict quantum phenomena, irreducibly discrete relative to each other. These predictions were and could only be probabilistic, which is, however, strictly in accord with what is observed in quantum experiments. While not originating in the IWI view, Gödel's incompleteness theorems and the undecidability of the continuum hypothesis may, as discussed in this chapter, be open to and may entail this view in mathematics. This is because it may not be possible to assume that the reality underlying what we try and cannot *mathematically* define as the continuum may not be either continuous or discrete, or anything else our thought can conceive of.

The argument of this chapter is, thus, governed by four main principles—the continuity principle, the unambiguity principle, the composition principle, and the RWR/IWI principle. Ultimately, the composition principle is most important for this argument. The creative, *poetical*, essence of art, philosophy, and mathematics and science, is composition. Their *political* essence is freedom, a freedom of thought in deciding what it aims to do, [69, pp. 13–14, 48–49]. What ultimately drives mathematics is the power of composition and freedom in creating new compositions, or using them, and not rules, logical or other, necessary as such rules are for mathematics as a disciplinary field. Mathematics gives human thought an extraordinary freedom, in spite of logical and other constraints it requires, and sometimes because of them, because they are not subject to coercive exterior ideologies, however dominant. Such ideologies could be and are resisted by philosophy and art as well, but mathematics and science, especially mathematics, might be argued to have more inherent resistance to such ideologies because they can only adopt an exterior ideology at the cost of compromising themselves as mathematics or science.

The following sections consider the relationships between geometry, or mathematics in general, represented by geometry, and respectively, philosophy (Sect. 12.2), physics (Sect. 12.3), geometry or mathematics with itself (Sect. 12.4), and art (Sect. 12.5), subtitled, "geometry before philosophy," "geometry before physics," "geometry before geometry," and "geometry before poetry," respectively. I use the word before in its double sense of preceding (logically rather than ontologically) and appearing in front of each of these field to be confronted by them, say, as a guest, or sometimes a ghost, when one opens the door of one's house, here that of philosophy, physics, or art.

12.2 Towers of Babel: Geometry Before Philosophy

Our experience is unique to each of us and at any given point of time is, in its entirety, unavailable in any outside observation or inference, which cannot be guaranteed to be correct even if it is only a partial one. Accordingly, insofar as mathematics is human, as it is assumed to be here (there are alternative views, discussed below), any mathematical object may be experienced differently by each human being. This is equally the case whether somebody, such as a mathematician or a physicist, with an expert knowledge in mathematics, or somebody with only a limited knowledge of it. Communicating and sharing our thinking, by one or another form of expression (linguistic, visual, mathematical, or other), is also human, however, and is unavoidable in and necessary for human life and culture. This sharing and, thus, continuity have both spatial and temporal aspects (broadly conceived so as to include social and cultural, such as institutional, workings of this continuity), local or across extended territories and periods. Spatial, contemporaneous, aspects are more prominent in communicating mathematics and science. Doing so, however, also involves temporal aspects, either in immediate terms, because in effect we always communicate from the past, however recent (possibly seconds ago, in more immediate exchanges), or in long terms, when we are reading something written some time ago, perhaps long ago, such as Euclid's *Elements*, [24]. A long-term communication tends to be equally split between mathematical cultures, as in the case of a present-day mathematician and the author or authors of the *Elements*, in this case unknown.

Grounding the unambiguity principle, mathematics and science capitalize on the possibility that this communication can be made unambiguous, at least sufficiently unambiguous, and in this sense *objective*. This is also true in experimental science in communicating the outcomes of experiments by means of general language, accompanied by technical terminology. The objectivity in the (realist) sense of representing the "object" considered, material in science or mental in mathematics, is a different matter, given a possible application of the RWR/IWI principle, which precludes any such representation, while allowing for the objectivity in the sense of unambiguous communication, and thus the unambiguity principle.

The functioning of geometry, or mathematics and science, depends on this reduction of ambiguity of our thinking, language, and knowledge. This functioning is always "institutional" in the broad sense of a community of practitioners, rather than only in its academic sense, which of course applies as well. Whether one sees the emergence of geometry, in a given form of it, as a single event, as Husserl did in "The Origin of Geometry" [37] (hereafter the *Origin*) or as a history combining several developments, which is the present view, this emergence was *instituting* geometry insofar as it entailed its collective, institutional character.[2]

Husserl based this reduction of ambiguity on what he called "the ideal objectivity," based on a more fundamental principle, defining his philosophy, the principle of the "phenomenological reduction" (also known as "philosophical epoché" or bracketing) of conscious experience to this type of ideality. More precisely, according to the principle, one ought to bracket the doctrinal content of all preexisting philosophy and conduct one's investigation within the framework define by this bracketing, in establishing what constitutes an ideal object, of which a properly geometrical object would be an example, too (e.g. [35, p. 10]). All extraneous elements of individual psychological experience are bracketed as well. Husserl's *Ideas I* also investigates the difficulties involved in the process, which, as Jacques Derrida shows in his analysis of the *Origin* and elsewhere are never quite resolved by Husserl, [35] [18, pp. 67–71].

[2] Husserl's essay, originally an untitled 1936 manuscript, was published by Eugen Fink in 1939. The translation used here is from the republication of the essay in [18]. The essay belongs to Husserl's final writings surrounding the *Crisis* [34].

The unambiguity principle is a less restrictive requirement, which allows for a more personal experience of mathematics and science (and as such does not entirely bracket this more personal experience), alongside the unambiguity of mathematical meanings and communications. The grounding of Husserl's argument for the ideal objectivity in the phenomenological reduction guides his argument for the origin of geometry, in his definition of geometry (explained below), in philosophy. Still, however, some geometrical mathematics preceded this origin and thus was confronted by philosophy before the geometry in Husserl's sense was developed, thus making the invention of geometry in Husserl's sense possible.

As indicated above, however, the unambiguity of mathematics and science or, for Husserl, even of the ideal objectivity, is dependent on general thinking and language, which allow for a general possibility of reducing the ambiguity of meaning. (For consistency, I shall hereafter use, as Husserl does, the term "general," although such proximate terms as "ordinary" or "daily" are also used with the same or related meaning by some authors cited here.) Physics is more dependent on general thinking and language. In experimental physics one must use general language in describing how experiments are performed, and classical physics, theoretical as well as experimental, may be seen as a conceptual and mathematical refinement of general thinking, vis-à-vis using more abstract mathematical concepts, adopted from modern mathematics, as in relativity and quantum theory [8, v2, p. 68] [33, p. 56]. Aristotle's physics may, by contrast, be seen as a philosophical, rather than mathematical, refinement of our general thinking, although, as noted, some of his concepts can be viewed more mathematically, [55, 56, 76]. The dependence on general thinking and language is, however, far from absent in mathematics, even beyond Euclidean geometry, which may be seen as a mathematical refinement of general thinking and language, similar to classical physics, the emergence of which depended on Euclidean geometry.

Modern mathematics, which abandoned the aim of representing natural objects and thus separated itself from physics, around 1800, broke radically with general thinking and language in the wake of set theory and its paradoxes. Quantum theory, too, moved toward a more radical departure from general language by using modern mathematics, including some developed around the same time, such as that of Hilbert spaces, noncommutative algebra, or group representation. Beginning with Heisenberg's invention of quantum mechanics in 1925, this break became complete in considering the ultimate reality responsible for quantum phenomena, at least in RWR interpretations. As explained above, these interpretations place this reality beyond any conception and hence expression, even a mathematical one, while using modern mathematics to predict what is experimentally observed as quantum phenomena. Quantum phenomena themselves are represented by classical physics, and as such retain closer connections to general thinking and language, also necessary for describing quantum experiments.

In any event, the fact that the possibility of the reduction of ambiguity is contained in general thinking and language, allows mathematics or science to capitalize on this possibility, for Husserl, in stronger terms of the ideal objectivity, including in the origin of geometry. According to Derrida:

> If we ask ourselves about the manner in which the subjective evidence of geometrical sense gains its ideal Objectivity, we must first note that (for Husserl) ideal Objectivity not only characterizes geometrical and scientific truths; it is the element of language (and thought) in general. "It is proper to a whole class of spiritual products of the cultural world, to which not only all scientific constructions and sciences themselves belong but also, for example, the constructions of fine literature", [37, p. 160], [18, p. 66][3]

This may be more than just an example, at least in the present view. "Fine literature" depends on an analogous, albeit not identical, *idealization* of general language into the language of literature or, as it is sometimes termed, "poetic language." (The concepts associated with this term vary but are sufficiently close to the one adopted here to put these variations aside for the present purposes.) This idealization need not imply, any more than in mathematics, idealism, such as Platonism, even though many mathematicians, philosophers, and literary authors, or artists in other domains, have adopted an idealist or Platonist view.

This view of literature was advanced, invoking a parallel with mathematics, by Mallarmé, one of the founders of literary modernism, without assuming idealism, including Platonism. For Mallarmé literature was, or was required for him to be, akin to mathematics in its "abstraction" from general language by means of literary compositions. Mallarmé's view echoes Aristotle's view of the language of poetry in *Poetics*, but it makes the difference between poetic and general language more radical by appealing to the abstract nature of poetic language and its proximity to mathematics. In this, Mallarmé was closer to Plato, in the absence of Platonism, than to Aristotle. This analogy and related connections between literature and mathematics have developed throughout the subsequent history of literary criticism, especially in literary structuralism, dominant in 1950s–1970s and in part inspired by mathematical structuralism, such as that of Bourbaki.

There are significant differences between literary or artistic abstraction and the ideal objectivity of mathematics or science or, in Bohr's terms (preferred here as well) their "refinement" of general thinking and language, beginning with the unambiguity principle in mathematics and science, [8, v2, p. 68]. Unlike the composition and continuity principles, the unambiguity principle is not applicable to literature or art, which retain the ambiguity found in general language and (often intentionally) enhances it in creating its compositions. Nevertheless, literature, more deliberately modernist literature, is not only a refinement of general language (which has been commonly recognized, from Aristotle's *Poetics* on), but also an abstraction from, or at least a transformation of, general language, especially in poetry. Fiction retains closer connections to general language, albeit not always, especially, again, in modernist fiction, such as, paradigmatically, that of James Joyce's *Finnegans Wake*, [40].

Literary compositions, especially in poetry, transform ordinary meanings of words, for example, by means of metaphors. Indeed, exemplifying Mallarmé's vision of poetry, manifested in his practice, a new poetic composition is a creation of

---

[3] The quotation of Husserl, modified by Derrida, follows [37, p. 160].

a new language. This language is different from mathematics by being linguistically connected much more closely to one or another general language, for example, French or a language so-called or perceived, since every "native" language is linguistically multiple. Nevertheless, compositionally, it is fundamentally different from general language. This compelled Mallarmé to view modernist poetry, including his own, on the model of "mathematical language," referring to Descartes's analytic geometry, which was one of the founding events of the history of modern (or at least proto-modern) mathematics in separating itself from general concepts and languages, [65]. The extension of this radical program of modernism involves such figures as James Joyce, Franz Kafka, Jorge Luis Borges, Samuel Beckett, Gertrude Stein, and William Faulkner. In its abstract nature (apart, again, from the unambiguity principle), modernist literature could be compared to the use, during the same period, of modern mathematics by Heisenberg and Dirac in quantum mechanics and quantum electrodynamics or the invention of algebraic topology, especially in the work, of Poincaré, and abstract algebra by Emmy Noether and Emil Artin in mathematics.

Husserl saw the origin of geometry as, along with Pythagorean arithmetic, the inaugural case of the reduction of ambiguity in mathematics and science, in his view, in terms of the ideal objectivity, rather than, as here in terms of the unambiguity principle. The latter is, again, a less stringent requirement, more permissive of elements that would be "purified" by Husserl's phenomenological reduction but that do not in fact inhibit the unambiguity necessary for the functioning of mathematics and science. It may be impossible to know which comes first, arithmetic or geometry, as Husserl was aware, even if without, as I shall do here, grounding this impossibility in the irreducibly multiple nature of any such origin. Husserl's first major book *The Philosophy of Arithmetic*, [38], preceding the *Origin* (1936) by four decades, was already concerned with the same problematic in the case of arithmetic. According to Derrida:

> *The Philosophy of Arithmetic*, Husserl's first important work, could have been entitled *The Origin of Arithmetic*. Despite a psychologistic inflection whose originality has often and justly been emphasized, it already concerns, as does "The Origin of Geometry," with the reactivation of the primordial sense of arithmetic's ideal unities by returning to the structure of perception and the acts of a concrete subjectivity. Husserl himself already proposed to account at once for the normative ideality of number (which is never an empirical fact accessible to a history in precisely the same style) and for its grounding in and through the lived act of its production, [18, p. 28].

"The lived act of production" and thus *construction* is a crucial aspect of Husserl's view of the origin of geometry, including against Platonism (which may be different from Plato's own thought), as well as Kant's view of geometry, to which Husserl's analysis is connected more closely. Kant gives geometry, while assuming it to be human, a form of preexistence before it is discovered. Both Derrida's and the present, more mathematically grounded, critique of Husserl are based in extending this constructivist aspect of Husserl's view beyond Husserl, in the present case following the IWI view. The latter, as explained below, is arguably more radical than that of Derrida, [63].[4] It is, however, worth considering Husserl's departure from Platonism and Kant a bit further to make clearer both the significance and limitations of Husserl's, as well as Kant's, contribution.

By Platonism I refer to the assumption of the existence of mental reality fully independent of an individual human thinking, which could only represent this reality or at least form a conception of it, which does not rise to confirmed representation. Mathematical Platonism, a twentieth-century trait, with several branches, in the philosophy of mathematics, is essentially based on the view of the existence of mathematical reality as fully independent of human thought and strictly discovered in the practice of mathematics. In Kant, by contrast, this reality, as an ideal objectivity, is something that belongs, universally, to human thought. At the same time, as against Husserl's view, geometry for Kant has a certain preexisting independent reality connected to the a priori nature of spatiality. As such it is discovered rather than created, as in Husserl by "the first geometer." Derrida emphasized this difference:

> For the inaugural mutation [in our thinking establishing geometry there] which interests Kant hands over geometry rather than creates it; it sets free a possibility, which is nothing less than historical, in order to hand it to us. At first this "revolution" is only a "revelation for" the first geometer. It is not produced by him. It is understood under a dative category, and the activity of the geometer to which the "happy thought" occurred is only the empirical unfolding of a profound reception. What is most often translated by "revelation" is the allusion to "a light that is given," to "a light dawns on" : "*Dem ersten . . . dem ging ein Licht auf.*", [18, pp. 39–40].

For Kant, it follows, geometry is established, as in Husserl, in a single event by a single person, as originally *revealed*, or *given*, defining its future potential development, making the practice of the first geometer and those to whom it is then handed unfold this revealed geometry by building its edifice. As against many, even most, subsequent (or preceding) Platonist views, however, it is revealed not from some exterior ideal reality but from within the first geometer's mind. In other words, it was a discovery made by a human subject, from within the interior of this subject thought, culturally and otherwise externally shaped as this interior may be. By the same token, geometry is not assumed to be determined in advance as concerns all its future content, which will only unfold in its history by continuously discovering new geometrical objects or new connections between existing objects.

Indeed, as against modern mathematical Platonism, Plato might have agreed with this view of geometry as a mathematical field. Plato, however, still assumed, at least in most interpretations, the ultimate (mental) reality as eternally existing, fully formed, independently of the human mind and thus preexisting whatever the

[4] Focusing on Husserl's argument for the origin of geometry strictly from philosophy, Derrida's analysis does not consider geometry as such. Derrida's critique of Platonism in his subsequent work mostly bypasses mathematics as well, although in several of his works he uses the concept of undecidable proposition, but only by analogy or metaphorically, in fact with a different meaning than in mathematical logic.

latter can discover. While, however, modelled on mathematics, this reality cannot be grasped by mathematics but only by philosophy [66, pp. 12, 86].

While Husserl agrees with Kant on the universality of geometry (or mathematics), for Husserl geometry is not "a revelation" of something preexisting even in the mind, but a creation of the first geometer. In stressing the difference between Husserl and Kant, Derrida says:

> Undoubtedly, Husserl' s production (*Leistung*) also involves a stratum of receptive intuition. But what matters here is that this Husserlian intuition, as it concerns the ideal objects of mathematics, is absolutely constitutive and creative: the objects or objectivities that it intends did *not* exist before it; and this "before" of the ideal objectivity marks more than the chronological event of a fact: it marks a transcendental prehistory. In the Kantian revelation, by contrast, the first geometer merely becomes conscious that it suffices for his mathematical activity to remain within a concept that it already possesses. The "construction" to which he gives himself, then, is only the explication of an already constituted concept that he encounters, as it were, in himself–a description which no doubt for Husserl as well would be true of every noncreative geometrical act, and which teaches us about the sense of ready-made geometry as such, but not about geometry in the act of being instituted, [18, p. 40]

Nor does any new geometrical invention (a concept, a theorem, or a theory) appears in the act of being instituted in Kant. In Kant, too, "the 'construction' is irreducible," but it follows a revelation of geometry:

> No doubt, once the geometrical concept has revealed its freedom with respect to empirical sensibility, the synthesis of the 'construction' is irreducible. And indeed it is an ideal history [just as it is in Husserl]. But [unlike in Husserl] it is the history of an operation, and not of a founding. [Kant's history] unfolds explicative gestures in the space of a possibility already open to the geometer. The moment geometry is established as such, the moment, that is, something can be said of it, then geometry already will be on the point of being revealed to the consciousness of the first geometer, who is not, as in [37], protogeometer, the primally instituting (*urstiftende*) geometer, [18, p. 41].

Thus, even if within a human mind, such as that of the first geometer, geometry is somehow already there, a priori. Not so in Husserl, for whom it must be invented from its prehistory. Kant makes geometry human, which is a crucial move, but he retains its pre-existent or a priori-like nature. Husserl makes the reality of geometry created, constructed in human history and, thus, as he suggests, by the history of culture, which moves his view further away from Platonism.

Husserl, however, still makes this construction governed by ideal objectivity, defined by phenomenological reduction, vs. *the unambiguity principle*. The latter only requires the possibility and necessity of "objectivity" in the sense of the unambiguous definition and communication of mathematical and scientific concepts, propositions, or theories, without unconditionally limiting their experience to phenomenological reduction. In other words, the unambiguity principle allows for more personal, subjective (and thus more open to ambiguity) aspects of experience, even when this experience is, or is related to, that of mathematical and scientific concepts. More personal, such as psychological or cultural, aspects of experience are found in any mathematical or scientific practice, no matter how abstract the mathematics used may be. While necessary for the collective or institutional functioning of mathematics and science as such, the applications of the unambiguity principle, has its limitations and cannot be guaranteed. As they do elsewhere, exterior circumstances can and often do shape the trajectories of mathematical or scientific thinking, especially innovative ones, which are rarely limited to mathematical or scientific factors, although the origins of such trajectories are often untraceable.

In the present view, the original invention of geometry or, it follows, the process of its emergence is the effect of this multifaceted complexity, at work from within and from without geometry. This view, thus, precludes the assumption that there could have been a single origin of geometry, any more than of arithmetic or any mathematical field, or of mathematics itself, for example, singularly, from philosophy. It is possible that a realization that certain mathematical objects could have been understood in terms of what became defined as, or (already in existence) called, geometry occurred for the first time to a single person or a group of persons. Husserl's comment on the subject in the *Crisis* may be seen as consistent with this possibility:

> Serving as methodical praxis of mathematicians, in this form of long understood acquisitions, are significations which are, so to speak, sedimented in their embodiments. And thus they make mental manipulation possible in the geometrical world of ideal objects. (Geometry represents for us here the whole mathematics of space-time.)
>
> But in this mathematical praxis we attain what is denied us in the empirical praxis: "exactness"; for there is the possibility of determining the ideal shapes in absolute identity, of recognizing them as substrates of absolutely identical and methodically, univocally determinable qualities. This occurs not only in particular cases, according to an everywhere similar method which, operating on sensible intuitable shapes chose at random, could carry out idealizations everywhere and originally create, in objective and univocal determinateness, the pure idealities which correspond to them. For this, (rather,) certain structures stand out, such as lines, triangles, circles. But it is possible— *this was a discovery which created geometry*—using these elementary shapes, single out in advance as universally available, and according to universal operations which can be carried out with them, to *construct* not only more and more shapes, which because of the method that produced them, are intersubjectively and universally determined. For, ultimately the possibility emerges of producing constructively and univocally, through an a priori, all-encompassing systematic method, *all* possibly *conceivable* ideal shapes, [36, pp. 26–27] - translation slightly modified; some emphasis added.

Husserl appears to be ambivalent concerning whether this "infinite" project was in place in the ancient Greek geometry, the view assumed in the *Origin*, or emerges only with modernity, the view that appears to be adopted in the *Crisis*. I shall comment on this ambivalence below. Either way, thus understood, geometry is not merely a praxis of mathematical manipulations in the spatial (or temporal) world of ideal objects. Instead, it is defined as the possibility of creating "all possibly *conceivable* ideal shapes," by a pre-established (and in this sense a priori) all-encompassing method, thus, providing this possibility of extending geometry infinitely into the future. Thus, vs. Kant, what is a priori is a method and not geometry that this method can construct. It is difficult to be certain that this project or only this project was uniformly understood or practiced as geometry either in ancient Greece or in modern mathematics. It may have existed, and in modern mathematics appears to have existed, as merely one of the multiple conceptions of geometry or forms of geometrical practice, without, at least in some of these practices, being concerned with dealing with "*all* possibly *conceivable* ideal shapes," or extending geometry infinitely into the future. In other words, geometry, ancient Greek or modern, continued to pursue practices and ways of understanding geometry preceding the discovery of this particular geometry, and to invent new such practices and alternative ways of understanding geometry after this discovery.

It is also difficult to be certain that this discovery, which created geometry, was a single event, as Husserl appears to claim, even if one that sees this "event" as extended to a process, possibly collective, although the thought grounding this discovery might have occurred to somebody. His elaboration just cited clearly tells us that some geometrical-like mathematics was already in place before this discovery. This prehistory also makes it difficult to claim that this discovery emerged from philosophy, even if one accepts Husserl's view that geometry should be seen as defined by "the possibility ... of producing *all* possibly *conceivable* ideal shapes," which is, again, far from self-evident.

It may be impossible to trace, except partially, the event of the invention of geometry of any kind (that defined by Husserl included), or the network of trajectories leading to it. Nevertheless, given the available evidence concerning ancient Greek mathematics and culture, it appears more plausible that the "origin" of geometry, again, of any kind, was a multi-trajectorial, rather than only philosophical or, philosophically mediated, emergence of it, beginning with the pre-Socratic mathematics, and preceded by and dependent on still others, in a long history, from the Babylonian or even earlier mathematics on. An annotated Babylonian clay tablet (between 1800 and 1600 BC), the diagonal shows an approximation of its length (the square root of 2 in our terms) in four sexagesimal figured to approximately six decimal figures. The evidence of previous mathematics, as in Egypt, is scarce, but it was there. So, the ancient Greeks are latecomers in this history, much as they shaped the subsequent European history of mathematics and culture. One can even speak, adopting Gary Tomlinson's metaphor concerning the emergence of music, of "a million years of mathematics," keeping in mind that concerning anything before, roughly 50,000 years ago, one can only invent evolutionary fables, akin to those concerning the origins of thinking, consciousness, language, logic, music, or art, which may be plausible and useful but are unlikely to ever be confirmed, [77].[5]

This long history is not inconsistent with Husserl's claim concerning the discovery of geometry in the sense of creating "*all* possibly *conceivable* ideal shapes," by a pre-established all-encompassing method. In fact, as is clear from the passage in the *Crisis* under discussion, Husserl might have accepted this history, even if not in the degree of multiplicity assumed in the present chapter. He was also right to have warned against projecting the European ("Greek") sense of mathematics (or philosophy) on to ancient one in India, Babylonia, or China.

[5] Tomlinson's excellent book confirms the last qualification, even if sometimes against its own grain.

The point here is, however, defined by the opposite question: How, historically, ancient Greek was the ancient Greek mathematics, or how philosophical it is in its origin? Husserl argued, more strongly in the *Origin*, that philosophy, in fact that of a single philosopher, necessarily interjects itself in this history to discover geometry thus defined. This is because otherwise one can, in principle, continue the technically geometrical mathematical practice preceding this discovery apart from this discovery, or alongside this discovery, making it multiple, which is what I argue here. The essentially philosophical nature of this discovery is the reason for Husserl's claim concerning the origin of geometry, thus defined, in philosophy. The loss of the connection between modern geometry (or geometrical or mathematical sciences), in their pursuit of this project and philosophy becomes the crisis of modern European science, which is also the crisis of modern European philosophy.

By contrast, I argue, that, while, again, still only one among others, even this conception of geometry might have emerged in mathematics itself, and then entered philosophy. In other words, geometry might have been there before philosophy, preceding it and standing in front of philosophy, inviting or demanding its attention. I am not claiming that such was necessarily the case, but only that it might have been the case, which is sufficient to question Husserl's unconditional claim of the origin geometry in this sense from philosophy. It is of course important to keep in mind that Husserl only claims the origin of geometry in this sense from philosophy, and not of geometry in general, that is, not of any possible form of geometry. Nevertheless, while geometry is before his philosophy, again, in both senses, Husserl wants to position ancient Greek philosophy before geometry, while it is possible that geometry was before philosophy, again, in both senses, even then.

Husserl, as noted, appears to be ambivalent concerning the provenance of this origin in ancient Greek philosophy. While ascertained in the *Origin* or the Vienna lecture, [34], in which the *Crisis* originated, the *Crisis* itself, just cited, associated this type of infinite project with modern and not ancient Greek geometry, making even the *Elements* short of this project. As Husserl says: "But Euclidean geometry, and ancient mathematics in general only knows finite tasks. Aristotelian syllogistics belongs here also as an a priori which takes precedence over all others. Antiquity goes this far, but never far enough to grasp the possibility of the infinite task which, for us [as modern], is linked as a matter of course with the concept of geometrical space and with the concept of geometry as the science belonging to it", [36, p. 22]. In other words, while ancient Greek in the Vienna lecture before the *Crisis* and then in the *Origin* after the *Crisis*, this project and thus, in the first place, the discovery, of geometry as creating "*all* possibly *conceivable* ideal shapes," is modern in the *Crisis*.

The existence of such a project is more easily historically supported, for example, by Descartes's and Galileo's work, during modernity. Husserl sees modern geometry or mathematics (of which geometry would then be part) as in accord with this type of project. The problem for Husserl is the divergence of modern mathematics or science from philosophy and then, in fact as a result, from each other. This divergence is the crisis of modern science. Philosophy is in a correlative crisis as well by losing mathematics and science as its constitutive parts. Modern science, including mathematics, is practically effective, but is ultimately detrimental to (European) humanity by separating its project from that of philosophy. On the other hand, the project of ancient Greek science, especially mathematics, geometry, was, in Husserl's view, unified with philosophy. This surmise may have been the main reason for Husserl's argument in the *Origin*, thus, offering a positive view of geometry there, in his definition of geometry as creating "*all* possibly *conceivable* ideal shapes," by a pre-established all-encompassing method. Accordingly, one might argue that Husserl prefers to see the origin of this project in the philosophy of ancient Greece.[6]

Ultimately, however, at stake in this chapter is our own mathematics and science, and whether one needs to ground the continuity and unambiguity principles in the ideal objectivity, as in Husserl, or not, which is the present view. By the same token, in the present view, the separation of mathematics and science (no longer necessarily defined on Husserl's model just considered), from philosophy is not seen as a crisis either. That does not mean that this separation is unconditional or always necessary in the present view. Such connections or disconnections are a matter of contingent alliances, sometimes desirable or even necessary, but not always so. The balance of mathematical and philosophical thinking, in either mathematics or philosophy, in mathematics was, arguably, contingent in both cases, as it is in modern mathematics and science or philosophy, although there is indeed more disciplinary separation between them in modernity.

Husserl is right about a greater degree of this separation. The question is whether it is quite as unconditional as he sees it, or whether the separation constituted a crisis, which it does not in the present view. Was there a philosophy at the time of Babylonian mathematics? While it is highly unlikely that there was anything akin to philosophy in ancient Greece, it is difficult to assume that thinking akin to philosophical thinking did not exist then. Hence, the present view of the relationships between mathematical and philosophical thinking might have been similar there, although our information concerning these relationships there is too scarce to make a reasonably reliable assessment.

Husserl refers to a biblical Babylonian tower of future mathematics in a version of the Vienna lecture, but not to Babylonian mathematics, as part of the early history of mathematics, of which he, again, was aware. This history, irreducible to a single event or even a single history, suggests that mathematics, including geometry, in its many projects, had emerged and has continued to exist in the course of building something akin to the tower of Babel. The one, be it ancient or modern, that Husserl defines, as the construction of "*all* possibly *conceivable* ideal shapes," is one such project, but, I argue, only one among others. The building of this tower is still ongoing, although, unlike the mythical tower of Babel of languages, the tower of Babel of mathematics obeys the unambiguity principle.

---

[6] Derrida also suggests that one can bring these two views closer to each other along related lines but arguing that "even in the spirit of the *Crisis*, the modern infinitization will mark less an authentic up surging than a kind of resurrection of geometry", [18, p. 130].

Husserl invokes the tower of Babel, under the condition of the single, and moreover, singularly, "*only*!," Greek origin geometry, and a homogeneity of its future in one version of the Vienna lecture: "Only Greek philosophy leads, by a specific development, to a science in the form of infinite theory, of which Greek geometry supplied us, for some millennia, the example and sovereign model. Mathematics—the idea of the infinite, of infinite tasks—is like a Babylonian tower: although unfinished, it remains a task full of sense opened onto the infinite. This infinity has for its correlate the new man of infinite ends", [34] cited in [18, p. 129].[7] While this claim is, again, in accord with the *Origin*, in the *Crisis* this project of geometry becomes modern, also breaking away from its unity with philosophy, which is the crisis of modern science. Either way, one can think of two alternative tower-of-Babel-like architectures of geometry or mathematics. Husserl appears to think that the continuity and the unambiguity of geometry entail his model of a *single* (potentially) infinite theory, which does not make his tower much of a tower of Babel. No wonder that both him and earlier Leibniz thought that one can defeat the tower of Babel of language by the infinite tower of logic, defined by its homogeneity, in principle universal, a view noted by Derrida, who cited Husserl's passage from the Vienna lecture, [18, pp. 100–101, note 108]. Leibniz's "universal characteristic," mentioned above, was a manifestation of the same philosophy, prominent in the present-day champions of (digital) AI and futurology based on it.

By contrast, the view adopted here implies that a heterogeneous tower of Babel of mathematics is possible, while assuming both the continuity and the unambiguity principle, both grounding, in their more rigid forms, Leibniz's and Husserl's view or that (often more rigid) of the present-day champions of AI. Husserl, in the *Crisis*, saw mathematics (especially when in harmony with philosophy) as a guidance for building a rational human society, and with it the European humanity, at the time at the state of the biblical tower of Babel, and in some of its aspects much worse. He was not wrong as concerns what he saw as the worst part of his contemporary world, especially in Germany then. That, however, might not have been the-tower-of-Babel problem, but the opposite problem, that of conformism. This is a separate subject. I would only add that that conformism is no friend of mathematics, which requires the maximal freedom and thus nonconformism of thought. In this regard, as I argued elsewhere, all new mathematics is minority mathematics, [69].

The unambiguity principle does distinguish the tower of Babel of mathematics from the tower of Babel of general languages, or that of literature and art. The latter also refines, even, in Mallarmé's view, as a form of abstraction, general thinking and language, but in the absence of the unambiguity principle, which absence they share with general thinking and language. I, nevertheless, argue that the unambiguity principle leaves room for multiplicity, even the irreducible, tower-of-Babel-like, multiplicity, in mathematics or science. Whatever unity one wants to ascribe to geometry, and one can and even must do so within certain limits (by the continuity

[7] Derrida uses a version of the Vienna lecture, translated by Paul Ricoeur, which is different from [34], published in [36].

principle), this unity is preceded by and proceeds from a multiplicity, even at the origin, in the present view, multi-trajectorial as well, of geometry.

Of course, there are situations in, or stages of development of, geometry or mathematics, or science, in which unities enter and take over multiplicities as well. The fundamental principle of these relationships is, however, that in general the multiple preceded the single, the complex preceded the simple, which principle is at the core of Derrida's critique of Husserl and defines Derrida's own philosophy, [19]. It equally defines the philosophy of such Derrida's contemporaries as Gilles Deleuze and Alain Badiou. This type of principle is also instructively considered by Imre Lakatos in the context of the twentieth-century axiomatization of mathematics, including geometry by Hilbert, and Gödel's incompleteness theorems, [46, pp. 3–23].

One might regret, and many have, that mathematicians in different fields or even in the same field, like geometry (there are so many geometries now!), no longer always understand each other. One might, however, and this author would, conversely, argue that there is nothing wrong or detrimental in living in this tower of Babel of mathematics, or in this world of mathematics, which would be my preferred and, in my view, a more accurate horizontal description. Given time, effort, and capability, one can learn, as many have, the language of a mathematical field different from one's own, without assuming that there is a single common encompassing language of mathematics. Just as in poetry, while a translation, even a good translation is possible, not everything can be translated. It may be tempting to think, and some, again, do, that there is only one mathematics, but this is not the case or at least may not be the case, although there are numerous overlapping and crossings between different areas and forms of mathematics. A fitting image would be a dynamical, ever expanding, Venn diagram with numerous and multiplying overlapping domains.

Husserl's argument was an important and influential step toward understanding this complexity in its departure from both Platonism (even if not that of Plato himself) and Kant's view of geometry in his first *Critique*, *The Critique of Pure Reason*, where geometry plays a key role, [41]. The most essential features of Husserl's analysis are, again, that (a) against Platonism, geometry or mathematics is, while objective, human; and (b) against both Platonism and Kant, that geometry or mathematics was created, constructed in a lived human experience, rather than discovered as anything that preexisted this experience. Of course, when we learn mathematics, we learn something that already existed in somebody's lived experienced and was recorded in the archive (sometimes immediately, such that of noted or even thought of those who teach us) of mathematical knowledge. This learning, however, takes place in our lived experience, and whatever we learn is constructed by this experience.

Husserl never moved toward thinking in terms of radical multiplicity advocated in this chapter, except, as noted, against his own grain (still in a limited way) in some of his writings, including the *Origin*. The RWR/IWI view does not appear to enter his argument even in this way. Certain developments of mathematics or physics which Husserl witnessed during his lifetime and some of which he mentioned might be seen as inviting one to do so. To follow this invitation would, however, require a very different philosophical orientation, uncommon in philosophy or foundational thinking in mathematics and physics during Husserl's lifetime. Bohr, born a quarter-century after Husserl, and initially Heisenberg, one generation younger than Bohr, were rare exceptions, and Heisenberg eventually retreated to a more realist view.

Husserl, who had a Ph.D. in mathematics, and studied, as a postdoc, with Leopold Kronecker and Karl Weierstrass (whose assistant Husserl was before switching to philosophy), considered modern mathematics, such as theories of manifolds (especially extending from Hermann Grassmann's work, but also Gauss, Riemann, and Cantor) in his earlier work. By contrast, his later works, such as the *Crisis* or (more understandably) the *Origin*, say little about key developments, many of them momentous, in contemporary or even nineteenth-century mathematics. The concept of manifold is important to the *Crisis*, and mentioned (once) in the *Origin*, [37, p. 178], some of its uses there echo modern, post-Riemannian, mathematical uses of the concept, but only echo. Neither Riemann nor Grassmann is mentioned in the *Crisis*. Also, the concept of manifold, *Mannigfaltigkeit*, has a long history in German philosophy as a philosophical concept (and came to mathematics, as in Riemann, from philosophy). Husserl's use of it also follows this history, along with modern mathematics. Foundational issues that emerge from the paradoxes of Cantor's set theory are not mentioned in the *Crisis* either. While Planck is mentioned in passing [36, p. 4], nothing is said about quantum physics, at the center of epistemological debate concerning physical reality then.

None of this is surprising given that the nature and the origin of the crisis at stake in Husserl has to do with the rise of modern mathematics and science. The crisis was a spiraling out-of-control divergence of mathematics or science from the unity established by philosophy, which was, correlatively, or even in the first place, also the crisis of modern philosophy. This joint crisis has remained part of European culture from the rise of modern science, and thus modernity, to Husserl's time. It is unlikely that Husserl, were he alive, would have seen any change in this regard in the present-day culture. As he says:

> Ultimately all modern sciences [including mathematics] drifted into a peculiar, increasingly, puzzling crisis with regard to the meaning of their original founding as branches of philosophy, a meaning which they continued to bear within themselves. This is a crisis which does not encroach upon the theoretical and practical successes of the special science; yet it shakes to the foundations the whole meaning of their truth. This is not just a matter of a special form of culture—"science" or "philosophy"—as one among others belonging to European mankind. For the primal establishment of the new philosophy is, according to what was said earlier, the primal establishment of modern European humanity itself—humanity which seeks to renew itself radically, as against the foregoing medieval and ancient age, precisely and only through philosophy. Thus the crisis of philosophy implies the crisis of all modern sciences as members of the philosophical universe: at first a latent, then a more and more prominent crisis of European humanity itself in respect to the total meaningfulness of its cultural life—its total *Existenz*, [36, p. 19].

In Husserl's view twentieth-century mathematics and science changed nothing in this situation and, if anything, have contributed to exacerbating it. One might, however, argue that the problem is instead Husserl's view of the situation, grounded, as is his view of the origin of geometry, in his assumption that the multiple follows and diverges from some primordial unity, defined in this above passage by, and as, philosophy. The present view is the opposite: The multiple, including mathematics and science, or literature and art (each already multiple), has preceded any unity, thus always limited or provisional, presumed in or claimed for, or by, philosophy, ancient Greek or modern. Philosophy is just part of one or another cultural multiplicity that preceded and continues to surround it. Husserl recognizes this, within certain limits, but in his view once created or imagined, philosophy, "a general ideal of universal philosophy," ancient Greek or modern, this unity becomes possible as beneficial for "modern European humanity itself."[8] Mathematics or science, whether geometry in ancient Greece or the mathematical-experimental physics of modernity, always derived or are born from this universal philosophy, in the *Crisis* interlinking ancient Greek and modern philosophy, under the heading of Renaissance. We use this term with some caution now, in part given the more complex and multiple nature of the history of modernity or ancient Greek culture in all its aspects, philosophy, mathematics, and science included.

In the present view, this multiplicity is irreducible, which still allows for both the continuity and unambiguity principle in mathematics, science, and philosophy. There is no philosophical universe of which modern science, including mathematics, is necessarily an organic part, or always needs to be. Philosophy, mathematics, and science, each already multiple, are all parts of a multiplicity, with contingently fluctuating connections or disconnections with each other, without ever amounting to a single whole, the One. Badiou, in this connection, speaks of the "multiple-without-One", [2, p. 35]. The unfolding multiplicity of this history is a multiplicity of multiplicities, and as such is not a "*uni*-verse" but a "*multi*-verse," to adopt a term from present-day cosmology (without adopting this cosmological hypothesis itself).

The trajectory of Husserl's thought remained outside, without much intersection with, this type of view, let alone the RWR/IWI thinking accompanying it here, even if, as indicated, his argumentation sometimes suggests the possibility

[8] While it would be impossible to deny a Eurocentrism of this argumentation, this Eurocentrism might be seen as tempered by Husserl's view that at stake here is a European cultural ideal, desirable but not necessarily required elsewhere. It is difficult to be definitive on this point, and I shall not insist on it. On the other hand, this ideal was manifestly defied in the Nazi Germany at the time, which would, accordingly, no longer qualify as European in this sense, as Husserl observed on several occasions. Husserl, who was Jewish, avoided the worst consequences of the Nazi regimes for the Jews, in part because he retired from the University of Freiburg, where he taught all his life, in 1926 and died in 1938. His life and work were nevertheless negatively affected during the Nazi years. While, however, the *Crisis* was shaped by the political situation in Germany and Europe at the time, one might argue that Husserl's personal circumstances and connections (such as those to Heidegger, who joined the Nazi party and became the Rector of the University of Freiburg in 1933), were less important than broader considerations connected to the problematics of the book. Among them were the role of contemporary mathematics or science and technology in the Nazi Germany, a role, which Husserl appears to have seen as a tragic manifestation of the separation of modern mathematics and science from "a general ideal of universal philosophy" as grounding "modern European humanity," incompatible with the Nazi ideology, expressly aimed at destroying this humanity.

of this multiplicity, against their own grain, but less so, if at all, suggesting RWR/IWI thinking. It might have been in part because Husserl bypassed exploring the mathematics or science that could have served as an additional impetus by suggesting both this multiple and RWR/IWI thinking, such as the developments in set theory culminating in Gödel's theorems in 1931, or quantum mechanics. Both were extensively discussed during Husserl's career, including at the time of his work on the *Crisis* in the 1930s. Husserl, who corresponded with Weyl (who was much attracted to phenomenology) was aware of these debates, although, unlike Göttingen or Berlin, the university of Freiburg, where Husserl was, was not the center of these debates. (Husserl turned down the invitation to move to Berlin in 1923.) One could have also found an inspiration in the type of thinking advocated here in modernist literature and art (some influenced by new mathematics and science), but Husserl does not appear to have been interested in them, similarly to Einstein and unlike Bohr.

It is also possible, however, and even likely that, as many others, such as Einstein, or less vehemently Weyl, Husserl rejected this type of thinking outright as inconsistent with realism, which defined his phenomenology, from *The Philosophy of Arithmetic* on and which he saw as an equally basic principle in either philosophy or mathematics and science. Einstein's position was closer to Kant than to Husserl, in contrast to Weyl, who, as I said, grounded physics, including, in his classic *Space-Time-Matter*, specifically relativity, in phenomenology [83]. For a while, Weyl was also attracted to intuitionism, which was close to and influenced by phenomenology. In sum, Husserl was strongly committed to realist thinking in mathematics and physics alike, as opposed to RWR or proximate thinking advocated by Bohr and those around him in quantum physics. It was, however, still rejected by many others working in quantum physics or the philosophy of quantum theory. Among the few philosophers during Husserl's time whose thinking had affinities (the degree varies) with RWR/IWI thinking, would be Friedrich Nietzsche, who was arguably the closest to RWR thinking, William James, and Ludwig Wittgenstein. It is difficult to extend this list. But then, as just explained, there were hardly many more mathematicians and scientists who even considered the possibility of an RWR/IWI reality. There were even fewer who had adopted the RWR/IWI view as fundamental (rather than a stop-gap measure). Among the exceptions were, again, Lebesgue, who entertained the possibility of this kind of reality already in 1905, or Bohr and a few around him, more open to the RWR/IWI view.

On the other hand, modernist literature and art have been impacted by Husserl's ideas, as have been the thinking of several of his contemporary mathematicians and physicists, such as Weyl (who, as noted, corresponded with Husserl), Kurt Gödel, and Fritz London, a quantum physicist. In fact, Husserl's phenomenology was had a significant impact on the development of realist thinking in mathematics and science. In philosophy, Husserl's influence has been massive, from Heidegger, his most famous student, to Derrida, whose philosophy benefited from Husserl's thinking, while more diverging from than following it. The present argument is different because rather than being influenced by Husserl, it uses his thinking, especially concerning geometry, as an example of a more sophisticated but still problematic case of realism, questioned here. The present argument also diverges from that of Derrida by virtue of giving a greater role to the composition principle, at most only implicit in Derrida, and using the RWR/IWI principle, not assumed by Derrida at least not in its strongest form which places the ultimate nature of the reality considered, material or mental, beyond the reach of thought.[9]

Husserl, as Kant, locates the origins of geometry in the famously mythical figure of Thales—the first philosopher, the first scientist, and the first mathematician. Whether it was Thales or somebody else, such a single figure and such a single event of the invention of geometry were assumed by both Kant, as a discovery, and by Husserl, as "the lived act of production," from philosophy. While assuming this event of the discovery of geometry as having occurred, Kant and Husserl were not aiming at tracing it as a historically recorded event or a definitive figure, which did not appear possible. It belongs instead to a different type of history, the history of concepts, in this case those comprising geometry.

This view is in accord with the idea of history defined by G. W. F. Hegel as *Geschichte*, vs. a history of recorded events, *Historie*. The concept of *Geschichte* was, the idea adopted by Husserl in considering the origin and history of geometry, [18, pp. 131–132]. (Kant, who died before Hegel's career took off, was less concerned with historical aspects of philosophy, made central by Hegel, whose historical orientation was widely influential, well beyond philosophy, including on Charles Darwin and Karl Marx.) Hegel's idea of the history of concepts is partially adopted here as well in considering mathematical concepts by assuming that every concept, however innovative, has a history. In the present view, however, this history is as part of the irreducibly multiple view of this history, which would preclude seeing any given concept, let alone a field, like geometry, as having a single origin of the type proposed by Husserl for geometry. Conceptual innovations are possible, but they are always an interplay of continuities and breaks with what preceded them. The question is what precedes an innovation, such as a new concept, theory, or field, such as geometry.[10]

In Husserl's view of the history of concepts, philosophy and its concepts precede geometry (or mathematics) and its concepts, derived from those of philosophy. Philosophy is interposed, first, into the (Western) cultural world, including as defined by general language, for geometry to be possible. This is a view that is

[9] This may be said even about his subsequent, philosophically more radical, works. (His introduction to Husserl's "The Origins of Geometry" was his first book.) As I have argued elsewhere, Derrida appears ultimately to settle for "the unnamable," which is in principle still thinkable, rather than for the unthinkable of the RWR/IWI view, [63, p. 960].

[10] Hegel's view of history may be seen as having a greater proximity to the present view, but still under the assumption of the ultimate unification, at least in principle, of all conceptually meaningful trajectories of history in what he sees as World History. This concept has been and remains debated ever since, not the least because it has been and remains the question what, or who, defines such trajectories as meaningful vs. those that are not and, as such, do not belong to World history. Hegel also asserts the ultimate superiority of philosophical thinking and concepts over all others forms of thinking and concepts, including mathematical and scientific ones.

especially challenged by the present argument, even more so than the origin of geometry as a single event. The latter view, as explained, could be accepted within the present framework, if suitably qualified, by assuming this event to be a local *process* of this invention, possibly with the corresponding conception of geometry, in accord with Husserl's definition considered above, first having occurred to a single person. As noted, Husserl might have accepted this qualification. The question here is the status of the project of geometry, specifically defined by Husserl, as "the possibility ... of producing constructively and univocally, through an a priori, all-encompassing systematic method, all possibly conceivable ideal shapes", [36, p. 27]. That is, the question is to what extent Husserl's definition of geometry thus discovered, and perhaps it was discovered as such in ancient Greece, possibly even by a philosopher, corresponded to the practice or a widespread view of geometry in ancient Greece, or in modernity.

It would be difficult to deny that philosophy, especially in ancient Greece, played a major role in the development of geometry and mathematics, specifically arithmetic, there. There is plenty of historical, archival, evidence to support this claim from the available archive of recorded events. Husserl's claim of the origin of geometry, *in his sense*, from philosophy in ancient Greece depends on this archive as well. As I argue here, however, given this archival evidence, it would be difficult to see the status of the project of geometry as defined by Husserl as more than a limited view of, let alone grounding the practice, of ancient Greek geometry, or for that matter modern geometry. In the latter case there were manifestly alternative views, let alone practices, of geometry and science emerging with modernity, marginal, as they might have been, for example, in the work of Girard Desargues, one of the founders of projective geometry.

Accordingly, while philosophy had a major role in the history of geometry in ancient Greece, it is likely that some trajectories of this emergence have extended from literary or other arts, for example, or that the emergence of philosophy itself, multiple in turn, proceeded in part from geometry or mathematics. The latter possibility was not considered either by Husserl or by Derrida, although Derrida's argument also suggests that this might well have been the case. So, Derrida would have been likely to agree with this possibility. Husserl is a different matter. In any event, I argue here that, while it is possible that the project of geometry as defined by Husserl emerged from philosophy, it is equally possible that the idea first occurred to a mathematician strictly mathematically and was then adopted and generalized by a philosopher, one Thales or another. In the present view, again, such a project would in any event have only been one project among others in ancient Greek mathematics.

In fact, this distinction between mathematicians and philosophers in ancient Greece is difficult to make, and this was of course well understood by Husserl, while, however, grounding his view, questioned here. Were the Pythagoreans, first philosophers and then mathematicians, or the other way around, or both all along? It may not be possible to be certain, but it would depend on the Pythagorean one considers, such as Theaetetus, whose mathematical work was so crucial for Plato's philosophy, [52]. Undoubtedly, poetry and art were part of the genealogy of Plato's (or pre-Socratic) philosophy, much as Plato fought against them, and they have also shaped the development or possibly even the emergence of mathematics, and were reciprocally shaped by mathematics and of course philosophy.

And then, how about ancient Greek physics, only barely and inconsequentially mentioned in the *Crisis* and not at all in the *Origin*? Were Thales and other pre-Socratics also not physicists and have they not been celebrated as such? Would not their physics have also shaped their mathematics, and the nature of the projects defining mathematics, including geometry, while their physics was in turn shaped by preceding mathematics, philosophy, or literature and art? It is difficult to assume that there ever was a time when physics and mathematics did not reciprocally shape each other, sometimes with physicists making mathematical discoveries and mathematicians physical ones. A broader intellectual and cultural multiplicity is at stake, especially insofar as concerns the role of literature and art as manifestations of the composition principle, not significantly if at all engaged and arguably underappreciated by Husserl.

In the present view, then, there is only a multiple history of the emergence of multiple geometries, from many trajectories—those of general thinking and language, literature and art, philosophy, physics, preceding forms of mathematics, and so forth. By the same token, these multiple geometries cannot be contained, either in their practice or in philosophically understanding them by the project of geometry as defined by Husserl, that "of producing constructively and univocally, through an a priori, all-encompassing systematic method, all possibly conceivable ideal shapes." This multiplicity, ultimately defined by the compositional nature of our thought, does not preclude the use of either the continuity principle or the unambiguity principle in mathematics and science, with the unambiguity principle ensuring the objective nature in both domains, which is part of the continuity of their history.

The unambiguous character of geometry, or mathematics and science, defined by the unambiguity principle, emerges in this history, in part because they exist in general language, thinking, and culture, which are more permissive of ambiguity. Husserl, again, acknowledges this pre-existence in the case of ideal objectivity. This character of mathematics and science does not eliminate more subjective, including more personal, and hence sometimes ambiguous, aspects of mathematical or scientific experience. The presence of these aspects in mathematics or in philosophy worried Husserl throughout his career as a philosopher, leading him to his idea of phenomenological reduction, which ideally curtailed these aspects.[11] In the present view, these aspects are ultimately irreducible in and in fact help mathematical and scientific creativity, a fact, again, underappreciated by Husserl.

Arguing, as I am here, for this multi-trajectorial, compositional becoming, in which the continuous and the discontinuous are in a complex interplay, radically breaks with Platonism in mathematics, or realism in physics. One equally breaks with Platonism, now also Plato's own, if one adopts the RWR/IWI principle in which

[11] The ultimate impossibility of this curtailment, indeed emerging from within Husserl's text, against its grain, was convincingly shown in several of Derrida's critiques of Husserl, especially in [20].

case the term becoming no longer applies either, any more than the term being, to the ultimate reality responsible for the phenomena considered in mathematics or physics. In sum, assuming, correlatively, both the irreducible multiplicity of the emergence of any concept, theory, or field, and the RWR/IWI principle, and thus combining the irreducible multiple and the irreducibly unthinkable, breaks with all Platonism, from Plato's own to mathematical Platonism or realism in physics, a materialist version of Platonism, of the twentieth and twenty-first century.

That is not to say that Platonist (or Kantian) views, on both counts—the *unconditional* reduction of the multiple to the univocal ideal objectivity and, even more commonly, rejecting the RWR/IWI view—are no longer around in foundational thinking in mathematics or physics. These views, in their various incarnations, persist and remain dominant. Indeed, the combination of this possibility of the radical break, thus defined, from Platonism or realism and yet their persistent dominance, which emerged in the early decades of the twentieth century, has defined and still does foundational thinking in and concerning mathematics and physics.

The advocacy of this radical departure has always been and remains a minority view. In reflecting on this situation in 1949, two decades after it emerged with the invention of quantum mechanics, Bohr spoke of "the dissent among the physicists themselves [as] the cause of skepticism about the necessity of going so far in renouncing customary demand as regards the explanation of natural phenomena," that is, as far as the RWR view of quantum phenomena as effects of a reality beyond the reach of thought, a view assumed by Bohr, [8, v2, p. 63]. Einstein was foremost on Bohr's mind, especially given that this comment was made in Bohr's tribute to Einstein, reflecting on their debates during these decades, [8, v2, pp. 32–66]. Einstein was the most prominent figure in, and became a symbol of, this dissent. But it was hardly Einstein alone. He was followed by many, event most, physicists and philosophers. Even most advocates of Bohr, including scholars of Bohr, stopped short of accepting or even perceiving more radical aspects of his view. Little, if anything, has changed since. The situation has been no different in the foundations of mathematics. There is little, if any evidence, for exploring the implications of Lebesgue's remarkable insight, mentioned at the outset and central for my argument here, that a reality beyond anything we can conceive might nevertheless exists.

12.3 "In Mathematical Language": Geometry Before Physics

More because than in spite of his radical, RWR, epistemological views, Bohr deeply understood and valued "the basic principles of science," including "the unambiguous logical representation of relations between experiences" and hence the possibility of unambiguously communicating them, [8, v2, p. 68]. Thus, his concept of complementarity (explained below) was expressly introduced to ensure the possibility of maintaining these principles in quantum physics, under an RWR interpretation, [5, p. 699]. Bohr's emphasis on this unambiguity in effect amounts to the unambiguity principle, without grounding it in Husserl's ideal objectivity, dispensed with by the present argument as well. In sum, Bohr accepted, as consistent with each other and with basic principles of science (including mathematics), the continuity principle, the unambiguity principle and the RWR principle. He did not expressly consider the composition principle, but clearly recognized the essential role of composition in the creative practice of physics, theoretical and experimental.

In contrast to Husserl's argument in the *Origin*, concerned with the origin of geometry and primarily addressing the historical and thus temporal implications of the continuity principle, Bohr was primarily concerned with the spatial functioning and implications of the principle in his contemporary mathematics and science. One of his aims was to argue that this principle and other basic principles of science are not threatened by the radical epistemological implications of quantum physics, especially in RWR interpretations, as many worried at the time and still do. As stated from the outset, however, the temporal aspects are unavoidable in the spatial functioning of the continuity and unambiguity principles, because a communication of anything is rarely, and rigorously never, strictly immediate, and often involves a manifested, possibly long, past.

Of course, what constitutes the basic principles of science may change and have changed, and could be, and has been, debated. At present, there is a nearly unanimous agreement on the unambiguity principle or logical consistency in mathematics and physics, or in physics, on the consistency of physical theories with the experimental evidence available or their capacity, in modern physics by means of mathematics, for predicting the outcomes of the experiments considered by them. (In general, a physical theory must be consistent with all available experimental evidence or confirmed physical laws rather than only those within its purview.) For Bohr, to be in accord with these requirements, as quantum theory was (and remains), it was sufficient to see it as conforming to "the basic principles of science." The present view is the same. On the other hand, many physicists and philosophers of science, Einstein, again, among them, saw a representation of how physical phenomena, quantum phenomena included, come about, in other words, realism, is a basic principle of science. Einstein's Princeton friend, Gödel similarly assumed Platonism to be such a principle in mathematics, his incompleteness theorems notwithstanding.[12]

It is not coincidental that, in stating that "when speaking of conceptual framework, we refer merely to the unambiguous logical representation of relations between experiences," rather experiments, as one might expect in dealing with physics, [8, v2, p. 68]. The reason is that Bohr makes a broader point, which, as will be seen presently, equally applies in mathematics. As stated from the outset, our experiences in mathematics and science combine individual and shared aspects, with

[12] Gödel also made an intriguing contribution to general relativity, by finding a solution of Einstein's equations, known as Gödel's metric, that allows for a retroaction in time and other controversial features, such as a rotation of the universe (never observed), [27]. This solution is not generally accepted as physically viable, although there are physicists and philosophers of physics who see a retroaction in time as possible. Of course, as allowing for time travel, this is a much beloved feature of the world of science fiction.

shared ones conforming to the unambiguity principle. "Individual" and "shared," are arguably better terms than "subjective" and "objective," although this depends on how one defines subjectivity and objectivity. In Husserl, the ideal objectivity is not without subjectivity because it is the objectivity of thought. Conversely, as Husserl realized as well, things are rarely entirely subjective, given the impact of the outside world, including human interactions, as thus intersubjectivity, on our thinking, however internalized this impact may be. Science and especially mathematics must maximally reduce the ambiguity of defining and communicating their findings, thus entailing the unambiguity principle. This is the definition of "objectivity" in Bohr, vs. the "objective" representation of the ultimate reality responsible for quantum phenomena, which representation is "*in principle* excluded" by Bohr's RWR interpretation, [8, v2, p. 62]. This representation, by means of classical physics, is possible and necessary in considering what is observed as quantum phenomena.

The uncertainty relations exemplify this situation. They are a defining feature of quantum phenomena, rather than of quantum mechanics, the predictions of which are, however, fully in accord with the uncertainty relations. This difference is important. While this is a major testimony to the accord between quantum mechanics and the experimental data it considers, the uncertainty relations remain an experimentally established fact, a law of nature, independent of any theory, and could in principle be predicted by an alternative theory. An uncertainty relation is represented by a formula, $\Delta q \Delta p \approx h$, where q is the coordinate, p is the momentum in the corresponding direction, which are measurable quantities observed in instruments, h is the Planck constant (strictly measurable as well, and Δ is the standard deviation, which measures the amount of variation of a set value. There is nothing unambiguous about the formula itself. It represents an experimentally confirmed law independent of any theory, although a physical interpretation of the uncertainty relations is subtle, in part in view of their statistical nature. It suffices to note here that the uncertainty relations are not a manifestation of the limited accuracy of measuring instruments, because they would be valid even if we had perfect measuring instruments. In classical and quantum physics alike, one can only measure or predict each variable within the capacity of our measuring instruments. In classical physics, however, one can in principle measure both variables simultaneously within the same experimental arrangement and improve the accuracy of this measurement by improving the capacity of our measuring instruments, in principle indefinitely. The uncertainty relations prevent us from doing so for both variables in quantum physics regardless of this capacity. Accordingly, what the uncertainty relations state, *unambiguously*, is that the simultaneous exact measurement or prediction of both variables (always possible, at least ideally and in principle, in classical physics) is not possible. On the other hand, it is always possible to measure or predict each variable ideally exactly. It is also possible to measure both inexactly. For Bohr, however, and others, especially those who hold RWR interpretations, the uncertainty relations also mean that one cannot unambiguously consider or even define both variables simultaneously in quantum physics, but only define one at each moment in time and thus make unambiguous statements about it.

This situation is understood by Bohr in terms of complementarity, a concept grounded in and supporting the unambiguity principle: the mutual exclusivity of certain measurements or predictions at any given point in time and yet the possibility, by a conscious decision, to measure either one or the other exactly at any given point in time, which, however, in principle preclude doing the same for the other, complementary, variable. The role of this conscious decision is crucial in quantum physics and is one of the key differences between it and classical physics or relativity, which do not contain either the uncertainty relations or complementarity, [64, pp. 209–218]. By the same token, one can only make, by means of quantum mechanics, the corresponding prediction, which is in general probabilistic, concerning a future measurement of one variable, while precluding one from doing so for the other. Complementarity, thus, helps to establish the possibility of strictly unambiguous definition or communication of everything involved in considering quantum phenomena, as, including, again, the uncertainty relations, equally unambiguously predicted by the formalism of quantum mechanics.

This formalism can also be unambiguously communicated, as can be all mathematics used in quantum theory, or any mathematics as such. Indeed, after defining in the passage cited above, "a conceptual framework" as "the unambiguous logical representation of relations between experiences," Bohr expressly comments on mathematics:

> A special role in physics is played by mathematics which has contributed so decisively to the developments of logical thinking, and which by its well-defined abstractions offers invaluable help in expressing harmonious relationships. Still, in our discussion ... we shall not consider pure mathematics as [entirely] separate branch of knowledge, but rather a *refinement* of general language [and concepts], supplementing it with appropriate tools to represent relations for which ordinary verbal expression is imprecise and cumbersome. In this connection, it may be stressed that, just by avoiding the reference to conscious subject which infiltrates daily language, the use of mathematical symbols secures the unambiguity of definition required for objective description." [8, v2, p. 68]-emphasis added.

Several qualifications are in order. First, I add concepts because it is difficult to dissociate words from concepts. A word, or as it is called in linguistics, a "signifier," can be "glued" or "re-glued" to a different concept, or what is, in linguistic, called a "signified," a concept applicable either in general language or in any semiotic systems, any system of signs, including in mathematics. There, for example, the word "curve" or "continuum," may have a meaning that has little to with its meaning in general language. Although there are exceptions, in general, it is only through a signified, as a concept, that a word, as a signifier, or a mathematical symbol, usually defined by its connections to other symbols, can refer to a concrete entity. A curve can, for example, refers a curve of a road, or a particular curve in mathematics, say, a parabola, $y = x^2$, which we can visualize just as we can visualize, or directly perceive, a curve of a road (which can in fact be parabolic). Or it can be a Peano curve, which we cannot visualize. It is, however, difficult to define a Peano curve entirely without using ordinary language either. One exception of a word without a concept, if one can still speak of a word in this case, is the word reality in "reality without realism," or ideality in "ideality without idealism." As beyond conception, either word is designed not to have a concept or signified associated with it. At the same reality without realism or ideality without idealism is a philosophical concept in the sense defined in the next section, a concept that refers to the impossibility of a concept or signified associated with the word reality or ideality. A reality without realism or ideality without idealism is, however, always part of the concept defined in the standard way from which such a reality or ideality is inferred, such as, in quantum physics, the concept of phenomena observed which are described by general language supplemented by mathematics, in this case that of classical physics.

Refinement, used by Bohr, is a term suggesting a capacious concept. It can, as noted, function, including, in accord with the unambiguity principle, in mathematics and science in the absence of the ideal objectivity in Husserl's sense of phenomenological reduction. It could also function beyond the domain of the unambiguity principle, but alongside the composition and continuity principles, as in literature and art. Indeed, it is a brilliant choice on Bohr's part, likely with a highly refined nature of the abstract mathematics used in quantum mechanics and quantum field theory, or other contemporary mathematics, or indeed mathematics in general.[13] Mathematics is refined as "a superior complement" of logic, just as poetry is "a superior complement" of language, according to Mallarmé.

As noted earlier, in a poem this refinement extends to the point of inventing its own language with each new line of poetry, giving a new meaning to each word. According to Mallarmé: "I say: a flower! And from the oblivion to which my voice relegates all contours, as something other than unmentioned calyces, musically arises the idea itself, and sweet, absent from all bouquets [of actual flowers]", [47, p. 851]. The flower, thus announced, enters a poetic composition (even if it only consists of one word) and loses its reference to any actual, specific flower, while retaining its singularity and concreteness as an idea, akin to a mathematical concept. Alain Badiou commented as follows: "If I say 'a flower,' I separate it from every bouquet. If I say 'let a sphere,' I separate it from every spherical object. At this point, matheme and poem are indiscernible", [2, p. 47].

This claim appear to me too unqualified. I would instead argue that they are analogous, rather than indiscernible, in this separation. A matheme is, at least in mathematics, still teachable (and the ancient Greek meaning of the word suggests) as something unambiguous and in this sense objective. This is not quite true about a poem, where the unambiguity principle does not apply, only the composition principle does. The flower of a poem is not a sphere in mathematics, although both are compositions, and refinement of the meanings of each word in general language, on superior supplement of logic and language respectively. Logic does require language, and language logic. This mutual dependence, however, still allows

[13] Bohr's brother Harald Bohr, with whom Bohr talked daily, was a world-renown mathematician, working in functional analysis and related fields, and the founder of the field of almost periodic functions. Niels Bohr reflected on mathematics, assuming the same view, in several of his works, especially in [7].

for the difference in question by virtue of the unambiguity principle in mathematics, but not in literature. What both share, as creative endeavors, is composition—the soul of both. Perhaps, then, we should (in addition to logic and calculations) teach mathematics as an art of composition, akin to poetry, as de Rham might have thought in reciting Mallarmé.

It is important to keep in mind that"avoiding the *reference* to conscious subject," invoked by Bohr as enabling mathematical symbols to secure the unambiguity of the definition of mathematical or (mathematized) physical concepts, is not the same as avoiding conscious subjects. Without consciousness neither the use of mathematical symbols (or concept) nor objective, unambiguous, description or communication of mathematics would be possible. This is Husserl's point as well, in his case at the cost of disregarding the unconscious, brought to the center stage by Sigmund Freud (Husserl's contemporary and a fellow student of Franz Brentano) or earlier Nietzsche. The unconscious is one of the primary sources of ambiguity, a fact profoundly understood by both Nietzsche and Freud, and explored by them and their followers. Nevertheless, the unconscious can still think mathematically and, hence, unambiguously. In fact, most mathematical thinking is unconscious, because most thinking is. On the other hand, mathematics or any thinking must make itself present to consciousness to function unambiguously, even for a single mathematician, but especially for mathematics to work collectively, institutionally.

This functioning is, accordingly, fully consistent with Bohr's argument, less stringent than that of Husserl and in principle open to the role of the unconscious. Indeed, Bohr's initial interests, before switching to physics, were in psychology, and he maintained interest in psychology throughout his life. Moreover, as noted, in considering complementarity, the role of conscious decisions, and thus consciousness, is crucial in quantum physics, and defines complementarity, while, again, allowing for the unambiguity principle, [64, pp. 207–218].

Finally, the "refinement" in question can take us very far from general language and concepts, a distance that is, again, manifested in modern mathematics as abstract mathematics or physics, such as relativity and quantum theory, that uses abstract mathematics. Nevertheless, it would still reflect the human nature of this refinement, including for Bohr, or under the exigency of the ideal objectivity, for Husserl. Is there any mathematics other than human, given that even the mathematics done by computers is still an extension of human mathematics? It is sometimes assumed that mathematics is something trans-human, for example, as existing in some independent ideal domain on lines of mathematical Platonism. It has also been suggested that mathematics is something that extraterrestrials would possess or could understand. This view will be put aside here. In Bohr's or the present view, as well as that of Husserl or Kant, mathematics is strictly human.

As such, it may not ultimately be suited to deal, because nothing human can, with the ultimate constitution—the ultimate character of reality—of matter or mind, either by representing or possibly even conceiving of this constitution, assuming that there is any way to verify any such conception apart from at least partially representing it. This is the question that was posed by quantum theory, especially with Bohr's 1913 theory and then quantum mechanics and quantum field theory in the 1920s, eventually, sometime by the late 1930s, leading Bohr to an RWR interpretations of it. By this time this question, which, as noted, shadowed foundations of mathematics from Cantor's set theory on, and was, as noted, expressly posed by Lebesgue in the 1900s, became more pronounced there with Gödel's incompleteness theorems (1931) and the undecidability of the continuum hypothesis, proved by Paul Cohen in 1960s. We now know that mathematics cannot represent itself and possibly disallows us even conceiving of its ultimate nature, which is nevertheless responsible for all that we conceive in mathematics.

This does not mean that we cannot continue to develop mathematics and invent new mathematics. Indeed, the development of twentieth and then twenty-first century mathematics, including its continuing inventions in and of geometries, was little affected by these questions. Neither was the development of physics based on mathematics, as all modern, mathematical-experimental physics, is. In fact, in RWR interpretations of quantum theory, the development of new mathematics, beginning with a new geometry, that of Hilbert spaces over $\mathbb{C}$, becomes the only way to develop new physical theories. This is because it may no longer be possible, and in RWR interpretations is strictly impossible, to use a mathematical representation of physical concepts developed first, as it was in classical physics or relativity. This mathematics is *before* physics, although it may be invented by a physicist, rather than being already available. Indeed, while it existed in mathematics, but was unknown to Heisenberg, the mathematics of quantum mechanics was reinvented by him, in a matrix version, rather than in (equivalent) terms of Hilbert space, introduced by von Neumann a bit later, [78].

According to Husserl, Galileo (who, along with Descartes, was central to Husserl's argument in the *Crisis*) "inherited" geometry as already established, and thus before his physics as well. Hence, Galileo had nothing to do with the origin of geometry. This "inheritance" was, of course, well recognized and gratefully acknowledged by Galileo himself. Galileo, however, gave geometry a cosmological meaning, and reciprocally the cosmos a mathematical, geometrical meaning, thus making the cosmos mathematical, or at least mathematically representable. It is not only or primarily a matter of considering the solar system geometrically, which was already the case in Copernicus, Descartes, and Kepler. Indeed, even apart from the fact that a version of the Copernican system was known to the ancient Greeks, the Ptolemean system was cosmological, too. Husserl notes these aspects of geometry as inherited by Galileo:"The relatively advanced geometry known to Galileo, already broadly applied not only to the earth but also in astronomy, was for him, accordingly, already pregiven by tradition as a guide to his thinking, which [then] related empirical matters to the mathematical ideas of limit", [36, p. 25]. It is the latter aspect of his thinking that is crucial here, as Husserl states in juxtaposing Galileo's modern thinking to the Platonism of the ancient Greeks (which one should, again, be careful to qualify in considering Plato's thought): "For Platonism, the real [*das Reale*] had a more or less perfect methexis of the ideal. This afforded ancient geometry possibilities of a primitive application to reality. [But] through Galileo's *mathematization of nature, nature itself* is idealized under the guidance of new mathematics; nature itself becomes—to express it in modern way—a mathematical manifold [*Mannigfaltigkeit*]", [36, p. 23].

In other words, Galileo replaced philosophy with mathematics, especially geometry (although he also uses functions, and thus algebra), in physics, thus bracketing philosophy, or at least the phenomenological part of it, from physics altogether. Accordingly, even though the unambiguity principle applied in both cases, this mathematical reduction was fundamentally different from the phenomenological reduction, primordially defining philosophical thought, in which and only in which mathematics and science would be grounded. In Galileo and in all modern physics after him, the primordial grounding of physics was mathematics, which was primordially grounded only in itself, rather than in philosophy. Admittedly, as Husserl notes in the *Crisis*, this grounding leaves open the question how, assuming that mathematics is human, this self-grounding of mathematics as human is grounded. Galileo was, I would argue, not unaware of this question, but bracketed it too, at least in its phenomenological aspects, which was possible for the purposes of mathematical physics. Husserl wanted to ground mathematics in the ideal objectivity of philosophy, which led to the origin of geometry. In the present view, geometry is grounded in the combination of the composition principle (not considered by Husserl) and the unambiguity principle, which need not require the ideal objectivity.

In any event, Galileo's modern cosmos is a mathematical cosmos, however far it extends, which is written as the book of nature in the language of geometry. On the other hand, the idea of the book of nature was borrowed by Galileo from theology. Thus, this new scientific cosmology emerged from two different (if already interactive) trajectories or technologies of thought, theological and mathematical, akin to the way the present-day smartphones have emerged from and merged two technologies, the telephone and the computer. Galileo's thinking was more multiply shaped, given philosophical trajectories of his thinking, from Plato and before on, or literary and artistic ones, such as Dante's poetry or Titian's paintings, mentioned by Galileo. In 1558, on the invitation by the Florentine Academy, Galileo also gave two lectures "On the Shape, Location, and Size of Dante's Inferno," which secured him a position of a lecturer in mathematics at the University of Pisa. (There are also more trajectories to the genealogy of the smartphone, some of which trajectories are not technological.) In Galileo's famous words:

> Philosophy is written in that great book which ever is before our eyes—I mean the universe—but we cannot understand it if we do not first learn the language and grasp the symbols in which it is written. The book is written in mathematical language, and the symbols are triangles, circles and other geometrical figures, without whose help it is impossible to comprehend a single word of it; without which one wanders in vain through a dark labyrinth. [26, pp. 183–184].

Galileo, thus, departs from Aristotle's physics both in making his physics that of nature alone (sometimes referred to as "the Galilean reduction") and, in this respect closer to Plato, in assuming a fundamentally mathematical character of the philosophy *written* in the book of nature the physics as conceived by him needed to *read*. For Plato, as noted, while mental and partially modeled on geometry, the ultimate nature of reality was not mathematical, at least not in the way it was for Galileo, but instead philosophical. In Galileo, on the contrary, a philosophy contained in the book of nature is written in the language of mathematics, geometry. Thus, as Husserl argues in the *Crisis*, Galileo does replace philosophy with mathematics in physics, bracketing philosophy altogether, at least phenomenology (or what became phenomenology in Husserl), in opposition to Husserl's phenomenological reduction.

The present argument agrees with this claim by Husserl, but not with Husserl's negative assessment of Galileo's move itself. This is not because the present argument is against philosophy. Neither was Galileo, quite the contrary; he merely bracketed, in Husserl's language, phenomenology or more accurately phenomenological reduction from physics, replacing with it a mathematization that still allowed for, in my terms here, the unambiguity principle. The present argument is philosophical as well, but without conforming to Husserl's phenomenology, while adopting and redelimiting some aspects of it. It does not position philosophy as uniquely grounding human thinking and culture, European or other, but only as one aspect of them, admittedly important, but no more unconditionally so than mathematics and science, or literature and art. Any one of these endeavors can, contingently, acquire a greater significance or dominance in a given human or cultural situation, especially when this endeavor is expressly pursued.

For the moment, more than anyone else before him, Galileo gave mathematics its defining role in modern physics, eventually codified by Newton in *Principia*, [53], and made modern physics experimental by making it mathematical, as Husserl and, following him, Heidegger argue, [31, p. 93]. As noted, this type of geometrical view was restored to cosmology by Einstein's general relativity, which made the physics of gravity a geometry by using modern mathematics (by then developed separately from physics), Riemannian geometry. Maxwell's electromagnetism, Einstein's main model, was a key precursor physically, but it was not as expressly geometrical was Einstein's general relativity, which Einstein had tried to bring together with electromagnetism for the rest on his life without ever succeeding.

It is possible that Galileo used "the book of nature" only as a metaphor, realizing its human provenance, thus also complicating Husserl's claim, [71]. Even if this was the case (which is not certain), the realist thinking implied in his statement has shaped modern physics and philosophy, unchallenged until quantum theory. It has persisted into our own time, notwithstanding this challenge, just as it did in mathematics and the philosophy of mathematics, notwithstanding the challenge posed by contemporaneous developments in mathematics, from the complexities of Cantor's set theory on.

Cantor's set theory and his investigations of the nature of mathematical infinity and his continuum hypothesis gave new and more radical dimensions to the question and, by 1900, to the crisis of continuity in modern mathematics, by separating the nature of continuity in mathematics from our phenomenal intuition of continuity especially radically and even irrevocably. Quantum theory, beginning with Planck's discovery of it in 1900, contributed to this crisis in relation to our representation, mathematical in modern physics, of nature, eventually, with quantum mechanics and quantum field theory, establishing entirely new relationships between mathematical continuity and physical discontinuity. They did so by relating their continuous mathematics (a complex combination of geometry and algebra) to the irreducibly discrete quantum phenomena in terms of probabilistic predictions. At the same time, they, at least in RWR interpretations, precluded a representation or even conception of how these phenomena come about, and thus placing mathematics before, in this case, also as strictly prior, to physics.

12.4 A "complément supérieur" of a "complément supérieur": Geometry Before Geometry

The divorce of abstract mathematics from both the representation of natural objects and our general phenomenal intuition, especially in the case of continuity, was emphasized by Weyl in his 1918 *The continuum: A critical reexamination of the foundation of analysis*, [82]. The book was followed closely by his 1918 classic *Time-space-matter*, [83], by which point quantum theory (although not yet quantum mechanics) entered his thinking concerning continuity and its role in physics, [83, p. 23]. According to Weyl: "the conceptual world of mathematics is so foreign to what the intuitive continuum presents to us that the demand for coincidence between the two must be dismissed as absurd", [82, p. 108]. "Coincidence" is not the same as "relations," which may be unavoidable, at least insofar as it is difficult to think of continuity, spatially or temporally, apart from our phenomenal intuition.[14] On the other hand, it is possible to define the continuum algebraically, arguably more in accord with the idea of time especially as it is used, as a continuous parameter, in physics.[15] At the time of the rise of modern mathematics, this parallel compelled Sir Rowan Hamilton (of the quaternions and Hamiltonian mechanics fame) to speak of "algebra as the science of pure time," vs. geometry as the science of pure space, in a work, under this title, inspired by his reading of Kant, [30].

The problem of the continuum was made more dramatic by Gödel's incompleteness theorems of 1931 and then Cohen's proof of the undecidability of the continuum hypothesis in the 1960s. Gödel made a key earlier contribution in 1940, by proving that the negation of the continuum hypothesis (the existence of a set with intermediate cardinality) could not be proven in standard set theory. These findings made us realize not only that we do not know but also and more radically that we cannot know how a continuous line, straight or curved (which does not matter topologically), is constituted by its points. In other words, there is not, thus

[14] Weyl's position concerning the nature of mathematical reality is a different matter. As is clear from his philosophical writings (e.g., [84]), Weyl was ultimately a realist (although not a Platonist) in mathematics and physics alike.

[15] This view of time and, correlatively, a retroaction in time are challenged by quantum physics, in part in view of complementarity, as considered in [64, pp. 207–218], [68].

far, and there may not be a rigorous mathematical concept of continuity, at least of a continuous set, that corresponds to our phenomenal intuition of continuity, in contrast to discrete mathematical entities at least some of which are in accord with our phenomenal intuition of discreteness.

These findings rekindled the question of the nature of mathematical reality in the foundations of mathematics, along with discomforts this question, as potentially implying the ultimately unthinkable nature of this reality, has caused in mathematical logic and the philosophy of mathematics. Both fields have been haunted by the *ghost* of the unthinkable from at least Cantor's set theory on. Neither, however, especially of the analytic philosophy of mathematics (as most philosophy of mathematics is), ever viewed the unthinkable as a positive or enabling condition of foundational thinking in mathematics, the possibility reflected in the IWI principle.

Gödel, as explained, exemplified this negative attitude in his strong adherence to Platonism, as much as his friend Einstein exemplified realism, a kind of Platonism of matter, in physics. Anti-Platonist aspects of Husserl's thinking appear to have had little effect on Gödel, who extensively studied Husserl's phenomenology in his later years. On the other hand, it might have been unsurprising because Gödel aimed at making Husserl's phenomenology mathematically grounded (without, it appears, any success) and perhaps also, and correlatively, at grounding this phenomenology in some form of Platonism, or at re-grounding his Platonism in phenomenology, [80]. Lebesgue was, again, an exception and, in early stages of this history, possibly the only exception, in realizing that the unthinkable might be possible in mathematics, just as Bohr realized that was possible and even necessary in quantum physics.

The nature of mathematical reality has of course been debated since the Pythagoreans. It was at stake in Plato, whose ghost has overshadowed this debate in modern mathematics, along with the ghost of the Pythagoreans, the ghost of the incommensurable. The discovery of incommensurable magnitudes, famously, the side and the diagonal of the square, led to the crisis of Pythagorean mathematics and to replacing their arithmetic (defined by commensurable magnitudes, proportions) with geometry as the dominant form of mathematics. Plato responded by attempting to avoid or overcome the unthinkable or the alogon in the logos of philosophy, a logos that, again, was modeled on mathematics, but was not a form of mathematics, [66]. The task of this overcoming in mathematics was left to mathematicians. Mathematicians and then mathematical logicians and philosophers of mathematics have attempted to do so ever since, periodically fending the unthinkable off, until Gödel's incompleteness theorems and the proof of the undecidability of the continuum hypothesis brought even greater difficulties into this task and possibly made it insurmountable. The hope, however, has never died, any more than that for realism in physics, notwithstanding the problems posed for it by quantum physics.

Mathematical Platonism has, however, been debated in turn throughout the twentieth and twenty-first centuries, although primarily as concerns the claim that a mathematical reality, even the ultimate mathematical reality, may exist independently of human thinking, rather than realism. Sometimes (Hilbert held this view), the consistency of a given definition of an entity is identified with the existence of this entity, an identification that becomes a form of mathematical Platonism if this existence is assumed outside human thought, but that allows for alternative views. Several alternative views have emerged (some preceding mathematical Platonism), such as logicism, formalism, intuitionism, conventionalism, or constructivism. Some of these positions are more open to the unthinkable, although no one so far has, to my knowledge, advanced an IWI interpretation of mathematical reality, except the present author in [66], which I follow in this section.

I shall not enter these debates, apart from noting that the question of the domain, "location," of this mathematical reality, as exterior to human thought, has never been adequately answered by mathematical Platonists. It is difficult to think of that which is not material and yet is outside human thought, unless it is divine, which, however, is not a common assumption among mathematical Platonists. More recently, various virtual-reality type possibilities (including those making ourselves and thus our thinking a form of virtual reality created by some extra-terrestrial beings) were proposed that would allow for containing all possible mathematics as part of virtual reality. These arguments will not be discussed here, but most, virtually all of them, appear to be species of the Platonism of the virtual.

My assumption here that the reality of human thought, different for each of us, is the only domain in relation to which one can speak of the reality of mathematical objects or concepts, in juxtaposition to the material reality of nature considered in physics. As emphasized from the outset, the existence of this physical reality is still a human assumption and thus is a product of our thought. In part given the shared neurological constitution of our brains, it is possible to recast the Platonist mathematical reality as a potentiality—the same, even if multiply branched, potentiality—in principle realizable by human thought. This view would be closer to Husserl's view of geometry, discussed above, a view implying this type of potentiality, in "producing constructively and univocally, through an a priori, all-encompassing systematic method, *all* possibly *conceivable* ideal shapes [of geometry]", [36, p. 27]. Kant, as explained, while assuming mathematical reality to be human, sees it as pre-existing, which brings his view close to Platonism, but not mathematical Platonism, where this reality is assumed to exist independently of human thought. I shall comment on this recasting of the Platonist mathematical reality as potentiality below. Most mathematical Platonists assume more than such a potentiality: they assume an actual existence of this reality, only to be discovered rather than eventually constructed.

What could be claimed to exist in our thought without much controversy are mathematical specifications, from strict definitions to partial and indirect characterizations, possibly implying more complete specifications. Such specifications would involve concepts, structures, logical propositions, and so forth, either already in existence in previous mathematics (that is, in the thought of others, transmitted in one way or another) or new mathematical compositions. The question, then, becomes whether anything else mathematical exists in our thought beyond such specifications and the local (mental) mathematical ontologies they define, at least in our conscious thought. We can unconsciously think of other properties of a given object or assemblage of objects, which may either be eventually made conscious, "discovered" by consciousness, or never become conscious. The latter possibility does not change the nature of the situation, because one could either claim that the unconscious could only contain such specifications or that the object or broader reality in question, as different from or exceeding such specifications, somehow exists there. Some unconscious specifications may, again, never become conscious. It is even possible that some mathematics, even very great mathematics, just as poetry (more commonly considered in this connection), has never left the unconscious. There are accounts by mathematicians or poets of dreaming of mathematics or poetry which could not be remembered as what it was, although sometimes it can be remembered or recalled. Only our memory of dreams is conscious, but never dreams themselves. Anything in our unconscious can, however, enter our consciousness. Our unconscious does most of our mathematical thinking, as it does most of our thinking in general, and some of this thinking may never be realized by our conscious thought.

The present view is that nothing mathematical can exist beyond what can in principle be so specified by thought, possibly in one's unconscious. This, IWI (ideality without idealism), view differentiates the situation from that of quantum theory in the RWR (reality without realism) view, which assumes the independent nature of this unthinkable material reality. In this view, quantum phenomena, observed by our consciousness, are described by classical physics and thus specified by thought, and quantum theories are specified by thought mathematically. On the one hand one assumes the existence of a stratum of physical reality, responsible, through its interaction with measuring technology, for these phenomena, to be beyond representation or even conception and thus beyond any specification by thought. This stratum is strictly beyond the reach of thought, even though it is inferred as such on the basis of these phenomena, only possible as phenomena because of our existence as humans. By contrast, in the present view, even if some form or part of mathematical reality may be shared and must be by the continuity principle, it cannot be assumed to exist independently of thought, somebody's thought. Only physical matter is assumed to exist in this independent way, although this existence is, again, still an assumption that belongs to thought. As noted, this assumption has been challenged, even to the point of denying that the ultimate nature of reality is material rather than mental. It is, again, possible that there are real things that exist in our unconscious that will never become conscious. It is equally possible, however, that they will enter our consciousness at one point or another. By contrast, the ultimate constitution of nature, in RWR interpretations, is not assumed to become available, at least as things stand now and possibly ever.

The unthinkable, of the RWR or IWI type may be part of a given theory even if it does not designate something that actually exists. A theory may define entities that it cannot account for, without being able to know whether they exist or not, while offering an effective account of other objects it considers. Thus, applying the power axiom to a countable infinite set implies an "entity" cannot be proven to correspond to the continuum (say, as defined numerically as the set of real numbers), in view of the undecidability of Cantor's continuum hypothesis. This fact introduced an unthinkable element in set theory as initially axiomatically defined, say, by the Zermelo–Fraenkel (ZFC) axiomatic (which assumes the axiom of choice), a theory that is still workable otherwise. Assuming, constructing, the domain in which this correspondence obtains or, as is equally possible because either the continuum hypothesis or its negation can be added as an axiom to ZFC set theory, the alternative domain where this negation applies, defines two alternative theories, either of which is consistent if and only if ZFC is consistent. Either of these theories will, however, inevitably contain, by Gödel's first theorem, at least one undecidable proposition, which will lead to the same type of situation, a process that could in principle continue interminably. Nevertheless, unless one assumes some form of mathematical Platonism, either domain defined at each such point, or possibly both as independent of each other, can only exist in an individual human mind.

Our mathematical specifications could be made more complete or modified by new concepts, structures, and logical propositions, which would change the objects or concepts in question, as say, a Riemann surface, as it developed during over, by now, a long period. However, in the present view, it is not possible to see such changes as referring to the same mathematical object (which can, again, be a broad and multiple entity), approached by our evolving concepts. Instead, they create new objects or concepts. Thus, new concrete multiplicities of Riemann surfaces are created by each modification of the concept. There are no Riemann surfaces existing by themselves and in themselves somewhere; there are only objects, so defined, when we think about them—consciously, at a given point of time, or unconsciously, either at a given point in time or over longer intervals.

This strictly constructivist view of mathematical reality is historically traceable to Kant's view of mathematics as the synthetic construction of mathematical concepts by thought and then Husserl's understanding of geometry as discussed here, keeping in mind that Kant still sees mathematical reality as somehow preexisting and then discovered by means of these synthetic concepts. The present view is more radical, on both counts, that of the nature of constructivism and the IWI view. By the same token, the "constructivism" assumed here is not the same as the trend known as "constructivism" in the foundational philosophy of mathematics, from intuitionism on (e.g., constructive set theory, constructive recursive mathematics, or topos theory). Mathematics is only what can actually be thought—be created, composed by (conscious or unconscious) thought. It can either be something new or something adopted from one record or another, and then expressed, communicated, and so forth, in accord with the continuity and the unambiguity principles.

It is not uncommon to encounter a situation in which a mathematical entity that cannot be given, at the time of its introduction or possibly ever, an adequate mathematical specification, and is only specified partially. In other words, such a situation could only be consistently related to indirectly, by means of a properly specified mathematical concept or set of concepts. In the present view, however, any mathematical entity can only be assumed to exist or be real if it is specified in some way, possibly by using ordinary concepts and language or philosophical concepts, partially supplemented by mathematical concepts, which may be unconscious.

In sum, it cannot be assigned existence or reality as independent of thought, *somebody's* thought.

As noted, in view of Gödel's theorems, mathematics, at least if it is rich enough to contain arithmetic, cannot completely represent and thus fully specify itself: it cannot mathematically formalize all its concepts, propositions, or structures, and ultimately itself so as to guarantee its consistency. This fact is not merely a matter of the inexhaustibility of mathematics within any formal systems, something that has been invoked well before (and of course after) Gödel, for example, by Luitzen E. J. Brouwer, the founder on intuitionism, with whose view on this, Gödel, who refers to Brouwer in his dissertation, was familiar. Unlike this type of inexact philosophical assessment, Gödel's are rigorous mathematical findings beginning with constructing within a formal system (such as Peano's arithmetic) specific mathematical propositions that are undecidable in this system. This type of exactitude may be unattainable in philosophical discourse, which has been striving to achieve it from the pre-Socratics on, although it can come close at its best.

It follows that the ultimate nature of human thought cannot be mathematically formalized either. This is, however, a separate issue, in part, related to that of AI, the possible capacity for consciousness and thinking on the part of digital computers, digital AI, and so forth, which will be put aside here. On the other hand, it does not follow that there actually exists some corresponding unspecifiable reality. One can make this assumption, as Gödel did on Platonist lines, while claiming that there is, at least for now, no human means, mathematical or other, to specify this reality. This assumption is, however, not a consequence of Gödel's incompleteness theorems. These theorems only imply that it is impossible to prove that all possible specifications, within any such system, are consistent. As such, they imply the unthinkable within the logic of any such system, without, again, assuming that there exists some independent mathematical or otherwise mental reality that, in each such case, is responsible for the mathematical structures considered, in the way the ultimate constitution of physical reality is in physics. (Gödel's theorems do allow that a given system could be proven to be inconsistent.)

What may be inconceivable is *why* this is the case, what is the nature of the overall reality responsible for this situation. This is, however, not a mathematical or meta-mathematical question, any more than is the question why our interaction with nature by means of quantum physics enables us to make correct probabilistic predictions. These questions concern the neurological nature of our thought, although they may not ultimately be answerable by neuroscience. That is, the ultimate reasons why we think and for the way in which we think may also be beyond the reach of our thought, which, again, need not mean that we cannot advance neuroscience under this condition.

One could, in considering such as yet unspecified mathematical objects or concepts, speak in terms of a hypothetical *potentiality* defined by the assumption that a mathematical object or concept of a certain type would come into existence as real. The question becomes then that of how one determines this possibility, specifically either in terms of necessity (in this case in a greater proximity to Platonism, even if this potentially only concerns what will come into existence in a human mind) or in terms of probability, which is the present view.[16] In this view such potentialities are only partially, probabilistically, determined by what is sometimes called plausible reasoning, very important in mathematical thinking, [11].

There are different and possibly incompatible ways in which this potentially may or may not become reality. Consider the equation $x^2 + 1 = 0$ and complex numbers. While this equation (which may be assumed to exist in our thought) had no real-number solution, one could have and some had envisioned that it should have a solution and that a mathematical entity or a multiple of such entities, a new type of number, should exist. This possibility came to be realized, although not all such possibilities have been. Complex numbers eventually became a mathematical reality, similarly to the way irrational numbers, which, while beyond the reach of the thought of the Pythagoreans, eventually, after two millennia, became part of a mathematical reality. In the present view, however, neither type of number was a mathematical reality before they were correspondingly specified in somebody's thought, in somebody's unconscious. In the case of complex numbers, it happened to a sufficient degree (vis-á-vis earlier approaches) by the time of Gauss, a crucial figure in giving complex numbers a mathematical reality. The existence of this reality is not denied here. Complex numbers or any mathematical entities have a mathematical reality each time one thinks of them, consciously or unconsciously (in the latter case possibly as long as one is alive), but not otherwise.

In sum, in the present view, all mathematical reality is constructed, and this construction may, ontologically, involve multiple mathematics. This multiplicity is also a consequence of Gödel's undecidability, as exemplified by Cantor's continuum hypothesis, mentioned above. This hypothesis was crucial not only for the question of continuity but also for the question of Cantor's hierarchical order of infinities (the infinity of which was one of his discoveries) and thus for the whole edifice of Cantor's set theory. The hypothesis was, as noted, proven undecidable by Cohen in 1961, with Gödel contributing a crucial part earlier.

It follows that one can extend classical arithmetic in two ways by considering Cantor's hypothesis as either true or false, that is, by assuming either that there is no such intermediate infinity or that there is. This allows one, by decisions of thought, to extend arithmetic into mutually incompatible systems, ultimately infinitely many such systems, because each of them will contain at least one undecidable proposition. Gödel's incompleteness theorems may not be a necessary condition of the multiplicity of possible mathematics. The existence of Cantorian and non-Cantorian set theories, or the existence of Euclidean and the various non-Euclidean geometries established earlier, does not depend on these theorems. Gödel's theorems, however, imply the inevitability of such a multiplicity in principle, on mathematical grounds,

[16] A view of quantum probabilities in terms of potentiality, via Aristotle's concept of potentia, was entertained by Heisenberg in his later works, under the assumption that the ultimate nature of reality in quantum physics is in principle representable mathematically, via such potentiality, the concept used throughout [33].

whether one pursues such alternatives or not, in the way say mathematics has pursued both Euclidean and non-Euclidean geometry. In this case, however, one could still unify them with the same broader framework, say, as defined, following Riemann, by the concept of manifold, one of the most remarkable mathematical compositions. In the case of undecidable propositions, the only thing that these alternative forms of mathematics share is the fact that they are forms of mathematics, which forms are, however, incompatible with each other.

The role of a decision of thought, especially in the case of Gödel's theorems, is a remarkable aspect of this situation and thus of contemporary mathematics. Allowing one, by such a decision, to extend arithmetic or any system containing it into two mutually incompatible systems, and (by iteration) to infinitely many such systems has often been seen as troubling and even intolerable. One might, however, find an appeal in it, as Badiou does:

> As we have known since Paul Cohen's theorem, the Continuum Hypothesis is ...undecidable. Many believe Cohen's discovery has driven the set-theoretic project into ruin. Or at least it has "pluralized" what was once presented as a unified construct. ...[M]y point of view on this matter is ...the opposite. What the undecidability of the Continuum Hypothesis does is complete Set Theory as a Platonist orientation. It indicates its line of flight, the aporia of immanent wandering in which thought experiences itself as an unfounded confrontation with the undecidable. Or, to use Gödel's lexicon: as a continuous recourse to intuition, that is, to decision. [2, p. 99]

By a Platonist *orientation* Badiou means the association of mathematics with the movement of thought, its orientation toward the new and toward the multiple, rather than any form of mathematical Platonism, even that of Plato himself, insofar as Badiou's Platonism is a Platonism of the multiple, while Plato's was not. On the other hand, in the "immanent wandering" that Gödel's theorems bring about, a decision concerning which new mathematical system (made equally possible by an undecidable proposition arising in a given mathematical system, containing arithmetic) one follows is not fully mathematical or possibly even logical in nature. Nor is the potential future course of mathematics, a history of events in its "immanent wandering," defined by the sequence, in principle open, of such decisions, fully defined by mathematical considerations. The nature of such events can only be analyzed historically or philosophically, as Badiou argues, even though he sees topos theory as a mathematical concept that can be used in this analysis e.g. [2, pp. 166–167], [3]. There is a question of how much freedom of choice one has in such a decision, a question that I put aside for the moment, merely noting that there are always factors that limit our freedom of choice. This makes the concept of decision preferable to that of choice, insofar as a decision is defined by a combination of the freedom of choice, chance, and necessity. I shall return to this aspect of the situation in the next section.[17]

---

[17] The situation, as concerns both, the mutually exclusive nature of the two mathematics thus emerging, and the role of the (sufficiently) free decision of thought, which one to pursue, is thus parallel to Bohr's complementarity. This is an intriguing parallel, but one must be cautious as concerns its limits. The subject is, however, beyond my scope.

It is, as I said, in principle possible to assume that all such possible constructions of mathematical systems form a single, if multiply branching, potentiality, ultimately realizable in principle. I shall not assess this view except by noting that even if one adopted it, one would still only allow for a vast constructible mathematical potentiality rather than independent mathematical reality. Would this view be in practice equivalent to mathematical Platonism? While it may be in practice, the difference in principle remains important, both in general and because it would be impossible to assume that, being infinite, this potentiality could ever be realized. In any event this is not a view assumed here.

The present view is that of the development of mathematics as defined by the history of contingent decisions in the absence of a single reality or potentiality, while assuming an IWI type reality, each time different at such junctures, even if still each time inconceivable, unthinkable. This view extends beyond the situation defined by undecidable propositions. Mathematics creates new and multiple mathematical realities and, with them, new and multiple mathematical potentialities all the time, apart from any undecidable propositions. The ultimate nature of reality, whether mental or material or, interactively, both, that makes this (multiple) process possible, may be an IWI or RWR, or IWI/RWR reality. This process will only end when mathematics is no longer with us, and one day it might not be, although it would be difficult to argue that something mathematical-like will not remain in our thought, making it possible to bring mathematics back, old or new, perhaps entirely new and even now unimaginable mathematics.

As indicated, one could envision eventually downloading all possible mathematics, created by us or our computer, in some, say, silicon-based, media, and future mathematicians rediscovering it, or similarly, imagine that all possible mathematics is so downloaded by some extra-terrestrials and that we have been discovering it all along. The first possibility would not in any way challenge the present argument, because this mathematics would still be human and not anything that emerged independently of us. Nor, in principle, would the second, in any event not essentially. There is, however, no special reason to assume, although some do, that anything existing in the universe other than us has mathematics, any more than any other human attributes or capacities. Mathematics is just one of them: it is something that is human and only human. It may not be enough to understand the ultimate workings of nature, dead or alive, or of thought itself, either as the product of the brain or something considered in its own accord, which is not the same as *existing* independently of our bodies.

It is not so easy either and perhaps ultimately impossible to understand how it is possible to be human or what it ultimately means. That, however, still leaves us plenty to think about and to understand, and in the first place to create, in mathematics, science, philosophy, and art. To be human is not a deficiency of mathematics, any more than that of science, philosophy, and art. It is their virtue and our fortune because they are refined products of our thought—"*compléments supérieurs*," as Mallarmé defined them. Indeed, a new, more abstract, mathematics may and often does, as, in the case of geometry, with Riemann, or algebra, with Galois, become a "*complément supérieur*" of preceding mathematics. Thus, already being a "*complément supérieur*" of thinking and language, a new, more abstract, mathematics becomes a "*complément supérieur*" of a "*complément supérieur*." The same process is also found in physics, in modern physics, defined by mathematics and its refinements, or in poetry, independently of mathematics, which can still serve as a model, as in Mallarmé. In each field, this is only possible by means of inventing new compositions, in accordance with the composition principle, for which the primary model is poetry, in the broad sense of poetics, thus including all art.

12.5 The Chaosmos of Compositions: Geometry Before Poetry

In the present view, the poetics of mathematics is the poetics of concepts. The term "poetics," again, refers to both: (a) the practice of the composition of concepts, or theories or fields, which (while they contain other constituents) are conglomerates and thus *compositions* of concepts; and (b) the study, as in this chapter, of this practice. Concepts are, I argue, especially crucial in defining the creative character of mathematics or other sciences, mathematical, like physics, or not. Concepts are, however, important in other aspects of its practice, including pedagogical practice, which requires understanding concepts and thus, first, constructing them by thought. We do not always focus, including in teaching, on the role of concepts in mathematics, as opposed to the logical or calculational aspects of mathematics, which are of course important in mathematics, including in enabling the continuity and unambiguity principles there. Plato's *Meno* provides arguably the earliest known and still paradigmatic example of this conventional teaching of geometry as logic and calculation, including by implicitly appealing to both the continuity principle, in using the already acquired knowledge to acquire a new knowledge, and the unambiguity principle in communicating geometrical knowledge. It is, however, possible to teach mathematics by considering its conceptual nature and to rewrite *Meno*, as such figures as Abel, Galois, or Riemann invite us to do, [15, pp. 179–180], [69, pp. 21–23]. If one adopts this view, mathematical education becomes equally the pedagogy of concepts, a pedagogy implicit in any introduction of a new concept in any domain, as Gilles Deleuze and Félix Guattari argue, [16, p. 112].

One needs, however, first to define concepts in terms of their compositional nature, as Deleuze and Guattari do as well. They reserve this compositional concept of concept primarily and even exclusively to philosophy, but as I argue here, this concept of concept is extendable to mathematics and science. They also juxtapose the concept as a creation of thought to the concept of concept, defined conventionally (as a generalization from particulars), in commercial professional training, shaped, I would argue, primarily by procedural, "algorithmic," learning, based in calculations, [16, p. 112]. Accordingly, huge corporate investments into AI and AI-based training are hardly surprising. That university administrations are enamored with it, even in the humanities, is not surprising either, in part because it attracts corporate investments. It is, however, more distressing, in the absence of equal attention to creative thinking, such that based in the invention of new concepts.

Nobody is much encouraged either to use poetry in teaching or understanding mathematics. Manin's writings are among exceptions of the first use. De Rham's students, such as de la Harpe, were lucky to hear him reciting Mallarmé, and thus to learn from it. These students, I surmise, were mathematicians, among whom reading poetry was not uncommon, although it appears less common among those of the younger generation now. If I am permitted a personal recollection, when I studied mathematics and theoretical physics at the University of Leningrad (now St. Petersburg), reading and discussing contemporary poetry was ubiquitous. Many students of Manin (he was in Moscow) would not be surprised to encounter using poetry in his classes. Athanase Papadopoulos, to whom this chapter is dedicated, is among those for whom literature, including I might add, Russian poetry, and other arts are not only an important part of their life but also part of the culture of mathematics.

It is even more unusual to bring mathematics to bear on teaching or scholarly understanding of literature, beyond explicating references to mathematics in literary works, which is sometimes done by mathematicians in writing about literature. As noted, Galileo was a pioneer in this regard as well by giving his lectures on the geometry of Dante's *Inferno*. This is, however, hardly sufficient for understanding the deeper mathematical aspects of literature, such as those found in structural and epistemological parallel between literature and mathematics, rather than in references to mathematics in literary works. As Bohr said about such interconnections: "We are not dealing here with more or less vague analogies, but with an investigation of the conditions of the proper use of our conceptual means of expression", [8, v2, p. 2]. Admittedly, Bohr is referring here to other science or philosophy, because literature and art rarely *investigate* these conditions in the way mathematics, science, and philosophy do. But literature and art can stage these conditions, which mathematics, science, and philosophy can then explore, including by means of concepts, which is why Bohr speaks of *conceptual* means of expression.

The very term "concept" is often used without being defined or by assuming some common understanding of it in mathematical and scientific or even philosophical literature.[18] There is, admittedly, no single definition of concept, no single concept of concept. The concept of concept I shall adopt here builds on my earlier arguments, but expressly considered here in terms of the composition principle, only implied in these earlier arguments. This concept of concepts in part follows that of Deleuze and Guattari's concept of a philosophical concept, but also departs from their argument concerning concepts, [16], [66, pp. 52–76], [69]. They juxtapose their concept of concept, as a *philosophical* concept, to mathematical and scientific concepts, and the creative practice of philosophy as the invention of concepts to that of mathematics and science. In fact, while casually mentioned sometimes,

[18] The subject discussed in detail in [66, pp. 52–76].

mathematical or scientific concepts are not really defined by Deleuze and Guattari either. These concepts appear to be seen by them in accord with a common use of concepts, as generalization from particulars defining specific representatives of a given concept, say, the concept of a tree from particularities defining actual trees. Deleuze and Guattari even deny that mathematics or science have concepts in their sense. "The concept," they say, "belongs to philosophy and only to philosophy", [16, pp. 11–12, 33–34]. They see mathematics and science as fundamentally defined by logical propositions and formal elements, such as functions, which are composed of elements without being concepts in the same sense.

This is a surprisingly conventional view for these famously unconventional philosophers. As noted from the outset, logic and calculations are indeed essential to mathematics or science, including for maintaining the unambiguity and continuity principles there. I argue, however, that concepts have the primary role in *creative* thinking in mathematics or science, manifesting the composition principle there.[19] Moreover, as noted, the roles of concepts and compositional thinking are not limited to creative aspects of mathematics, because working with already established concepts and teaching them are indispensable in all mathematical or scientific practice. Calculations, too, require concepts, making the unconditional opposition between them impossible to maintain, even in considering figures, such as Galois, Riemann, or Richard Dedekind, who gave concepts a dominant role in their work. They were perfectly capable of difficult calculations. Conversely, calculations may lead to new concepts, as they did in the case of Abel, or Kronecker, famous for his insistence on the primacy of calculations in mathematics.

In the present understanding, a concept is not merely a generalization from particulars (which commonly defines concepts in linguistics, cognitive sciences, or analytic philosophy) or a general or abstract idea, although a concept may contain such ideas, especially abstract mathematical ideas, where they may become concepts. A concept is a multicomponent structure, defined by the organization, *composition*, of its components, and some of these components may be concepts in turn. In other words, the invention of concepts is grounded in the composition principle. Of course, as emphasized from the outset, every concept, no matter how innovative, has a history defined by earlier concepts or other elements. Just as a theory or a field, such as geometry, a concept always emerges in this history, and hence has some continuity with preceding concepts, it would not be possible otherwise. A new concept or a new theory is a product of an interplay of continuities with and breaks from preceding concepts or theories. Continuity is more prevalent in the case of theories or fields, such as geometry, even while new concepts are invented and change some aspects of the theory or field to which they belong.

[19] There are important new approaches to logic in mathematics, such as homotopy type theory, [12, 79]. Concepts, however, still play an essential role in them. There are other aspects of mathematical thinking, for example, a narrative one, the role of which, as constitutive rather than merely auxiliary (which has always been recognized), has received a considerable amount of attention during recent decades. For a representative collection of essays on the subject, see [23]. The present author was among the contributors, [62].

Some concepts, such as those of a group by Galois, a manifold by Riemann, or a topos by Grothendieck, were more radically innovative. Still, they had connections with earlier concepts, such as (putting aside their earlier history) those of Adrien-Marie Legendre in the case of a group, Gauss's extraordinary theorem (*theorema egrerium*), which states that the curvature of a surface is intrinsic to the surface, in the case of a manifold, and those of cohomology and category theories in the case of a topos.

The definition of a concept just given is only a general, *abstract*, definition. What meaningfully establishes a concept is the specific character of its composition, defined by both the nature of each component and the character of their relations within this composition, which is unique to a new concept, and which makes this concept new. The creative essence of concepts is their compositional individuality, defined by the specificity of this composition. A group is composed of its elements, but in accordance to very specific rules, which give it its structure and role in mathematics in group theory and beyond, and then it multiplies further to still more specific concepts of a group, a finite group, a Lie group, an algebraic group, and so forth, and then into further concepts within each of these. A single-component concept is a product of a provisional cut-off of its multicomponent organization. In fact, there are always cut-offs in delineating a concept, which result from assuming some of the components to be primitive entities whose structure is not specified. These components could, however, be specified by alternative delineations, leading to a new concept, containing new primitive (unspecified) components.

The understanding of the nature and workings of concepts just outlined may be illustrated by the concept of space, defining the history of geometry beginning with its emergence. Emerging at the intersection of mathematics, physics, philosophy, and general thinking, and defined by such constitutive concept-components as point, line, plane, or distance, the concept of space and each of these components have a long history of definitions, modification, and transformations. A phenomenal sense of space is arguably as old as human consciousness. Kant did not see space, or time, as a concept, but rather as an a priori form of phenomenal intuition. He had a point, even if one rejects (as many have done since Kant's own time) his a priori view of space and time, according to which these phenomenal forms precede any concept of space and time. As a geometrically representable physical concept, space is at least as old as ancient Greek geometry, preceding Euclid, who did not define, but only implied, the concept of (three-dimensional) space in the *Elements*. The history of the concept of space, as a mathematizable physical and eventually purely mathematical concept, extends from the Pythagoreans to Euclid to Descartes (a coordinate space) to Riemann (a space defined as a manifold) to F. Hausdorff (topological space) to Hilbert (Hilbert spaces in functional analysis) to Grothendieck (topos), with further versions within each, and others in between and after them. The earlier concepts of space just mentioned, including that of Riemann, were physical, mathematically represented, but after Riemann, in part following him, purely mathematical concepts of space became common, the last three just mentioned among them. Such mathematical concepts could be returned to physics, as were, *representationally*, Riemann spaces in Einstein's relativity, or Hilbert spaces over $\mathbb{C}$ in quantum theory, without, in RWR interpretations, using these spaces to represent the ultimate reality responsible for quantum phenomena, but only to predict, in general probabilistically, the outcomes of quantum experiments.

Grothendieck's topos remains arguably the most abstract concept of spatiality in mathematics thus far, with a rich and remarkable genealogy. This genealogy is initiated by Riemann's concept of covering space and Galois theory, applied to the fundamental group, and absorbs and transforms a formidable array of the concepts of modern mathematics. Grothendieck compared his contribution to the idea of space in mathematics, as defined by the concept of topos to that of Einstein and Schrödinger in physics, [28, p. 68]. One can easily understand his reference to Einstein, whose general relativity used Riemannian geometry in redefining gravity as the curvature of the space considered, and thus made general relativity geometrical. On the other hand, unless Grothendieck refers to Schrödinger's work on general relativity (which does not seem likely), a reference to Schrödinger may appear strange. In fact, it makes sense or (as I am not sure that Grothendieck had it in mind) can be given a sense, which is as follows.

Quantum mechanics, established a new type of relationship, fundamentally probabilistic in nature, between purely mathematical spaces, Hilbert spaces over $\mathbb{C}$, in Schrödinger's case infinite-dimensional ones, and three-dimensional physical space, in which quantum phenomena are registered by experiments, mathematically represented as a real manifold. These predictions require rules added to the formalism, such as Born's rule, which relate complex quantities of the formalism to real numbers corresponding to the probabilities of quantum events. One takes the square moduli of the eigenvalues of the operators associated with quantum variables, such as position, momentum, or energy, or equivalently, multiply these eigenvalues by their complex conjugates. Doing so gives one real numbers, corresponding, once suitably normalized, to the probability densities of observed events. In the simplest case, when ψ is a wave function for a point particle in position space, the probability density function $p(x, y, z)$ for predicting a measurement of the position at time t_1 equals $\|\psi(x, y, z, t_1)\|^2$. Integrating over this density gives the probability or (if one repeats the experiment many times) statistics of finding the particle in a given area. Born's rule, thus, allows one to move from complex quantities of the formalism to probabilities, as real numbers between zero and one, by means of a purely mathematical procedure, which has no representational, only probabilistic, relation to physical reality any more than does the formalism itself.

The only representational relations in quantum mechanics, especially in RWR interpretations, are those to quantum phenomena observed in measuring instruments as represented (along with the observable parts of these instruments), by classical physics, defined over $\mathbb{R}$, rather than by quantum mechanics or quantum field theory. As defined over $\mathbb{C}$, either theory cannot represent anything actually observed in quantum experiments, any more than how what is so observed comes about. It can only predict the probabilities of the data obtained in these experiments. This is, however, strictly in accord with the experimental evidence, because no predictions other than probabilistic are in general possible in the case of quantum phenomena, at least as things stand now. Technically, it was Heisenberg who first defined this architecture of quantum mechanics in his discovery of it, based in RWR-type thinking, vs. Schrödinger's thinking in his discovery. Schrödinger initially thought that his wave function would represent quantum processes, as wave-processes, in space and time. In addition to the problem of discontinuity of quantum phenomena (which brought major difficulties to his approach from the outset), this type of representation became virtually impossible when he had to introduce a complex wave function in his time-dependent equation, the defining equation of quantum mechanics. Schrödinger never reconciled himself with the RWR view of Heisenberg and Bohr, and he came to see quantum mechanics as merely a convenient method of calculation rather than as a proper fundamental theory, [73, p. 167], [64, pp. 145–166]. However Schrödinger himself viewed the situation, his equation contained a new form of mathematical spatiality in physics, that of Hilbert spaces over $\mathbb{C}$, and new relationships between the (mathematically) continuous and the (physically) discrete entities. The continuous entities are defined by the formalism of the theory and the discrete ones by the phenomena observed and, probabilistically, predicted by this formalism. The same situation is found in quantum field theory.

This situation may also be seen as a physical version of Grothendieck's reversal, on the same occasion, of the standard relationships between the discrete and the continuous in mathematics. Grothendieck suggested that one might, contrary to the standard view, see the continuous as the approximation of the discrete, attributing the idea to Riemann: "[Riemann] pointed out that it could well be that the ultimate structure of space is discrete, while the continuous representations that we make of it constitute perhaps a simplification (perhaps excessive, in the long run ...) of *a more complex reality*; That for the human mind, 'the continuous' was easier to grasp than 'the discontinuous', and that it serves us, therefore, as an 'approximation' to apprehend the discontinuous", [28, p. 68]. This is not what Riemann actually said on that occasion, suggesting instead that "the reality underlying space" may be either continuous or discrete, [72, p. 33], [66, pp. 217–219]. Grothendieck didn't explain his meaning, although he might have had in mind in that the continuous algebraic topology or geometry (dealing with manifolds over $\mathbb{R}$ or $\mathbb{C}$) is a kind of approximation of his topos theory (and étale cohomology) over a finite field. As he noted, probably with these connections in mind: "Mathematical developments of recent decades have, moreover, shown a much more intimate symbiosis between continuous and discontinuous structures than was imagined, even in the first half of this century", [28, p. 68]. This is equally true about twentieth-century physics, especially quantum mechanics and quantum field theory.

Indeed, in quantum field theory, assuming the continuity of space or time may be related and (possibly lead to difficulties not found in quantum mechanics, such as the appearance of mathematically divergent integrals, handled by the so-called renormalization procedure, which is not fully rigorously defined mathematically. These difficulties have made the view that, to return to Riemann's apt phrase, "the reality underlying space", [72, p. 33], might be discrete and, it follows, at least locally finite, to have periodically reemerged from the rise of quantum field theory to our own time. (Under this assumption the divergences inherent in the current formulation of quantum field theory would disappear.) This is the case apart from the problem of quantum gravity in hypothetical approaches to which this view is even more common. From this perspective, a physical version of Grothendieck's reversal would be that the continuous conception of space and time are approximations of the discrete ultimate reality, an approximation adopted by current quantum theories, mathematically continuous in turn. Quantum field theory remains, however, a viable theory in handling all other fundamental forces of nature, even if, not in the way many would prefer, especially, again, in view of the necessity of renormalization, [66, pp. 271–274], [70]. There is, thus far, no effective discrete theory or, in the first place, rigorous concept of the discrete ultimate reality for doing so.

Quantum mechanics and quantum field theory established the new relationships between the mathematical continuity of their formalism in terms of Hilbert spaces over $\mathbb{C}$ (or some mathematically equivalent formalism) and the *physical discontinuity* of quantum phenomena, always discrete relative to each other, with each represented in continuous space and time, defined as a manifold over $\mathbb{R}$. In RWR interpretations, this situation appears in a different light because the ultimate reality responsible for discrete quantum phenomena, predicted by this continuous formalism, is beyond conception. Hence, it cannot be assumed to be either continuous or discrete, any more than either spatial or temporal, or conforming to any concept we can conceive of. Accordingly, such concepts as a particle, which is discrete, or a wave, or a field, which is continuous, cannot apply to the ultimate constituents of nature, even though these constituents are referred to as elementary particles or quantum fields.[20] All that is discrete now are quantum phenomena observed in space and in measuring instruments. From this perspective, Hilbert's invention of a Hilbert space may be seen as his greatest contribution to geometry, tremendous as his other contributions to geometry or foundations of geometry were.

The historical sketch here offered is too fragmentary and does not do justice to the complexities of the history of the concept of space in mathematics and physics. It also puts the history of the concept of space in philosophy aside. Thus, as noted above, Thom argued that something like topological space is already found in Aristotle and hence in his physics, and certainty the idea of continuity, eventually leading Cantor to the continuum of set theory, extends from Plato, or the pre-Socratics, [55, 56, 76]. On the other hand, the ideas related to discontinuous spatiality, to "the reality underlying space" as possibly discrete are much more recent, although older that one might think, certainly preceding Riemann. What belongs only to the twentieth century is the RWR/IWI view, according to which the ultimate nature of reality, material or mental, may be beyond conception and hence not be either continuous or discontinuous, or for that matter, spatial or temporal, any more than conform to any other concepts. But that, as Lebesgue argued in connection with the problem of the continuum in Cantor's set theory, does not mean that what we cannot conceive does not exist. Quantum theory brought this possibility into physics and to our understanding of material reality.

[20] A discussion of this subject from the RWR perspective is offered in, [64, pp. 273–306].

A concept, then, is an assemblage of its components connected by its composition, which define its specific individual, and at the time of its invention, unique, nature. In this respect, as stated from the outset, a concept is akin to a work of art, always defined by its composition. "Composition," Deleuze and Guattari say, "is the sole definition of art", [16, p. 190]. Arguably, however, this is the single definition (to the degree there can be one) of a creative production anywhere, including mathematics and science, in accord with the composition principle. It may be argued that, insofar as art represents the compositional capacity of human thinking, and transforms this capacity into thought, art ultimately governs thought in all human endeavors. Art, as a culturally defined practice, especially manifests this capacity. This capacity, however, also shapes philosophy, mathematics and science, indeed all creative human endeavors, or the reception of such compositions, which, while not the same as the original creation, involves constructive and, thus, creative aspects.

Components of a composition, conceptual or artistic, may be new, too, but need not be, and it is arguably impossible for all of them to be new. Their arrangement must, however, be new for a conceptual or artistic composition to be new, a new mathematical or scientific, or philosophical, concept or theory, a new work of art, or type of art, or field of art. A theory, or a disciplinary field, mathematical, scientific, or philosophical, becomes an organized assemblage of concepts, accompanied by other elements. It is a product of a broader *poiein*, which in mathematics or science is commonly collective, although a theory may be initiated, by a single new concept, such as a Galois theory, initiated by the concepts of a Galois group, or Riemannian geometry by the concept of a manifold, and thus a theory initially practiced by a single person. In the present view, a creative conceptual mimesis (which may borrow, "imitate," elements of previous concepts but combines them in a new way) is a technical mimesis of previous concepts, rather than a Platonist mimesis of something that belongs to a preexisting independent reality. The same view would apply to philosophical concepts or theory or works of literature and art. As Percy Bysshe Shelley argued in his preface to *Prometheus Unbound*:

> As to imitation, poetry is a mimetic art. It creates, but it creates by combination and representation. Poetical abstractions are beautiful and new, not because the portions of which they are composed had no previous existence in the mind of man or in Nature, but because the whole produced by their combination has some intelligible and beautiful analogy with those sources of emotion and thought and with the contemporary condition of them. One great poet is a masterpiece of Nature which another not only ought to study but must study. He might as wisely and as easily determine that his mind should no longer be the mirror of all that is lovely in the visible universe as exclude from his contemplation the beautiful which exists in the writings of a great contemporary. The pretence of doing it would be a presumption in any but the greatest; the effect, even in him, would be strained, unnatural and ineffectual. A poet is the combined product of such internal powers as modify the nature of others, and of such external influences as excite and sustain these powers; he is not one, but both. Every man's mind is, in this respect, modified by all the objects of Nature and art; by every word and every suggestion which he ever admitted to act upon his consciousness; it is the mirror upon which all forms are reflected and in which they compose one form. Poets, not otherwise than philosophers, painters, sculptors and musicians, are, in one sense, the creators, and, in another, the creations, of their age. From this subjection

> the loftiest do not escape. There is a similarity between Homer and Hesiod, between Aeschylus and Euripides, between Virgil and Horace, between Dante and Petrarch, between Shakespeare and Fletcher, between Dryden and Pope; each has a generic resemblance under which their specific distinctions are arranged. If this similarity be the result of imitation, I am willing to confess that I have imitated. [75, pp. 134–135]

The use of "abstraction" is notable here. It is reminiscent of Mallarmé's view in making poetical abstraction not only or even primarily fictional, but also a refined form of thinking and language, for Mallarmé in affinity with mathematics. Just as that of Mallarmé, however, Shelley's argument, need not and, I would argue, does not imply Platonism or idealism, notwithstanding Shelley's invocations of ideality or idealism, such as "the beautiful idealisms of moral excellence," represented in the figure of Prometheus in the poem, [75, p. 135]. Shelley's use of plural, "idealisms," is indicative of his divergence from Platonism or Idealism, in which there is only one ultimate idealism of moral excellence. Mallarmé invokes such terms, too, similarly representing an ethical or artistic attitude rather Platonism or similar forms of idealism. In the case of Shelley, this invocation can be read, along the lines of skepticism of David Hume's philosophy, a well-established connection in Shelley's thinking. One is not dealing with a Platonist mimesis of primordial or eternal original forms, but only with a partial technical mimesis of preceding compositions. At the same time, as against the representational use, dominant in general language, a "poetical abstraction" is a transformational refinement of general language, and as such is new and unique.

It is instructive to consider from this perspective an iconic modernist text, Jorge Louis Borges's "Pierre Menard, the author of Don Quixote," in which Cervantes's "original," copied word by word, becomes a new, *Pierre Menard's* novel. Borges's famous example manifesting this fact is:

> It is a revelation to compare Menard's Don Quixote with Cervantes'. The latter, for example, wrote (part one, chapter nine):
>
> . . . truth, whose mother is history, rival of time, depository of deeds, witness of the past, exemplar and adviser to the present, and the future's counselor.
>
> Written in the seventeenth century, written by the "lay genius" Cervantes, this enumeration is a mere rhetorical praise of history. Menard, on the other hand, writes:
>
> . . . truth, whose mother is history, rival of time, depository of deeds, witness of the past, exemplar and adviser to the present, and the future's counselor.
>
> History, the mother of truth: the idea is astounding. Menard, a contemporary of William James, does not define history as an inquiry into reality but as its origin. Historical truth, for him, is not what has happened; it is what we judge to have happened. [9, p. 43]

This is not a matter of a correct reading of Cervantes's novel, if it is possible to speak of a (single) correct reading of it. Instead, it is an illustration of the uniqueness of each literary composition, in this case, clearly allowing for and even implying an ambiguity. Menard arrived at his decision to strictly copy the text of *Don Quixote*, as against Cervantes's novel that cannot be copied as *Cervantes's* novel, a novel that will always remain unique to Cervantes. Every reading of a text, or every (re)construction of the author's experience that creates it, make both unique, a unique event of experience. The novel, reauthored by us, becomes unique each time, as does any work of literature or art.

This situation is also a manifestation of the essential ambiguity inherent in general language and is retained in the compositional abstractions or refinements of literature, as against, mathematics and science, which require the unambiguity principle. Insofar as it represents a new mathematical concept or theory, a mathematical text does not, in general, repeat an earlier text. It can of course repeat some elements of an earlier concept or text and recombine them in a new composition, a new truth of mathematics, without, again, assuming its Platonist origin in some preexistent, primordial domain. By the unambiguity principle, however, a mathematical or scientific text must have the same mathematical or scientific meaning, which disregards a different additional meaning it can have for an individual person. There can be no Pierre Menard, the author of the *Elements*, as a mathematical text. There are always personal components, some related to mathematics, to our experience of it, but they are bracketed in the mathematical functioning of the *Elements*. Conversely, there may be shared elements in our experience of literature and art, but they coexist, on equal footing, with the unique aspects of our experience. The unambiguity principle is not required for the functioning of literature and art, and instead might inhibit this functioning.

According to James: "Truth HAPPENS to an idea. It BECOMES true, is MADE true by events", [39, p. 73]. This sense of event applies to that of each composition, rather than only in a radical transformation or a creation of a new theory or way of thinking, a revolution, although such a revolution can also happen in literature and art, or mathematics and science. Shelley was a political revolutionary, too, as Galois was, and he thought of poetry in revolutionary and even (broadly) political terms of transforming the world. One can find possible suggestions of this view in Galois as concerns mathematics, but it is difficult to be certain, while Shelley is explicit, or at least more readily allows for this reading.

For Shelley as for Mallarmé, every poetic composition was such a transformative, revolutionary event. This is one of the meanings of Mallarmé's title, "*Crise de Vers*," a crisis enacted in each poetic composition, by "a throw of dice launched ... from the depth of a shipwreck," launched by a poetic thought, as his *Coup de Dés* famously says, adding and even more famous: "a throw of dice will never abolish a chance", [48, p. II]. I would not attempt to give this statement a definitive meaning, which may not even exit, as opposed to many possible meanings. One way to read it, however, is to see it as applied, not to the game of dice as such, but to a new poetic composition, such as *Coup de Dés* itself, the future fate of which was part of Mallarmé's meaning. Creating a new composition and "throwing" it into the world only begins the "game," that of betting on its future survival and success, rather than ends this game, as a throw of dice does in the game of chance. A new composition is a bet on its future, never certain at the time of its creation.

The same singularity and the probabilistic bet of the future define the birth, event, of a mathematical composition, in mathematics or physics and makes them akin to poetry or art. The concepts of Galois group and Riemann's manifold were among the greatest such bets ever, at the time, responding to the crisis of mathematics, albeit it more hidden than realized, except of course by Galois or Riemann. On the other hand, the idea of the different order of infinities by Cantor or that of the undecidable propositions by Gödel led to a crisis, a shipwreck of the foundations of mathematics. Heisenberg's discovery of quantum mechanics was "a throw of dice launched ... from the depth of a shipwreck" of quantum theory in crisis at the time, which rescued quantum theory. Quantum mechanics is of course itself a theory in which the role of randomness and probability is irreducible, and no other theory of quantum phenomena is possible on experimental grounds, as things stand now. Quantum physics (both quantum phenomena and quantum mechanics or quantum field theory) only contains randomness, however, rather than is entirely random, because it allows for probabilistic or statistical predictions (purely random events do not) and contains the so-called quantum correlations. Probability is not randomness, which brings an element of disorder in our interactions with the world. By contrast, probability restores a degree of order to this interaction in the interplay of randomness and order.

The life of a composition in literature and art, philosophy, or mathematics and science is not merely random either. It is partially ordered and hence probabilistic as concerns the author's or a culture's bet on it. This make this life *chaosmic*, in Joyce's coinage (itself a micro composition of chaos and cosmos), reflecting the interplay of randomness or order, or chance and necessity, [40, p. 118]. This interplay is Democritean in its ancient Greek genealogy. In the present view, however, the outcome of this interplay is no longer fully calculable in advance, as it would have been, by either human or divine means, for Democritus and for the subsequent history of philosophy, or physics, such as classical statistical physics, before quantum theory entered the stage, eventually leading to RWR interpretations of it. The latter disallow one to see the recourse to probability as resulting from the insufficient but, at least in principle, possible knowledge of how the phenomena observed come about, as in classical physics, such as in classical statistical mechanics or in just the case of a coin toss. While impossible in practice, in principle, if one has suitable measurement devices, tracking all movements involved, and computers, one could calculate exactly how the coin will fall in a toss. In quantum physics, by contrast, the recourse to probability is the result of the impossibility, in principle, of knowing or even forming a conception of how what is observed comes about. According to Bohr:

> [I]t is most important to realize that the recourse to probability laws under such circumstances is essentially different in aim from the familiar application of statistical considerations as practical means of accounting for the properties of mechanical systems of great structural complexity. In fact, in quantum physics we are presented not with intricacies of this kind, but with the inability of the classical frame of concepts to comprise the peculiar feature of indivisibility, or "individuality," characterizing the elementary [quantum] processes. [8, v2, p. 34]

Bohr adds later in the same article: "[I]n quantum mechanics, we are not dealing with an arbitrary renunciation of a more detailed analysis of atomic phenomena but with a recognition that such an analysis is *in principle* excluded", [8, v2, p. 62]. This statement expressly defines the nature of quantum chaosmos in RWR interpretations. While it is not clear (and a matter of interpretation in turn) whether Joyce himself subscribed to the RWR view of his concept of "*chaosmos*," he expressly refers to quantum theory in the novel. "I am working out a quantum theory about it, because this is the most tantumising [a play on tantalizing] state of affairs" he says, [40, p. 149]. The sentence also describes the novel itself, which scholars have tried to "calculate" ever since. In this case, unlike (luckily for us) in quantum theory, there is no mathematics to do so. There are, however, other means to understand the chaosmic reality of Joyce's text.

In any event, one can never fully count on a success of our bets, as Joyce was aware, in general and as concerns *Finnegans Wake* itself, or any of his works. It is possible that very great mathematical or scientific ideas, or poetry, even greater than anything we know (by whatever criteria) have been discarded or lost and may never be recovered even partially. Thus, Abel's 1923 paper "A general representation of the possibility to integrate all differential formulas" was lost, while being reviewed. We are lucky to have most of Abel's work, containing his major ideas. We may not always be so lucky, because some important ideas have never become part of the record, archive, in which they could in principle be found one day. But we can only build on what this archive contains.

While, however, the role of composition in art, philosophy, and mathematics and science, is fundamental, the differences between their compositional thinking and practice remain essential. Thus, a work of art, as a composition, may but need not contain concepts in the present sense, although so-called conceptual art does. Arguably most significant is, however, the defining role of the unambiguity principle in mathematics, science, and philosophy, but not in literature and art. By the same token, a concept or a theory, while created as a new composition in mathematics and science, can function directly or with minor modifications in, in Kuhn's terms, the normal practice of mathematics, science, and philosophy, transformed by this new concept or theory. The degree of such a transformation may vary, thus, as stated from the outset, creating a complex interplay of continuity and discontinuity with the preceding configuration of thought. Philosophy, at least some philosophy, may be associated with or even defined as (as it is in Deleuze and Guattari), a revolutionary practice of the invention of concepts, which becomes its normal practice. This practice is closer to that of art, which, while it may use a tradition, is compelled to be innovative in each work. The nature of philosophical and artistic compositions remains different, by virtue of both the role of concepts and especially the unambiguity principle in philosophy.

As explained earlier, literary compositions also retain closer connections to language, vis-á-vis compositions in mathematics or mathematical sciences. Because of this connection, a general-language functioning can be returned to a literary composition by a reading, but at the cost of losing the compositional innovation of the work. A similar loss is sometimes found in philosophy, when a philosophical concept, expressed in language, is given a common-sensical meaning and thus misunderstood. This often happens to Bohr's "complementarity," as both a physical and philosophical concept, about which Bohr said: "In the last resort an artificial word like 'complementarity' ... does not belong to our daily concepts serves only briefly to remind us of the epistemological situation ... encountered [in quantum physics]", [6, p. 87]. Complementarity does more than this, but in considering it, one must take into account this epistemological situation, the RWR situation, which is not something we encounter in daily life. Otherwise, the concept loses its proper meaning.

Mathematics allows us to break with, or free ourselves from, general, or even philosophical and physical, concepts, which can inhibit or even preclude dealing with the reality we must confront in mathematics and physics. In quantum physics, mathematics allows us to dispense with classical physical concepts, in dealing with the reality ultimately responsible for quantum phenomena. As Heisenberg said, shortly after the introduction of quantum mechanics, at the time assuming an RWR interpretation, rather than the mathematical realism, with a Platonist bent, of his later view:

> It is not surprising that our language [or conceptuality] should be incapable of describing processes occurring within atoms, for . . . it was invented to describe the experiences of daily life, and these consist only of processes involving exceedingly large numbers of atoms. Furthermore, it is very difficult to modify our language so that it will be able to describe these atomic processes, for words can only describe things of which we can form mental pictures, and this ability, too, is a result of daily experience. Fortunately, mathematics is not subject to this limitation, and it has been possible to invent a mathematical scheme—the quantum theory [QM]—which seems entirely adequate for the treatment of atomic processes. [32, p. 11]

Words can do more, and they can describe things of which we cannot form mental pictures, including mathematical symbols. This qualification does not, however, undermine Heisenberg's main point, amounting to the fact that our language and thought are the product of our evolutionary biological and neurological development defined by the experiences consisting only of processes dealing with huge numbers of atoms. Accordingly, there is no special reason to assume that they will be able to describe nature on the atomic or still smaller scales, or conversely, on very large, say, cosmological scales. One might instead doubt that one can do so even mathematically, as Heisenberg eventually assumed, given that mathematics, too, is only human and as such is the product of the same evolutionary development. We are fortunate to be capable of using it to predict quantum phenomena.

Mathematics, however, especially abstract mathematics, such as that used in quantum mechanics and quantum field theory, does have a great, even if not complete, degree of freedom from the limitations of daily language and thinking. This freedom also ensures the application of the unambiguity principle within much broader limits, as Bohr noted in his comment, cited earlier, to the effect that "by avoiding the reference to conscious subject which infiltrates daily language, the use of mathematical symbols secures the unambiguity of definition required for objective description", [8, v2, p. 68].

In the RWR/IWI view, in physics or in mathematics itself, mathematics enables us to relate to things in nature and mind which are beyond the reach of thinking, [66, pp. 83–95]. In quantum theory it does so by enabling us to estimate probabilities or statistics of quantum events, to which Heisenberg refers above by speaking of his scheme, quantum mechanics, as "entirely adequate for the treatment of atomic processes." As noted earlier, not everyone, beginning, again, with Einstein, would see this, merely probabilistically predictive, treatment, let alone the RWR view of a fundamental physical theory, as "entirely adequate." In any event, Heisenberg, while modestly avoiding mentioning himself as the inventor of the scheme, is right to say that it was fortunate for us to be able to invent, compose, the mathematical scheme of quantum mechanics. In fact, it is equally fortunate that nature allows us to use any "mathematical scheme" in physics, because the freedom of mathematics from the limitations of common language and concepts does not guarantee that any such scheme will work. Indeed, that of quantum theory might not work beyond the scales now handled by quantum field theory, just as the scheme of classical physics no longer worked in quantum theory. In fact, while most hope otherwise, it is not clear that any mathematical scheme will work beyond these scales.

It is not too far off the mark, however, to say that Heisenberg's quantum mechanics was a kind of abstract composition of concepts, divorced from representing physical reality, akin to those of Kandinsky or Piet Mondrian. Such a painting is abstract vis-à-vis a representational painting (putting aside for the moment the complexities involved in this opposition, which is not unconditional), but is concrete in its composition of lines, shapes, colors, and so forth, and it is still related historically to other paintings, abstract or representational. In commenting on continuous manifoldness, Riemann observed: "opportunities for creating concepts whose instances form a continuous manifold occur so seldom in everyday life that color and the position of sensible objects are perhaps the only simple concepts whose instances form a multiply extended manifold. More frequent opportunities for creating and developing these concepts first occur in higher mathematics", [72, p. 24]. Abstract art appeared half a century after Riemann's death. His first sentence, however, could have inspired Kandinsky, whose abstract art in part emerged through converting color and objects of our sensible perception into abstract compositions. There is no evidence that he knew of Riemann's lecture, although he might have through both his Russian and German cultural connections. He certainly knew of non-Euclidean geometry, and relativity and its geometry, including the idea of four-dimensional space, much discussed during the time when he was developing his abstract art, during the first decade of the twentieth century.

There were, of course, other, well-known, trajectories that had led to his abstract paintings, for example, those extending from music, the main source of his use of the term "composition." I am not making any strong claim concerning such historical influences. My point is a comparison between the nature of abstraction in mathematics, including that used in physics, specifically, in quantum theory, and abstract painting, a comparison that is hardly controversial or especially new. What is new is an emphasis of the compositional nature of these abstractions in both domains. Geometry, however, from its most basic aspects to some sophisticate contemporary ideas, was part of the genealogy of abstract art.

Did the latter, conversely, have influence on modern mathematics? It would be difficult to think that it did not in some cases. I am aware of no evidence of any influence of modern art on Heisenberg's thinking, unlike, it appears, that of Bohr. Eventually, but well after quantum mechanics was in place, Bohr bought a cubist painting by Jean Metzinger ("A woman on a horse"), admittedly a cubist and hence not technically abstract painting, but still not a realist one either. I am not definitively claiming that abstract or other modernist art had decisively impacted Bohr's thinking, or that of Heisenberg or others in physics and mathematics, such as Emmy Noether, [66, pp. 194–195]. Bohr's receptiveness to modernist art may have been more a reflection of his way of thinking about the nature of reality and its representation, or as the case may be, the impossibility thereof, rather than of the influence of modernist art in shaping his thinking, originating in considering the nature of reality in quantum physics.

The invention of quantum mechanics and abstract art required a tremendous imagination, as would any invention in any field that would transform it so radically. While, however, along with the composition principle, the continuity principle applies in both domains, regardless of how radical a break with the preceding tradition may be, the unambiguity principle brings special aspects and some difficulties to the use of imagination in mathematics and science. It is fitting to consider in this connection Richard Feynman's comment on scientific imagination, equally applicable to mathematical imagination (essential in modern physics as well), in relation to both, expressly to the continuity principle and, implicitly but unmistakably, to the unambiguity principle:

> Our science makes terrific demands on imagination. The degree of imagination that is much more extreme than required for some of the ancient ideas. The modern ideas are much harder to imagine. We use a lot of tools, though. We use mathematical equations and rules, and make a lot of pictures. [25, Book 2, 20-10-1]

While such pictures are primarily heuristic, as Feynman confirms in elaborating on this point, mathematics provides abstract tools, including those that allow us to deal with what we cannot picture. This becomes essential in quantum mechanics, especially in RWR interpretations, and in much of modern mathematics itself, many, even most of the objects of which, such as infinite-dimensional Hilbert spaces, used in quantum mechanics, are beyond out capacity to visualize. They can only be formally represented mathematically, thus, in a certain sense, only algebraically. Feynman continues his reflections as follows, in effect bringing both the continuity and unambiguity principles to bear on the situation:

> The whole question of imagination in science is often misunderstood by people in other disciplines. They trust to test imagination in the following way. They say: "Here is a picture of some people in a situation. What do you imagine will happen next?" When we say, "I can't imagine," they may think we have a weak imagination. They overlook the fact that whatever we are *allowed* to imagine in science must be consistent with everything else we know: that the electric field and waves we talk about are not just some happy thoughts which we are free to make as we wish, but ideas that must be *consistent with all the laws of physics we know*. We can't allow ourselves to seriously imagine things which are obviously in contradiction to the known laws of nature. One has to have imagination to think that has never been seen before, never been heard before. At the same time the thoughts are restricted in a strait jacket, so to speak, limited by the conditions that come from our knowledge of the way nature really is. The problem of creating something which is new, but which is consistent with everything which has been seen before, is one of extreme difficulty. [25, Book 2, 20-10-11]

From the present (RWR/IWI) perspective, because we don't know or possibly cannot imagine the way nature really is, we only deal with what we have "seen before" in our experiments and theories, thus in accord with the continuity and unambiguity principles. Both are implied by Feynman in connection with the mathematical nature of our theories, beginning with the mathematics of classical electromagnetism, where the concept of the field already poses difficulties for general language and thinking. As discussed here, these difficulties become just about insurmountable in quantum physics. By virtue of, as Heisenberg said, being "free from these limitations," mathematics can come to the rescue on both counts: that of maintaining the unambiguous nature of mathematical and scientific theories at any given time and that of the continuity of physics as a mathematical experimental science. Both principles reflect the necessity for our imagination and the compositions it creates to remain consistent "with everything which has been known before" in order for mathematics and science, often with the help of mathematics, to function as mathematics and science. Otherwise, artistic imagination is not so different, and it is no less difficult, even though, in the absence of the unambiguity principle, one does not need to be strictly consistent with the knowledge of what (in art) has been seen before. One still relates, however, to one or another artistic tradition, and in this sense to something that has been known.

Intriguingly, Feynman next shifts to an aesthetic question, that of beauty, but a "beauty that we cannot see:" "While I am on this subject [of imagination in science], I want to talk about whether it will ever be possible to imagine beauty that we can't see. It is an interesting subject", [25, 20-11]. As emphasized from the outset, my main concern in this article is compositional creativity, "poetics," rather than "aesthetics." I would like, however, to reflect on Feynman's comments, because the aesthetics in question depends on composition, through which and only through which our mind "perceives" the beautiful. The question is also that of the meaning of the verb "see," by which Feynman, initially, refers to our immediate sense perception, but only initially. How about seeing in our thought, in our mind's eye? This is, I would argue, ultimately Feynman's question, which pertains to mathematics as well, including geometry, which we primarily and in a certain sense always seen in our mind's eye, even when we look on a geometrical figure represented in one media or another. But then, everything we see we always see in our mind eye, in the vision constructed by our brain, even in our immediate sense perception, or what so appears. This perception is never immediate, rigorously speaking: it is always mediated by the neurological processes in the brain and, thus, by thinking.

Feynman brings up a rainbow, a phenomenon that is actually connected to electromagnetism, but is also a paradigmatic example of the beautiful in daily life. Feynman's comments, thus, connect both domains. He says: "When we look at a rainbow, it looks beautiful to us. Everybody, says, 'Ooh, a rainbow.'" He adds:

> (You see how scientific I am. I am afraid to say something is beautiful unless I have an experimental way of defining it.) But how would we describe a rainbow if we were blind? [And] we are blind when we measure the infrared reflection coefficient of sodium chloride, or when we talk about the frequency of waves that are coming from some galaxy we can't

> see—we make a diagram, we make a plot. For instance, for the rainbow, such a plot would be the intensity of radiation vs. wavelength measured with a spectrometer for each direction in the sky. [25, 20-11]

Feynman then sketches the graph of electromagnetic waves as a function of wavelength for three angles, and asks: "Now, do we find [this] graph beautiful? It contains much more detail than we apprehend when look at a rainbow, because our eyes cannot see the exact details in the shape of the spectrum. The eye, however, finds the rainbow beautiful. Do we have enough imagination to see in the spectrum curves [in the graph] the same beauty we see when we look directly at a rainbow? I don't know", [25, 20-11].

Actually, he did know, and he also had enough imagination to see this beauty in the spectrum curves, which is perhaps not quite the same but not fundamentally different. This is the problem of our phenomenal experience, as Kant realized in defining the beautiful or the sublime, sometimes used in describing a rainbow as well, on which I shall comment below, [42]. In the case of the spectrum curves, mathematics becomes part of our experience, as something we can see in our mind's eye or on paper, such as a diagram, a graph, a figure, or an equation, which a mathematical physicist, like Feynman, or a mathematician can *experience* as beautiful. Even in this case, however, we still see them in our mind's eye. The objects of such experience may not be as common as rainbows, but the nature of the experience itself is not so different, beginning with the fact that this experience is strictly individual. We may agree on the beauty of some objects, and in Kant's view even must do so to define them (in Kant, only as objects in nature) as universally and purely beautiful. (This view has further complexities and has been challenged, which is, however, secondary for the moment.) Nevertheless, as the common saying goes, beauty is always in the eye of the beholder, or as the case may be, in the blindness of the beholder, blindness toward the rest of the world. In Keats's words, defining what he called "negative capability" (one of his most famous ideas): "with a great poet the sense of Beauty overcomes every other consideration, or rather obliterates all consideration", [43, p. 193].

That could not quite be said about a great mathematician or scientist, although in this case, too, the sense of mathematical or scientific composition, guiding creativity, tends to dominate, even if not, obliterate all other considerations. There is, however, an interior experience of the beautiful in mathematics and science (as opposed to being blind to the frequencies defining color, noted by Feynman, which are not part of experience). As Feynman adds: "There is plenty of intellectual beauty associated with the equations"[25, 20-11-15]. He then proceeds to a discussion of wave-equations, undoubtedly seen by him as intellectually beautiful. One needs to know a great deal of mathematics to experience these equations as beautiful. "The laws of physics must have mathematical beauty," Dirac famously said, [22, p. 124].

Perhaps this is so, although it is not clear why the laws of physics must have this beauty, or even must be mathematical, although they are mathematical in modern physics. They can have this beauty and do for some of us, such as Dirac or Feynman, and many others, but not for all. Bohr never appeals to the idea of beauty in his arguments. And then, what does mathematical beauty mean here? The idea of elegant simplicity is often invoked, in general and in interpreting Dirac's statement, which appears consistent, given his other statements and assessments of physical theories with what Dirac had in mind, although Dirac himself has never explained it.[21] On the other hand, he commented on the lack of mathematical simplicity or elegance in the renormalization of QED, a theory he created. Consider Dirac's equation, which he undoubtedly saw as beautiful. In the elegant simplicity of its compact symbolic form, reproduced on the plate in Westminster Abbey commemorating Dirac, it is:

$$i\gamma \cdot \delta\psi = m\psi.$$

This elegant simplicity is similar to that of Einstein's field equations in general relativity.

$$G_{\mu\nu} + \Lambda_{\mu\nu} = \kappa T_{\mu\nu}$$

($G_{\mu\nu}$ is the Einstein tensor; $G_{\mu\nu} = R_{\mu\nu} - \frac{1}{2}Rg_{\mu\nu}R_{\mu\nu}$is the Ricci tensor, a symmetric second-degree tensor that depends only the metric tensor and its first and second derivatives; R is the scalar curvature; $T_{\mu\nu}$ is the stress energy tensor; Λ is the cosmological constant, and $\kappa = \frac{8\pi G}{c^4} \approx 2.077 \times 10^{-43} N^{-1}$ is the Einstein gravitation constant, and G is the Newtonian gravitational constant.)

This elegant simplicity is, however, equally deceptive in both case. As Feynman himself noted, in commenting on "Dirac's matrices," defining his equation, but making a more general point: "Why should the fundamental laws of Nature be so that one cannot explain them to a high-school student—but only to a quite advanced graduate student? And we claim they are simple? In what sense are they simple? Because we can write them in one line. But it takes 8 years of college education to understand the symbols. Are there any simple ideas in these laws?" [A letter to T. A. Welton, February 10, 1947; cited in [74, p. 408].

Einstein's equation unfolds into a system of ten nonlinear, hyperbolic-elliptic partial differential equations. This is a far cry from anything based in our daily concepts. Infinitesimally, the spaces considered are metrically Euclidean in accordance with Riemann's concept of manifold, mathematically representing physical space. SR applies in a local or infinitesimal neighborhood of each point, which gives the corresponding local spacetime the Minkowskian metric. Physically, this metric corresponds to the fact that all reference frames that are in free fall are equivalent. The concept of spacetime was introduced by Hermann Minkowski in 1908. It was

[21] Elegance or simplicity is not the same as beauty according to Kant. Kant does not speak of mathematical beauty, arguably because it is difficult to avoid concepts in dealing with mathematical objects, while Kant defines the experience of the beautiful or the subline as strictly associated with feelings in the absence of concepts, [42]. Kant does have a concept of the mathematical sublime, but this concept has to do with the idea of the numerically infinite escaping our capacity to grasp it.

not part of special relativity as originally developed by Einstein, and was initially accepted by Einstein with some reluctance, but it proved useful in developing GR. While uncontroversial as a mathematical tool, as a physical concept, spacetime involves complexities and its status as such has been debated, unlike, or much more than, the status of Riemannian spaces of GR as representational concepts. Their representational status has been questioned as well.

Similarly, the "one line" of Dirac's equation encodes a complex mathematical machinery. As introduced by Dirac, the equation was:

$$\left(\beta mc^2 + \sum_{k=1}^{3} \alpha_k p_k c\right) \psi(x,t) = i\hbar \frac{\partial \psi(x,t)}{\partial t}$$

$$\alpha_i^2 = \beta^2 = I_4 (I_4 \text{ is the identity matrix})$$

$$\alpha_i \beta + \beta \alpha_i = 0$$

$$\alpha_i \alpha_j + \alpha_j \alpha_i = 0$$

This equation unfolds into four coupled linear first-order partial differential equations for the four quantities that make up the wave function. P is the momentum operator. The wave function $\psi(t, x)$ takes value in a Hilbert space $X = C^4$ (Dirac's spinors are elements of X). For each t, (t, x) is an element of $H = L^2(R^3) \otimes C^4$. This architecture allows one to predict the probabilities of high-energy (QED) events associated with the relativistic electron, just as Schrödinger's equation does for the nonrelativistic electron in QM. An analogous, if more complex, situation is in the Yang–Mills equations of quantum field theory. In the latter case, rather than particles, one begins with quantum fields, a complex and still not entirely settled concept, including as concerns its relation to the concept of a particle.

In all these cases we deal, as in poetry, with the creation of new mathematical compositions, under the conditions of the unambiguity (not necessary in poetry) principle. The sense of beauty is, on the other hand, a matter of individual experience of these compositions, related most essentially to how we feel about them. Hence, as I said, it is doubtful that, unlike in poetry according to Keats, in creating new mathematical compositions "the sense of Beauty overcomes every other consideration, or rather obliterates all considerations." This may, however, happen sometimes, and the search for simplicity is often a good guidance for developing a new concept or a theory.

The concept of "intellectual beauty" goes back to Plotinus, who expressly used it, or Plato, where it has mathematical overtones, even if, as discussed earlier, not strictly mathematical. It was, with this genealogy in mind, memorably described by Shelley in his "Hymn to Intellectual Beauty." The poem associates intellectual beauty with "the awful shadow of some unseen Power," in line 1, repeating the word "unseen" in line 2, and ultimate "a dark reality" (l. 2): "It visits with inconstant glance/Each human heart and countenance" (ll. 3-4). As that of Plato's and Plotinus's philosophy, that intellectual beauty of Shelley's poem is beyond the world perceived by our senses or more immediately conceived by our thinking. Intellectual Beauty only casts a shadow over this world and leaves a trace in it. Our mind, however, our mind's eye, must still be able to see Intellectual Beauty itself, to conceive of it, from its trace, as real, or assume it to be real, beyond this shadow, thus, again, echoing Plato, but only echoing. As I said, there is also a skeptical, Humean, side to Shelley's thinking. This side would complicate this ontology of Intellectual Beauty, to the point of questioning this concept itself, as something that can be glimpsed from its trace. It is even possible to read Plotinus, against his own grain, as making the trace [*ichnos*] of intellectual beauty a trace of something formless, if not entirely unthinkable ("*To gar ichnos tou amorphou morphē*", [59, *Enneades* 6.7.33.30] rather than a form.[22] This subject would, however, require a separate analysis. In any event, the (material) RWR-type reality or (mental) IWI-type reality cannot be beautiful because it cannot be "seen" even in our mind's eye or traced otherwise. It is beyond thought.

Feynman might not have been aware of this history of the idea and used the expression "intellectual beauty" in its common sense of something perceived by one's mind as beautiful in its appeal, in this case as something mathematical. The Intellectual Beauty of Shelley's poem, or in Plato and Plotinus, is not mathematical. The phrase "intellectual beauty," however, echoes this idea, as something at a distance that can be reached or is unreachable by thought. The nature of the reality ultimately responsible for quantum phenomena is not at stake in Feynman's concept, as Feynman only refers to mathematics, in fact that of classical electromagnetism. The mathematics of quantum theory, including that of Feynman's path integrals, must have been on his mind, however. Feynman might not have been aware either (although it is possible that he was) of Keats's appeal to the rainbow in his poem *Lamia*, as something beautiful or rather sublime, "awful," in Kant's sense of "absolutely great," and thus, while visible, escaping being captured or framed by imagination, [42]. I have discussed this passage on an earlier occasion, [60], but would like, in closing, to revisit it, revising this earlier discussion from the perspective of this chapter.

12.6 Conclusion: Of Mysteries, Dreams, and Divine Madness

Keats's *Lamia* contains a famous passage about a rainbow, the beautiful (or sublime) experience of which, while it can be captured by poetry, is threatened by "cold philosophy." This makes the poem one of the forerunners of the "two-cultures" confrontation, which has overshadowed our culture since the 1950s, when C. P. Snow introduced the concept. The concept, much debated since as well, is too

[22] This aspect of Plotinus's thinking has been considered by Derrida (who cites this sentence), in yet another analysis of Husserl's phenomenology, [17, p. 157].

reductive, given that these two cultures, science, including mathematics (seen as significantly responsible for the split), and the humanities, are interactive and that each of them is already comprised of several cultures. Nevertheless, if suitably qualified, the distinction is not out of place, and in some situations is significant. One might, accordingly, see science and the humanities, each already multiple, as interactively heterogeneous, creating multiple relationships between themselves as they develop. Keats says:

> . . . Do not all charms fly
> At the mere touch of cold philosophy?
> There was an awful rainbow once in heaven:
> We know her woof, her texture; she is given
> In the dull catalogue of common things.
> Philosophy will clip an Angel's wings,
> Conquer all mysteries by rule and line,
> Empty the haunted air, and gnomed mine—
> Unweave a rainbow . . .
> [*Lamia*, Part II, ll.229–238]

Given that Keats has both Descartes and Newton in mind, philosophy here is also natural philosophy or what we now call physics. Keats may be right about fleeing charms, at least some of them. Kant, incidentally, distinguished the beautiful from merely charming, [42]. Would the "beautiful," which, in creating poetry "obliterates all [other] considerations" for Keats, would also fly "at the mere touch of cold philosophy"? Probably, yes, for Keats. The remainder of the passage, however, has proven to be among the least prophetic of Keats's insights. Or rather, he was only partly right: philosophy or mathematics and science can do so and have done so, but they do not have to do so. Keats, were he alive, might have welcomed this more poetic capacity of mathematics and science enabling them to create things akin to a rainbow in their beauty. I am less certain if he would have wanted any "cold philosophy" ever to arrive, however slowly, where poetry gets so quickly and with so much beauty on its wings, "the viewless wings of Poesy," invoked by Keats in "Ode to a Nightingale" (l. 33). Curiously, these wings are "viewless," the wings of music and language, refined by the compositions of poetry. Philosophy, or mathematics and science, could, however, do this as well. They have their wings.

The "cold philosophy" of quantum theory, our best theory of light now, tells us that the ultimate unweaving of the rainbow may not be possible. The reality that, in interacting with the reality of our bodies and minds, and thus through them, creates a rainbow un-unweavable: there is no unweaving that can reach it. Hence, it cannot be catalogued in terms of common or any other things either. It cannot be reduced to the known, elemental constituents, for example, of the kind that Democritean atomism or, conversely, classical wave optics would envision. If one adopts an RWR interpretation of quantum QED (necessary for dealing with photons), the ultimate nature of light cannot be envisioned: while we would not be able to see anything without it, light itself cannot be *seen*, or even thought of in its ultimate nature, either as particle or as waves. What we call photons is neither in RWR interpretations. The mystery of light cannot be conquered by rule or line, the great instruments of "cold philosophy," converted into new algebra and geometry of quantum theory. Light, in its ultimate nature, can only be handled by the mathematical compositions of quantum theory. Quantum theory, however, makes this mystery a mystery without mysticism. A mysticism would assume some divine or divine-like agency responsible for the reality considered, even if this agency cannot be assigned any human attributes, as in the so-called mystical or negative theory. Even when not so named but still referring to something responsible for what we encounter in the world, "agency" is of course still a human attribute.

It is, however, the responsibility of all "cold philosophy," not the least when it becomes mathematics and science, to confront the unthinkable and make it part of their thinking if it becomes necessary given the nature of reality considered. Doing so need not prevent and might instead enhance the advancement of thought and knowledge, allowing us to use our "rules" and "lines," and to invent new ones. Nevertheless, whatever our mathematics might create in this confrontation, or in its confrontations with reality in mathematics itself, might still only be a dream, even if at its best drive by the imperative of advancing thought, where even a small step, even half a step counts. In Osip Mandelstam's remarkable lines in his poem on the death of Andrei Bely, one of the founding figures of Russian modernism, best known in the West for his pioneering modernist novel *Peterburg* (1913) [4]: "A dream within a dream, inside which one is dreaming to move ahead at least by a half a step" ["10 January 1934," ll. 26–28]. By have a step at the time, Bely nevertheless momentously advanced Russian and European literature, just as did Mandelstam, one of the greatest Russian poets ever.

In *Lamia*, also a poem about dreams, Keats says: "Real are the dreams of Gods" (Lamia, Part I, l. 126). But, which was also Keats's point in the poem, we are not gods and our dreams—our artistic, philosophical, and mathematical or scientific compositions, or our ordinary dreams, which are compositions, too, sometimes quite extraordinary—must deal with things that exist, are real, but are beyond what we can see, or, as this chapter argued, possibly beyond what we can in principle think. For gods, there would not be such a distinction, as they would be able to see and, thus, to think and know everything that exists. Dreams, however, are messengers of our thoughts, are our thought, and desires. Here is a problem of and for digital AI: if computers cannot dream, they cannot think, and as far as we know they do not dream. Will they ever dream?

For us, the reality of the unthinkable, as a reality without realism, or beyond all realism, may define the ultimate constitution of nature. Because our thought, the creation of our brains, is the evolutionary product of the ultimate constitution of nature, this constitution, here conceived as an RWR, is also ultimately responsible for our thought. That includes the unthinkable, IWI, within thought, possibly revealed by twentieth-century mathematics or philosophy. A reality without realism, RWR, in nature is responsible for an ideality without idealism, IWI, in thought. It is, however, thought and only thought that defines any concept of reality, even when it places this reality, material or mental, beyond its own reach. It is also thought that, in its interaction with the world, natural and human, defines the fields of inquiry in which such a concept of reality is created, because it, in the first place, it defines all human endeavors, beginning with our daily life.

Our daily life is an endeavor, too, and as such is the invention and use of compositions and a freedom (and sometime pressure) to do so. Elevating our compositions to heights that may sometimes seem almost divine, art, philosophy, and mathematics and science are, nevertheless, human endeavors, born from and continuously shaped by our daily life, which makes everything that we do human.

The origin of geometry, which origin is understood here as its emergence through many interacting paths by the work of many minds, would not have been possible otherwise either, any more than the emergence of any mathematics. When discovered, however, geometry might well have appeared as something far above anything in daily life, at a divine height and was perceived by many as something divine, just as poetry was, compelling Plato to famously speak of poetry as a "*divine madness*," $\theta\epsilon\iota\alpha\ \mu\alpha\nu\iota\alpha$, [57, 58]. He would not see geometry as madness or even as divine. But if the greatest poetry is divine madness, the best geometry must be, too. This divine madness, however, only means a remarkable, difficult to believe to be possible, but perfectly human nature of their best compositions.

Plato, a poet and a mathematician himself (unlike his teacher Socrates, who was neither), perhaps knew or sensed this vertiginous yet human greatness of both, mathematics and art. He was, however, reluctant to accept this greatness, fearing to surrender to either poetry or mathematics the logos, the logical logos, of philosophy, for which he also reserved a capacity to reach the divine. But then—and Plato might have sensed this, too—the best philosophy is only possible by the divine madness of its compositions, and not logic, logical as these compositions are. Poetry, mathematics, and philosophy may not be divine, but, thanks to the work of many more minds than those, already numerous, that made their emergence possible, all three have retained their power to elevate their compositions to divine heights, and will, hopefully, keep this capacity alive as long as they exist.

Acknowledgments I am grateful to the editors of this volume, celebrating Athanase Papadopoulos, for inviting me to contribute. It a great honor to do so. I am grateful to Athanase for invaluable exchanges on the subjects considered in this chapter.

References

1. Aristotle, *Poetics*. Trans. A. Kenny (Oxford University Press, Oxford, 2013)
2. A. Badiou, *Briefings on Existence*. Trans. N. Madarasz (SUNY Press, Albany, 2006)
3. A. Badiou, *Logics of Worlds*. Trans. A. Toscano (Continuum, London, 2009)
4. A. Bely, *Peterburg*. Trans. J. Elsworth (Pushkin Press, London, 2013)
5. N. Bohr, Can quantum-mechanical description of physical reality be considered complete? Phys. Rev. **48**, 696–702 (1935)
6. N. Bohr, Causality and complementarity, in *The Philosophical Writings of Niels Bohr, Volume 4: Causality and Complementarity, Supplementary Papers*, ed. by J. Faye, H.J. Folse (Ox Bow Press, Woodbridge, 1994 (1937)), pp. 83–91

7. N. Bohr, Mathematics and natural philosophy, in *The Philosophical Writings of Niels Bohr, Volume 4: Causality and Complementarity, Supplementary Papers*, ed. by J. Faye, H.J. Folse (Ox Bow Press, Woodbridge, 1994 (1956)), pp. 164–169
8. N. Bohr, *The Philosophical Writings of Niels Bohr*, 3 vols. (Ox Bow Press, Woodbridge, 1987)
9. J.L. Borges, *Labyrinths* (New Directions, New York, 1964)
10. A. Connes, *Noncommutative Geometry*. Trans. S. K. Berberian (Academic Press, San Diego, 1994)
11. D. Corfield, *Towards a Philosophy of Real Mathematics* (Cambridge University Press, Cambridge, 2006)
12. D. Corfield, *Modal Homotopy Type Theory: The Prospect for a New Logic of Philosophy* (Oxford University Press, Oxford, 2020)
13. J.F. Dauben, *Georg Cantor: His Mathematics and Philosophy of the Infinite* (Princeton University Press, Princeton, 1990)
14. P. De la Harpe, *Topics in Geometric Group Theory* (University of Chicago Press, Chicago, 2000)
15. G. Deleuze, *Difference and Repetition*. Trans. P. Patton (Columbia University Press, New York, 1995)
16. G. Deleuze, F. Guattari, *What Is Philosophy?*. Trans. H. Tomlinson, G. Burchell (Columbia University Press, New York, 1996)
17. J. Derrida, *Margins of Philosophy*. Trans. A. Bass (University of Chicago Press, Chicago, 1975)
18. J. Derrida, *Edmund Husserl's "Origin of Geometry": An Introduction*. Trans. J. Leavy (University of Nebraska Press, Lincoln, 1989)
19. J. Derrida, *The Problem of Genesis in Husserl's Philosophy*. Trans. M. Hobson (University of Chicago Press, Chicago, 2003)
20. J. Derrida, *Speech and Phenomena: And Other Essays on Husserl's Theory of Signs*. Trans. D. Allison (Northwestern University Press, Evanston, 2023)
21. R. Descartes, *The Geometry of René Descartes*. Trans. D. E. Smith, M. L. Latham (Dover, Mineola, 1954)
22. P.A.M. Dirac, The relation between mathematics and physics. Proc. R. Soc. (Edinburgh) **59**, 122–129 (1939)
23. A. Doxiadis, B. Mazur (eds.), *Circles Disturbed: The Interplay of Mathematics and Narrative* (Princeton University Press, Princeton, 2012)
24. Euclid, *Thirteen Books of the Elements*, 3 vols. T.L. Heath (ed.) (Dover, Mineola, 1989)
25. R. Feynman, R.B. Leighton, M. Sands, *The Feynman Lectures in Physics*, 3 vols. (Basic Books, New York, 2011)
26. G. Galilei, Assayer, in *The Controversy on the Comets of 1618*, ed. by S. Drake, C.D. O'Malley (University of Pennsylvania Press, Philadelphia, 1966), pp. 131–355
27. K. Gödel, An example of a new type of cosmological solutions of Einstein's field equations of gravitation. Rev. Mod. Phys. **21**, 447 (1949)
28. A. Grothendieck, *Récoltes et Semailles, I, II. Réflexions et Témoignage sur un Passé de Mathématicien* (Gallimard, Paris, 2022)
29. J. Hadamard, *Cinq Lettres sur la Théorie des Ensembles*. Bull. S. M. F. t. 33, 261–273 (1905)
30. W.R. Hamilton, Theory of conjugate functions, or algebraic couples, with a preliminary and elementary essay on algebra as the science of pure time, in *Mathematical Papers of W. R. Hamilton, Vol. III. Algebra, Sect.I.*, ed. by H. Halberstram, R.E. Ingram (Cambridge Univ Press, Cambridge, 1967 (1835)), pp. 3–96
31. M. Heidegger, *What is a Thing?*. Trans. V. Deutsch (Gateway, South Bend, 1967)
32. W. Heisenberg, *The Physical Principles of the Quantum Theory*. Trans. K. Eckhart, F. C. Hoyt (Dover, Mineola, rpt. 1949 (1930))
33. W. Heisenberg, *Physics and Philosophy: The Revolution in Modern Science* (Harper & Row, New York, 1962)
34. E. Husserl, Philosophy and the crisis of European humanity, in *The Crisis of European Science and Transcendental Phenomenology: Introduction to Phenomenological Philosophy*. Trans. D. Carr (Northwestern University Press, Evanston, 1970 (1935)), pp. 269–300

35. E. Husserl, *Ideen zu einer reinen Phänomenologie und phänomenologischen Philosophie, Vol. III of Husserliana* (M. Nijhoff, The Hague, 1952)
36. E. Husserl, *The Crisis of European Science and Transcendental Phenomenology: Introduction to Phenomenological Philosophy*. Trans. D. Carr (Northwestern University Press, Evanston, 1970)
37. E. Husserl, The origin of geometry, in *Edmund Husserl's "Origin of Geometry": An Introduction*, ed. by J. Derrida. Trans. J. Leavy (University of Nebraska Press, Lincoln, 1989), pp. 155–180
38. E. Husserl, *Philosophy of Arithmetic: Psychological and Logical Investigations with Supplementary Texts from 1887–1901*. Trans. D. Willard (Springer, Heidelberg, 2003)
39. W. James, *The Meaning of Truth* (Harvard University Press, Cambridge, 1978)
40. J. Joyce, *Finnegans Wake* (Oxford University Press, Oxford, 2012)
41. I. Kant, *Critique of Pure Reason*. Trans. P. Guyer, A. W. Wood (Cambridge University Press, Cambridge, 1997)
42. I. Kant, *Critique of Judgement*. Trans. N. Walker, J. C. Meredith (Oxford University Press, Oxford, 2009)
43. J. Keats, *The Letters of John Keats, Vol. 1*, H.E. Rollins (ed.) (Harvard University Press, Cambridge, 1958)
44. T. Kuhn, *Black Body Theory and the Quantum Discontinuity, 1894–2012* (University of Chicago Press, Chicago, 1987)
45. T. Kuhn, *The Structure of Scientific Revolutions* (University of Chicago Press, Chicago, 2012)
46. I. Lakatos, *Mathematics, Science and Epistemology: Philosophical Papers, Volume 2*, ed. by J. Worral, G. Currie (Cambridge University Press, Cambridge, 1980)
47. S. Mallarmé, Crisis in poetry, in *The Norton Anthology of Theory and Criticism*, ed. by V. Leitch, et al. (W. W. Norton, New York, 2010), pp. 847–851
48. S. Mallarmé, Coup de Dés [A Throw of Dice], in *The Number and the Siren*, ed. by Q. Meillassoux. Trans. R. Mackay (Sequence Press, New York, 2012), pp. 225–271
49. Yu.I. Manin, Georg Cantor and his heritage. arXiv.math.AG/0209244 v1 (2002)
50. Yu.I. Manin, What then? Plato's Ghost: The Modernist Transformation of Mathematics. Not. Am. Math. Soc. **57**(2), 239–243 (2010)
51. Yu.I. Manin, Time and periodicity from Ptolemy to Schrödinger: paradigm shift vs. continuity in history of mathematics, in *Geometry in History*, ed. by S.G. Dani, A. Papadopoulos (Springer Nature, Switzerland AG, 2019), pp. 129–138
52. S. Negrepontis, D. Kalisperi, The mystery of Plato's Receptacle in the Timaeus resolved, in *Handbook of the History and Philosophy of Mathematical Practice*, ed. by B. Sriraman (Springer Nature, Switzerland AG, 2021)
53. S.I. Newton, *The Principia: Mathematical Principles of Natural Philosophy*. Trans. I. B. Cohen, A. Whitman (University of California Press, Berkeley, 1999)
54. A. Papadopoulos, Looking backward: from Euler to Riemann, in *From Riemann to Differential Geometry and Relativity*, ed. by L. Ji, A. Papadopoulos, S. Yamada (Springer Nature, Switzerland AG, 2017), pp. 1–94
55. A. Papadopoulos, Topology and biology: from Aristotle to Thom, in *Geometry in History*, ed. by S.G. Dani, A. Papadopoulos (Springer Nature, Switzerland AG, 2019), pp. 89–128
56. A. Papadopoulos, René Thom: from mathematics to philosophy, in *Handbook of the History and Philosophy of Mathematical Practice*, ed. by B. Sriraman (Springer Nature, Switzerland AG, 2024), pp. 385-429
57. Plato, *Phaedrus*. Trans. A. Nahemas, P. Woodruff (Hackett, London, 1995)
58. Plato, *Ion*. Trans. B. Jowett (Polit Press, New York, 2010)
59. Plotinus, *The Essential Plotinus*. Trans. E. O'Brian (Hackett, London, 1975)
60. A. Plotnitsky, *Epistemology and Probability: Bohr, Heisenberg, Schrödinger, and the Nature of Quantum-Theoretical Thinking* (Springer Nature, New York, 2009)
61. A. Plotnitsky, The spacetimes of Nympheas: matter and multiplicity in Einstein, Monet, and Deleuze and Guattari, in *Revisiting Normativity with Deleuze*, ed. by R. Braidotti, P. Pisters (Bloomsbury, London, 2012), pp. 37–50

62. A. Plotnitsky, The disaster of the diagonal, in *Circles Disturbed: The Interplay of Mathematics and Narrative*, ed. by A. Doxiadis, B. Mazur (Princeton University Press, Princeton, 2012), pp. 407–446
63. A. Plotnitsky, Structure, sign, and play and the discourse of the natural sciences: after the Hyppolite-Derrida exchange. Mod. Lang. Notes **134**(5), 953–966 (2019). https://doi.org/10.1353/mln.2019.0105
64. A. Plotnitsky, *Reality Without Realism: Matter, Thought, and Technology in Quantum Physics* (Springer Nature, Heidelberg, 2021)
65. A. Plotnitsky, Modernist literature and modernist mathematics I: mathematics and composition, with Mallarmé, Heisenberg, and Derrida, in *The Palgrave Handbook of Literature and Mathematics*, ed. by Tubbs et al. (2021). https://doi.org/10.1007/978-3-030-55478-1
66. A. Plotnitsky, *Logos and Alogon: Thinkable and the Unthinkable in Mathematics, from the Pythagoreans to the Moderns* (Springer Nature, Heidelberg, 2022)
67. A. Plotnitsky, A toss without a coin: information, discontinuity, and mathematics in quantum theory. Entropy **24**, 532 (2022). https://doi.org/10.3390/e24040532
68. A. Plotnitsky, *t* is not time: reality, causality, and the arrow of events in quantum theory. Eur. Phys. J. (2023). https://doi.org/10.1140/epjs/s11734-023-00979-2
69. A. Plotnitsky, 'That which a minority construct': Abelian mathematics in Abel, Galois, Noether, and Grothendieck, in *Handbook in the History and Philosophy of Mathematical Practice*, ed. by B. Sriraman (Springer Nature, Cham, 2024)
70. A. Plotnitsky, Mathematical physics, mathematics against physics, in *Topology and Geometry*, ed. by A. Papadopoulos (Springer Nature, Cham, forthcoming, 2024)
71. A. Plotnitsky, D. Reed, Discourse, mathematics, demonstration, and science in Galileo's *Discourses Concerning Two New Sciences*. Configurations, vol. 9.1 (2001), pp. 27–64
72. B. Riemann, On the hypotheses that lie at the foundations of geometry, in *Beyond Geometry: Classic Papers from Riemann to Einstein*, ed. by P. Pesic (Dover, Mineola, 2007 (1854)), pp. 23–40
73. E. Schrödinger, The present situation in quantum mechanics, in *Quantum Theory and Measurement*, ed. by J.A. Wheeler, W.H. Zurek (Princeton University Press, Princeton, 1983 (1935)), pp. 152–167
74. S.S. Schweber, *QED and the Men Who Made It: Dyson, Feynman, Schwinger, and Tomonaga* (Princeton University Press, Princeton, 1984)
75. P.B. Shelley, *Shelley's Poetry and Prose*, D.H. Reiman, H.B. Powers, (eds.) (W. W. Norton, New York, 1977)
76. R. Thom, Les Intuitions Topologiques Primordiales de l'Aristotélisme. Rev. Thomiste No.3 **88**, 393–409 (1988)
77. G. Tomlinson, *Million Years of Music: The Emergence of Human Modernity* (MIT Press, Cambridge, 2015)
78. J. Von Neumann, *Mathematical Foundations of Quantum Mechanics*. Trans. R.T. Beyer (Princeton University Press, Princeton, 2018)
79. V. Voevodsky, et al., *Homotopy Type Theory: Univalent Foundations of Mathematics* (Univalent Foundations Program, Princeton, rpt. 1983, 2013)
80. H. Wang, *A Logical Journey: From Gödel to Philosophy* (MIT Press, Cambridge, 2016)
81. A. Weil, *Foundations of Algebraic Geometry, 2nd edn.* (American Mathematical Society, Providence, 1962)
82. H. Weyl, *The Continuum: A Critical Examination of the Foundation of Analysis*. Trans. S. Pollard, T. Bole (Dover, Mineola, rpt. 1994 (1928))
83. H. Weyl, *Space-Time-Matter*. Trans. H. L. Brose (Dover, Mineola, 1952)
84. H. Weyl, *Philosophy of Mathematics and Natural Science* (Princeton University Press, Princeton, 2009)

Chapter 13
Pythagorean Book II of the *Elements* Restored and Pythagorean Incommensurabilities Reconstructed

Stelios Negrepontis and Vassiliki Farmaki

To Athanase in friendship and admiration:
ἄνθρωπος Θαλῆς *[the man is a Thales]*
Aristophanes, Birds. Line 1009

Abstract Unquestionably the greatest discovery of the Pythagoreans is the existence of incommensurable magnitudes, most probably the incommensurability of the diameter to the side of a square, but there is no agreement among historians of Greek mathematics on their method of proof. In this chapter, we present novel arguments not only for an anthyphairetic reconstruction of the original Pythagorean proof of incommensurability but also in favor of one that employs the Pythagorean Application of Areas in Excess and, in fact, Geometric Algebra. The main tool for this reconstruction is the restoration of Book II of the *Elements* to its original Pythagorean form. Thus, we place the Pythagorean theorem at the position II.4/5 (namely immediately after Proposition II.4 of the *Elements*), the Application of Areas in Defect at II. 5/6, in Excess at II.6/7, the Elegant Theorem(s) at (II.9/10 and) II.10/11.a, and the proof of the Pell property of the side and diameter numbers by mathematical induction at II.10/11.b. We then naturally expect that the original anthyphairetic Pythagorean proof of incommensurability is placed *before* Propositions II.9 & 10 on the side and diameter numbers and *after* the tools needed for its proof, the Application of Areas II.6/7 (and II.8), hence, *precisely* at the position II.8/9.

This reconstruction is strongly confirmed by

(a) the interpretation of the Pythagorean philosophical principles Infinite and Finite, in terms of the Pythagorean incommensurability proof, and

S. Negrepontis (✉) · V. Farmaki
Department of Mathematics, National and Kapodistrian University of Athens, Athens, Greece
e-mail: vfarmaki@math.uoa.gr

A. M. Uludağ, A. Zeytin (eds.), *Essays on Geometry*,
https://doi.org/10.1007/978-3-031-76257-4_13

(b) the Pythagorean derivation of the three kinds of angles from these principles, of the acute and obtuse angles from the Infinite, of the right angle from the Finite;

and in turn, the reconstruction confirms and explains

(a) the use of Geometric Algebra for its proof,
(b) the use of Mathematical Induction for the proof of Pell's property of the side and diameter numbers by means of Proposition II.10 of the *Elements*,
(c) the early dating of the Pythagorean incommensurability discovery, in connection with our study of Zeno's arguments and paradoxes, undermining the basic Burkert [16] tenet that Zenonian ontology is prior to mathematics,
(d) the presence of Proposition II.8 of the *Elements*, and
(e) the late, in fact, after Proposition II.10, appearance of the converse to the Pythagorean theorem, Propositions II.12 and 13.

13.1 Introduction

In the present chapter, we offer new arguments that help resolve some of the most enduring questions on Pythagorean Mathematics and Philosophy. The greatest mathematical discovery of the Pythagoreans, perhaps the greatest ever, is the discovery of incommensurability (Sect. 13.2). There are ancient proofs of incommensurability: an elementary arithmetical one, a later anonymous addition to the *Elements* as "Proposition X.117", a hint to this arithmetical proof by Aristotle in *Analytics Prior* 41a which probably refers to a more general proof by Archytas and, as it is not generally realized because it is couched in philosophical language, an anthyphairetic proof in Plato's *Meno* 80e-86e, 97a-98b [38] probably due to Theaetetus, the oldest one preserved. In addition, there are modern reconstructions; a modified arithmetical one by Knorr [32] and anthyphairetic ones by Chrystal [17], Tennenbaum, 1950, Fowler [21] that turn out to be consequences of Propositions II.9 or 10 and could well be ancient ones.

There are at least two strong but still *tentative* arguments according to which an *arithmetical* proof must be *rejected* as the original Pythagorean incommensurability proof in favor of an *anthyphairetic* proof, namely in favor of a proof that employs Proposition X.2 of the *Elements*, according to which an *infinite anthyphairesis* of two magnitudes is a criterion of their *incommensurability*.

One of these arguments is the Pythagorean stories about a disrespectful Pythagorean, probably Hippasus, which according to our interpretation correlate *the incommensurability with the Infinite*; the second argument is based on the fact that the Pythagoreans had introduced and studied *the side and diameter numbers* whose anthyphairesis is an initial finite segment of the infinite anthyphairesis of the diameter to the side of a square, the modern "convergents" of the continued fraction of the square root of 2 (Sect. 13.3).

The anthyphairetic proof of the incommensurability of the diameter to the side of a square by Theaetetus, which is the model for *Meno*'s theory of knowledge as recollection, employs the theory of ratios of magnitudes, a theory created by Theaetetus, and thus certainly is not the original Pythagorean proof (Sect. 13.4).

At this point a heuristic discussion intends to point out that Propositions II.9 & 10 of the *Elements* are in close connection with the side and diameter numbers and, through them, with the infinite anthyphairesis of the diameter to the side of a square, suggesting that the restoration of Book II of the *Elements* to its original Pythagorean form may lead us, *not only* to arguments, rejecting each of the three modern anthyphairetic reconstructions mentioned above as the original Pythagorean proof of incommensurability, *but, even more significantly,* to the original Pythagorean incommensurability proof and to the tools needed for its proof (Sect. 13.5).

So, we take our starting point from now on to be Book II of the *Elements* (Sect. 13.6).

The first partial restoration of Book II to its original Pythagorean form consists in adding the Pythagorean theorem, as Proposition II.4/5, and with the Pythagorean Application of Areas in Defect, as Proposition II.5/6, and in Excess, as Proposition II.6/7 (Sect. 13.7).

The additions of these Propositions on Application of Areas in Book II are related to the controversy of Geometric Algebra and are not universally accepted; thus these additions should be regarded as *tentative* (Sect. 13.8).

The second partial restoration of Book II to its original Pythagorean form consists in adding as corollaries to Proposition II.10, according to Proclus' *Commentary to Plato's Republic* [49] the so called *Elegant theorem* as Proposition II.10/11.a, and the Proposition on *the Pell property of the side and diameter numbers* as Proposition II.10/11.b; it is then natural to add a corresponding *Subtractive Elegant theorem*, as Proposition II.9/10, as the corresponding corollary to Proposition II.9, a proposition twin to Proposition II.10. Proclus, in his *Commentary to Plato's Republic* [49], describes also a *bona fide* proof of the Pell property of the side and diameter numbers by mathematical induction based on Proposition II.10. While solid and valid arguments for these additions are provided, we still choose to consider them as not definitive, hence the additions themselves should also be regarded as *tentative* (Sect. 13.9).

At this point we study the three anthyphairetic reconstructions, by Chrystal [17] adopted by Rademacher and Toeplitz [50] and by van der Waerden [61], by Tennenbaum, about 1950, and by Fowler [21] suggested as the original Pythagorean proof of incommensurability. Fowler's proof is based on the Elegant theorem, while Chrystal's and Tennenbaum proofs are found to be based on the Subtractive Elegant theorem (Knorr [33] related Chrystal's proof with Proposition II.9). The close association of these three proofs with Propositions II.9 & 10, both of which are closely related to the side and diameter numbers, which according to Proclus' comments, but also according to their nature as the "convergents" of the continued fraction of square root of 2 in modern language, were definitely *after the discovery and proof of the incommensurability*, lead us to reject them as the original Pythagorean proof (Sect. 13.10).

The structure of the Pythagorean Book II of the *Elements* now leads us to expect that the original incommensurability proof must be sought *before* the propositions related to the side and diameter numbers, namely at the position II.8/9 (Sect. 13.11).

Our expectation proves fruitful in that we obtain a proof of the incommensurability employing the emblematic Proposition II.4 and its consequences, namely the Pythagorean theorem II.4/5, the "completion of the square" II.6, the Application of Areas in Excess II.6/7, and finally Proposition II.8 (not having any known use till then). The Application of Areas in Excess is used in two basic ways for the incommensurability proof, of which one is typically school-algebraic, in the sense of van der Waerden, and provides the arithmetical value 2 for the second and third anthyphairetic quotients, while the second provides the preservation of the Gnomon, that shows that all the succeeding quotients, infinite in multitude, have also the same arithmetical value 2 (Sect. 13.12).

At this point we realize that Theaetetus' proof, given in Sect. 13.3, employed philosophically in the *Meno*, essentially transforms, with the help of Theaetetus' theory of ratios of magnitudes, the awkward Pythagorean Finitization of the preservation of Gnomons to the Logos Criterion for anthyphairetic periodicity.

The validity of our proof as the true reconstruction of the original one is *tentative* at this point, and must be supported by objective arguments in order to be considered the definitive one. We offer two such objective arguments.

The first is our interpretation of the two philosophic Pythagorean principles, the Infinite and the Finite, on the basis of the description of these principles by Aristotle, Simplicius, and Philoponus: the Infinite is the infinite anthyphairetic division of the diameter to the side of a square, while the Finite/Finitizer is the preservation of the Gnomons, providing full knowledge of an infinite and unknown entity (Sect. 13.13).

The certain dating of the discovery is subtle. The association of Hippasus in the Pythagorean stories does suggest an early dating but this is a rather weak evidence. Far stronger is the establishment that Zeno's arguments and paradoxes, in particular Zeno's *Fragment* B3, are closely connected with Pythagorean incommensurability (Sect. 13.14).

The second argument in favor of our reconstruction is our interpretation of the Pythagorean treatment of the three kinds of angles, mentioned in Plato's *Republic* 510c4-5 [46] and explained in detail by Proclus, in terms of the two Pythagorean principles, in which the Infinite, the infinite anthyphairetic division, is the cause of the acute and obtuse angle, and the Finite, the preservation of the Gnomons, the cause of the right angle. This second interpretation confirms not only our reconstruction of the original incommensurability proof, but also the use of Geometric Algebra (for the proof of the Incommensurability) and of Mathematical Induction (for the proof of the Pell property by the side and diameter numbers by means of Proposition II.10) by the Pythagoreans, and the reason why the converse to the Pythagorean theorem, Propositions II.12 and 13, appears late, and in fact after II.10 (Sect. 13.15).

Book II in its restored original Pythagorean form achieves a *unity* that is lacking in the *Elements*. It is no longer a Book consisting of disparate propositions with no discernible common theme, but it is transformed into a treatise aiming to develop the geometrical/algebraic tools, to use them in order to compute the infinite anthyphairesis of the diameter to the side of a square, and to establish the resulting incommensurability, and finally to examine the side and diameter numbers, the anthyphairetic approximations of the side and diameter numbers. Euclid destroyed this unity by moving the Pythagorean theorem to I.47, its converse to I.48, the Application of Areas to VI.28 & 29, and the proof of incommensurability to the problematic X.9.

13.2 Incommensurability, the Great Mathematical Discovery of the Pythagoreans, and the Question on the Method of Proof Employed by the Pythagoreans to Prove It

According to Definition X.1 of the *Elements*, two lines (or more generally magnitudes) a, b are called *commensurable* if there is a common measure, a line c, that measures exactly both lines a and b, namely that for some numbers m and n, we have $a = mc$ and $b = nc$. If a and b are not commensurable, then they are called *incommensurable*.

The greatest mathematical discovery of the Pythagoreans, arguably the greatest mathematical discovery ever, was their rigorous proof that there are lines a and b, such that a and b are incommensurable; in fact, they proved that the diameter and the side of a square are incommensurable. The ancient sources that mention this discovery include Proclus' *Synopsis of the History of Greek Mathematics*, in his *Commentary to the First Book of Euclid's Elements* 65,15–21 [48], a Synopsis that most likely goes back to Eudemus, and the *Anonymous Scholion In Euclidem X.1*, lines 21–47; and also the sources that recount persistently the story of a Pythagorean, most probably Hippasus of Metapontium, who disrespectfully made public to the "unworthy" non-Pythagoreans the secret discoveries of the Pythagoreans on incommensurability, and was punished by the divinities by drowning/perishing in the sea:

> Plutarch, *Nomas* 22,3,4-4,1 [47],
> Iamblichus, *On the Pythagorean Life* 246,10–247,1–7; 88,13–89,4 [30], and *On the Common Mathematical Science* 25, 27–39,
> Pappus, in his *Comments on Book X of Euclid's Elements* [44], and
> *Anonymous Scholion in Euclidem X.1*, 70–79

(Our interpretation of the story is outlined in Sect. 13.3.4, below).

Question 1 The greatest question concerning the mathematical achievements of the Pythagoreans is by what method the Pythagoreans obtained the original proof of incommensurability.

13.3 The Arithmetical Proof of Incommensurability, Hinted by Aristotle and Fully Described as "Proposition X.117" of the *Elements*, with Adherents from Heath [27] to Netz [42] Is Tentatively Rejected as the Original Pythagorean Proof in Favor of an Anthyphairetic Proof

13.3.1 An Elementary Arithmetical Method of Incommensurability Hinted in Aristotle's Analytics Prior *41a and Fully Preserved in "Proposition X.117", a Later Anonymous Addition to the* Elements

Some ancient sources, including Aristotle, *Analytics Prior* 41a [7] and "Proposition X.117" of Euclid's *Elements*, describe arithmetical proofs of the incommensurability of the diameter to the side of a square, and this has led a number of historians of Greek Mathematics to suggest that the original Pythagorean proof of incommensurability is to be found in the arithmetical proof of "Proposition X.117", in fact a later anonymous addition to Euclid's *Elements*.

13.3.2 Several Historians of Greek Mathematics Are Adherents to the Arithmetical Original Method of Proof

Among the adherents of such an original Pythagorean proof are Heath [27], Becker [12], Szabó [54], Zhmud [67], Netz [42]. It is rather amusing to follow Netz, pp. 80–81 [42] the last of the adherents, describing in his own words the proof of X.117:

> If so–already going beyond the scope of Plato's *Meno*–the diagonal of the square is what we would call $\sqrt{2}$ of the side of the square. It appears that Greeks were aware, perhaps very early on, that with the side of the square taken as the basis of measurement, the diagonal is irrational. How do we know that?
> This now becomes a little like Archytas's music theory.
> Assume the opposite–that the side and the diagonal are commensurable.
> This is tantamount to the claim that they are in the ratio of two integer numbers, say, $m : n$.
> Let us also assume that $m : n$ are the smallest such integers standing in this ratio–the key concern for Archytas's arithmetic!
> So, $m : n$ are not like 6 : 8, but like 3 : 4. They are $m : n$–the smallest such numbers.
> We also know, by Pythagoras's theorem (always the very same basic tool!), that $2m^2 = n^2$.
> Now, because n^2 is the double of something (it is the double of m^2), it is an even number. There is no way for a square number to be even without its root being even as well (odd times odd is odd, not even). So, n itself is an even number. Indeed, because n is an even number, it follows that n^2 is divisible by 4. (An even number is twice something; its square is therefore four times the something squared.) But we have $2m^2 = n^2$ or $m^2 = \frac{n^2}{2}$. Now, because n^2 is divisible by 4, it follows that $\frac{n^2}{2}$ is divisible by 2 or, in other words, is even.
> $m^2 = \frac{n^2}{2}$, and $\frac{n^2}{2}$ is even.

> We have learned, then, that m^2 is even. By the logic applied before, we've also learned that m itself is even. We have therefore established that both m and n are even, contradicting our preliminary hypothesis that m and n are the smallest integers of their ratio. (If they are both even, we just need to divide them both by 2 to get smaller integers with the same ratio. If they are both even, they are not like 3 : 4, but like 6 : 8.)
> Something is incurably wrong about our initial assumption, and in other words, no two integers can be found to display this ratio. The diagonal of the square is irrational.
> I do not know who was the first Greek to observe this. The observation is ingenious but simple; it could well have come about in the very first years of Greek mathematical writing.
> [...]
> We only know about this result now because it is tucked in as a kind of appendix at the end of Book X of Euclid's *Elements*.

13.3.3 Some Doubts About Whether Aristotle Was in Possession of the Simple X.117 Proof of Incommensurability

Van der Waerden, p. 110 [61] believed that the proof of incommensurability to which Aristotle refers in *Analytics Prior* 41a essentially coincides with the proof of Proposition X.117.

However, the detailed comments of Alexander Aphrodisieus on Aristotle's passage 260,7–261,28 describe a rather different proof of incommensurability of the diameter to the side of a square, the difference being principally its use of the strong Proposition VII.27.

The proof of incommensurability of the diameter that Alexander describes in his *Commentary to Aristotle's Analytics Prior 41a* [2] is as follows:

> Let $d^2 = 2s^2$. Suppose that d, s are commensurable. Then there are numbers m, n, such that $\frac{d}{s} = \frac{m}{n}$.
> By Propositions VII.20-22, we may assume that m, n are relatively prime.
> By the fundamental Proposition VII.27, m^2, n^2 are also relatively prime.
> By Proposition X.9, $\frac{d^2}{s^2} = \frac{m^2}{n^2} = \frac{2}{1}$.
> Hence m^2 is even. Hence $\frac{m^2}{2}$ is even, and since $\frac{m^2}{2} = n^2$, it follows that n^2 is even.
> Thus n^2 is both odd and even, a contradiction.

Alexander's proof is striking, in that it employs the unneeded 'heavy artillery' Proposition VII.27 to prove that n^2 is both odd and even, whereas Proposition X.117 reaches this conclusion with entirely elementary means. Thus, we must conclude that the elementary proof of X.117 was found after Alexander's time, possibly by Theon. A fortiori, then, the proof of X.117 was *not* available to Aristotle, and hence it is suggested that Aristotle, with his comments in *Analytics Prior* 41a, had in mind a proof similar to that described by Alexander. This suggests, in turn, that the proof that Aristotle had in mind was a specialization of a more general incommensurability theorem, involving VII.27 or one of its consequences, Proposition VIII.8, 14, 15, closely associated with Archytas. Thus, the elementary arithmetical proof of X.117 appears to be a final later product, not available yet in Aristotle's time, and thus not the original Pythagorean proof of incommensurability.

13.3.4 Our Interpretation of the Various Versions of the Pythagorean Story Involving the Disrespectful Pythagorean Correlate Incommensurability with the Infinite, Suggesting a Pythagorean Anthyphairetic Proof of Incommensurability by the Use of Proposition X.2

Pappus, in his *Comments on Book X of Euclid's Elements* [44] and the *Anonymous Scholion In Euclidem* X.1, lines 70–79, provides a fascinating explanation of this story by suggesting that the sea in which Hippasus was drowned symbolizes the Infinite, clearly referring to Plato's *Statesman* 273d4-e4 "infinite sea of dissimilarity".

καὶ τότ' ἤδη θεὸς ὁ κοσμήσας αὐτόν,
καθορῶν ἐν ἀπορίαις ὄντα,
κηδόμενος ἵνα μὴ χειμασθεὶς ὑπὸ ταραχῆς
διαλυθεὶς εἰς τὸν τῆς ἀνομοιότητος ἄπειρον ὄντα πόντον δύῃ,
πάλιν ἔφεδρος αὐτοῦ τῶν πηδαλίων γιγνόμενος,
τὰ νοσήσαντα καὶ λυθέντα ἐν τῇ καθ' ἑαυτὸν προτέρᾳ περιόδῳ στρέψας,
κοσμεῖ τε καὶ ἐπανορθῶν ἀθάνατον αὐτὸν καὶ ἀγήρων ἀπεργάζεται.

Therefore at that moment God, who brought order in the universe,
perceiving that it was in dire trouble,
and fearing that it might founder in the tempest of confusion
and sink in the infinite sea of dissimilarity,
he took again his place as its helmsman,
reversed in the previous period whatever had become unsound and unsettled
when the world was left to itself,
set the world in order,
restored it and made it immortal and ageless.
[Plato [45] with modifications by the authors]

It must be remarked that the Infinite [sea] in Plato's *Statesman* passage is menacing with chaos only if it is not periodic, while periodicity tames the Infinite by making it knowable. It is precisely this key to universal knowledge that the Pythagoreans revered and wanted to keep secret.

Pythagorean justice, as Aristotle describes it in *Nicomachean Ethics* 1132b21-23 [9] is reciprocal "anti-peponthos" justice, namely justice by a punishment corresponding and analogous to the affront; since Hippasus' punishment was drowning into the Infinite, we may conclude that the affront, the divulging of Incommensurability, was an affront to the Infinite, thus, connecting Incommensurability with the Infinite.

Proclus connects the principle of the Infinite with Incommensurability in his *Commentary to the First Book of Euclid's Elements* (=*In Euclidem*) 6,19–22 [48], and more specifically in 59,15–61,17 where he correlates incommensurability with an infinite decreasing sequence of Gnomons.

This connection of Incommensurability with the Infinite makes perfect sense if we consider that Euclid's only criterion for incommensurability is

Proposition X.2.
The infinity of the anthyphairesis of two lines implies the incommensurability of these two lines to each other.

By the *anthyphairesis* of two lines a and b, with $a > b$, the following process of successive reciprocal division is meant:

$$
\begin{aligned}
a &= k_0 b + c_1 \quad \text{with} \quad c_1 < b,\\
b &= k_1 c_1 + c_2 \quad \text{with} \quad c_2 < c_1,\\
c_1 &= k_2 c_2 + c_3 \quad \text{with} \quad c_3 < c_2,\\
c_2 &= k_3 c_3 + c_4 \quad \text{with} \quad c_4 < c_3,\\
c_3 &= k_4 c_4 + c_5 \quad \text{with} \quad c_5 < c_4,\\
&\dots\\
c_{n-1} &= k_n c_n + c_{n+1} \quad \text{with} \quad c_{n+1} < c_n,\\
c_n &= k_{n+1} c_{n+1} + c_{n+2} \quad \text{with} \quad c_{n+2} < c_{n+1},\\
&\dots
\end{aligned}
$$

where the lines $c_1, c_2, \dots, c_n, \dots$ are the *remainders*, and the numbers $k_0, k_1, k_2, k_3, k_4, \dots, k_n, \dots$ are the *quotients*. The anthyphairesis is either *infinite*, if it goes on forever, or *finite*, if there is a (final) step of the anthyphairesis in which a remainder divides the immediately previous remainder. A proof that the anthyphairesis of a to b is *infinite* thus provides a proof, an *anthyphairetic proof*, that the two magnitudes a and b are incommensurable.

13.3.5 The Pythagorean Infinite Musical Anthyphairesis in Philolaus Fragment 6 + Boethius, De Institutione Musica III.8 Implies a Musical Incommensurability

The Pythagorean correlation of the *Infinite* with the *Incommensurable* appears implicitly in the fascinating Philolaus *Fragment* 6. The *Fragment* sets up a "musical" anthyphairesis of musical intervals under the operation $*$ of composition, and provides the first three steps of the musical anthyphairesis of the octave $2/1$ to the fifth $3/2$:

octave = fifth $*$ fourth, fourth < fifth,
fifth = fourth $*$ tone, tone < fourth,
fourth = tone $*$ tone $*$ diesis, diesis < tone,

while the fourth step

tone = diesis $*$ diesis $*$ Pythagorean comma, Pythagorean comma < diesis

is preserved by Boethius, *De Institutione Musica* III.8 [14]. The form of the remainder, alternating between a power of 2 divided by a power of 3 or conversely, immediately shows that the remainder is never equal to the ratio $1/1$, and that the musical anthyphairesis is infinite, implying, in analogy to Proposition X.2, that there is no musical interval measuring (under composition) both the octave and the fifth.

13.3.6 The Pythagorean Association of Incommensurability with the Infinite Suggests an Anthyphairetic Original Proof of Incommensurability

In view of Proposition X.2 of the *Elements* and the corresponding musical incommensurability implied by Philolaus' Fragment 6, the Pythagorean stories, associating the Incommensurability with the Infinite, is a strong indication that the original Pythagorean proof of incommensurability was achieved by an *anthyphairetic* rather than an arithmetical proof, namely by employing Proposition X.2 of the *Elements*, according to which if two magnitudes possess infinite anthyphairesis, then these magnitudes are incommensurable.

13.3.7 The (Anthyphaireses of the) Side and Diameter Numbers Are Finite Initial Segments of the Infinite Anthyphairesis of the Diameter to the Side of a Square, an Indication that the Original Pythagorean Proof of Incommensurability Is Not Arithmetical but Anthyphairetic and the (Fundamental Pell Property of the) Side and Diameter Numbers Are Connected with Proposition II.10 of the Elements, an Indication that the Original Pythagorean Proof of Incommensurability Is Connected with Book II of the Elements

Another strong indication against the arithmetical and in favor of the anthyphairetic method of the original Pythagorean proof of incommensurability is the fact that the Pythagoreans defined and studied at depth *the side and diameter numbers after* their discovery of incommensurability as emphasized by Proclus.

The Recursive Definition of the Side and Diameter Numbers in Ancient Sources Based on the ancient sources for the side and diameter numbers which include:

Plato, *Republic* 546b-c [46]
Theon of Smyrna 43,5–43,15, 44,18–45,8 [56]
Iamblichus *Commentary to Nicomachus' Arithmetic* 91,21–92,3, 92,23–93,6 [31]
Proclus *Commentary to Plato's Republic* 2,24,16–25,13 and 2,27,1–29,4 [49],
we know that the Pythagoreans introduced and studied the double sequence of *the side p_n and diameter q_n numbers* defined *recursively* by

$$p_1 = q_1 = 1$$

$$p_{n+1} = p_n + q_n, \quad q_{n+1} = 2p_n + q_n \quad \text{for every natural number } n.$$

However, from the recursive definition given by the ancient sources, we would be at a loss to conceive a plausible way in which the Pythagoreans discovered these numbers. The recursive definition is usable on proving properties of these numbers but unenlightening on how the Pythagoreans thought of them.

Question 2 How did the Pythagoreans come up with the recursive definition of the side and diameter numbers?

Two Proclus' Passages, *In Republic* 2,27,1–11 and *In Euclidem* 61,12–17 clearly imply that the Pythagorean proof of incommensurability of the diameter to the side of a square *is prior to* the Pythagorean definition of the approximating side and diameter numbers.

Proclus informs us that the Pythagoreans introduced this double sequence after their discovery and proof of incommensurability:

(1) Proclus *In Euclidem* 61,12–17 [48]

	ὅπου δὲ τὸ σύνεγγυς ἀγαπῶμεν and when we are content with approximations
οἷον εὑρόντες ἐν γεωμετρίᾳ τετράγωνον τετραγώνου διπλάσιον, ἐν ἀριθμοῖς δὲ οὐκ ἔχοντες as having found in geometry a square double a given square, but not having such a square in numbers	
	ἑνὸς δέοντός φαμεν ἄλλον ἄλλου διπλάσιον ὑπάρχειν ὥσπερ τοῦ ἀπὸ τῆς πεντάδος ὁ ἀπὸ τῆς ἑπτάδος διπλάσιος ἑνὸς δέοντος. we say that a square number is the double of another square number short by one like the square of seven which is one less than the double of the square of five.

(2) Proclus *In Republic* 2,27,1–11 [49]

Ὅτι ἐπειδὴ *ἀδύνατον* ῥητὴν εἶναι τὴν διάμετρον τῆς πλευρᾶς οὔσης ῥητῆς (*οὐ γάρ ἐστιν* τετράγωνος ἀριθμὸς τετραγώνου διπλάσιος· ᾧ καὶ δῆλον ὅτι *ἀσύμμετρά* ἐστιν μεγέθη Since it is *impossible* for the diagonal to be rational when the side is rational (*for there is no* number squared that is twice a square – a fact that makes it clear that magnitudes are *incommensurable* – ...	
	ἐπενόησαν οὕτω λέγει οἱ Πυθαγόρειοι καὶ Πλάτων τῆς πλευρᾶς οὔσης ῥητῆς τὴν διάμετρον ῥητὴν οὐχ ἁπλῶς ἀλλ' ἐν οἷς δύνανται τετραγώνοις τοῦ διπλασίου λόγου ὃν δεῖ τὴν διάμετρον ποιεῖν ἢ μονάδι [δέ]ουσαν ἢ μονάδι πλεονάζουσαν· πλεονάζουσαν μὲν ὡς τοῦ Δ τὸν Θ δέουσαν δὲ ὡς τοῦ ΚΕ τὸν ΜΘ. the Pythagoreans and Plato contrived a way to say that when the side is rational the diagonal is not rational simpliciter but rather lies in the squares of the double ratio which it is requisite for the diagonal to make falling short or exceeding it by a unit – exceeding [by a unit] in the ratio of 4 to 9 but falling short in the ratio of 25 to 49.

The fact that the Pythagoreans introduced and studied the side p_n and diameter q_n numbers is a strong argument against the arithmetical and in favor of an anthyphairetic original Pythagorean proof of the incommensurability of the diameter d to the side s of a square because Anth$(d, s) = [1, 2, 2, 2, \ldots]$ and Anth$(q_n, p_n) = [1, 2, \ldots, 2$ (n-1 times)$]$

The fact that the Pythagoreans introduced the side and diameter numbers after their discovery of incommensurability as rational approximations of the diameter to the side of a square, as Proclus makes clear (Section 13.3.7.2), together with the fact that the following Proposition, not mentioned in ancient sources, reveals that this approximation is anthyphairetic. Noting also as Fowler [20], [23], pp. 240–262 shows that practically all approximating/simplifying expressions in Greek Mathematics were anthyphairetic make it certain that the initial Pythagorean definition of the side and diameter numbers was anthyphairetic and thus occurring *after* they had obtained the full knowledge of the infinite anthyphairesis of the diameter to the side of a square.

Proposition 13.1 *For every natural number n the pair of side p_n and diameter q_n numbers is unique such that: Anth$(q_n, p_n) = [1, 2, 2, \ldots, 2$ (n-1 times)$]$ and q_n and p_n are relatively prime.*

Proof We note that:

$$\mathrm{Anth}(q_1, p_1) = [1], \quad \mathrm{Anth}(q_2, p_2) = [1, 2].$$

We proceed by induction. Assume that

$$\mathrm{Anth}(q_n, p_n) = [1, 2, 2, \ldots, 2, \text{ (n-1 times)}]$$

and q_n, p_n are relatively prime. Thus we have:

$$\begin{aligned} q_n &= p_n + c_1 \\ p_n &= 2c_1 + c_2 \\ c_1 &= 2c_2 + c_3 \\ &\ldots \\ c_{n-3} &= 2c_{n-2} + c_{n-1} \\ c_{n-2} &= 2c_{n-1} \\ c_{n-1} &= 1 \end{aligned}$$

since q_n, p_n are assumed to be relatively prime. From the first two relations, it follows that

$$\text{Anth}(p_n, q_n - p_n) = [2, 2, \ldots, 2 \text{ (n-1 times)}]$$

and $p_n, q_n - p_n$ are relatively prime.

Then

$$\text{Anth}(p_n + q_n, p_n) = [2, 2, \ldots, 2 \text{ (n times)}]$$

and $p_n + q_n, p_n$ are relatively prime.

Then

$$\text{Anth}(q_n + 2p_n, q_n + p_n) = [1, 2, 2, \ldots, 2 \text{ (n times)}]$$

and $q_n + 2p_n, q_n + p_n$ are relatively prime, namely

$$\text{Anth}(q_{n+1}, p_{n+1}) = [1, 2, 2, \ldots, 2 \text{ (n times)}]$$

and q_{n+1}, p_{n+1}, q_n are relatively prime.

Schematically, the inductive argument from n to $n + 1$ proceeds in the following steps:

$\text{Anth}(q_n, p_n) =$ $[1, 2, 2, \ldots, 2$ (n-1 times)]				$\text{Anth}(q_{n+1}, p_{n+1}) =$ $[1, 2, 2, \ldots, 2$ (n times)]
			$2p_n + q_n =$ $(p_n + q_n) + p_n$	$q_{n+1} =$ $p_{n+1} + p_n$
$q_n =$ $p_n + c_1$		$p_n + q_n =$ $2p_n + (q_n - p_n)$	$p_n + q_n =$ $2p_n + (q_n - p_n)$	$p_{n+1} =$ $2p_n + (q_n - p_n)$
$p_n =$ $2c_1 + c_2$ $c_1 = 2c_2 + c_3$ … $c_{n-3} =$ $2c_{n-2} + c_{n-1}$ $c_{n-2} = 2c_{n-1}$	$p_n =$ $2(q_n - p_n) + c_2$ $c_1 = 2c_2 + c_3$ … $c_{n-3} =$ $2c_{n-2} + c_{n-1}$ $c_{n-2} = 2c_{n-1}$	$p_n =$ $2(q_n - p_n) + c_2$ $c_1 = 2c_2 + c_3$ … $c_{n-3} =$ $2c_{n-2} + c_{n-1}$ $c_{n-2} = 2c_{n-1}$	$p_n =$ $2(q_n - p_n) + c_2$ $c_1 = 2c_2 + c_3$ … $c_{n-3} =$ $2c_{n-2} + c_{n-1}$ $c_{n-2} = 2c_{n-1}$	$p_n =$ $2(q_n - p_n) + c_2$ $c_1 = 2c_2 + c_3$ … $c_{n-3} = 2c_{n-2} + c_{n-1}$ $c_{n-2} = 2c_{n-1}$

□

Thus, answering *Question 2*, this Proposition in connection with Proclus' passages considered in Section 13.3.7.2 reveals that the Pythagoreans introduced the side and diameter numbers as rational anthyphairetic approximations of the diameter to the side of a square as the ancient precursors of the convergents of modern continued fractions *after* they had already obtained full knowledge of the infinite anthyphairesis $[1, \text{period}(2)]$ of the diameter to the side of a square and thus had proved by Proposition X.2 the incommensurability of the diameter to the side of a square.

13.3.8 Tentative Rejection of the Arithmetical Reconstructions in Favor of an Anthyphairetic One

At this point, we have reached the tentative conclusion that we must reject the arithmetical reconstruction of the original Pythagorean proof of incommensurability in favor of an anthyphairetic one. We might say that Netz [42] is quite right in associating this arithmetical proof with Archytas, but quite wrong in thinking that this is the original Pythagorean proof of incommensurability. Additionally, Knorr's, p. 27 [32] reconstruction must also be rejected because it is a variation of the arithmetical one succeeding to correlate cleverly but rather artificially the incommensurability with the Infinite but still utterly unable to explain the generation of the side and diameter numbers.

13.4 The Anthyphairetic Proof of the Diameter to the Side of a Square by Theaetetus Which Is the Model for *Meno* 80e-86c, 97a-98b Employs the Theory of Ratios of Magnitudes and Is Certainly Not the Original Pythagorean Proof

It is not generally realized that Plato's *Meno* 80e1-86c3, 97a9-98b6 is based on an anthyphairetic proof of the incommensurability of the diameter to the side of a square as analyzed and interpreted in Negrepontis [38], a proof that must certainly be attributed to Theaetetus, as it makes use of his theory of proportion for magnitudes. We describe this proof in Proposition 3 below; the following elementary proposition is needed.

Proposition 13.2 *If a, b, c, d are line segments such that $ad = bc$ then $Anth(a, b) = Anth(c, d)$.*

Proof Let $\text{Anth}(a, b) = [k_0, k_1, \ldots]$. We proceed by induction. Thus $a = k_0 b + e_1$ with $e_1 < b$. Then $cb = ad = k_0 bd + e_1 d$. By Proposition I.44 of the *Elements*, there is a line segment f_1 such that $e_1 d = bf_1$. Since $e_1 < b$, it is clear that $f_1 < d$. Thus $cb = k_0 bd + e_1 d = k_0 bd + bf_1$; then $c = k_0 d + f_1$ with $f_1 < d$. Thus $\text{Anth}(a, b) = [k_0, k_1, \ldots] = [k_0, \text{Anth}(b, e_1)]$ while $\text{Anth}(c, d) = [k_0, \text{Anth}(d, f_1)]$ and at this point we have established for the tetrad b, e_1, d, f_1 the same condition $bf_1 = de_1$ that we had assumed for the original tetrad a, b, c, d and we are to prove that

$\text{Anth}(b, e_1) = \text{Anth}(d, f_1)$. We continue as in the first step and we finish the proof by induction. □

Proposition 13.3 (Pythagorean) *If b is a line and a is a line such that $a^2 = 2b^2$ then a is the diameter of the square with side b and* $\text{Anth}(a, b) = [1,\ \textit{period}(2)]$.

Proof (Theaetetus)

Part I
By the (isosceles) Pythagorean theorem line a is constructed as the diameter of the square with given side b. Clearly $a > b$. Since $a^2 = 2b^2$, it follows that $a < 2b$ (indeed if we had $a \geq 2b$ then $a^2 \geq 4b^2 > 2b^2 = a^2$, contradiction). Thus $b < a < 2b$. Hence,

$$a = b + c_1, \quad c_1 < b.$$

It follows that $b > 2c_1, \quad 3b > 2a$ (indeed if we had $b \leq 2c_1 = 2a - 2b$, then $3b \leq 2a$, hence $9b^2 \leq 4a^2 = 8b^2$, contradiction). Hence,

$$b = 2c_1 + c_2.$$

$$b > 2(a - b), \quad 3b > 2a.$$

Part II
Then $c_1 = a - b$, hence $c_2 = b - 2c_1 = b - 2(a - b) = 3b - 2a$, hence $bc_2 = b(3b - 2a) = 3b^2 - 2ab$, and $c_1^2 = (a - b)(a - b) = a^2 + b^2 - 2ab = 3b^2 - 2ab$. Hence $bc_2 = c_1^2$. Since $b > c_1$, and

$$bc_2 = c_1^2,$$

clearly $c_2 < c_1$, thus the second step is indeed anthyphairetic

$$b = 2c_1 + c_2 \quad c_2 < c_1.$$

By $bc_2 = c_1^2$ and Proposition 13.4.1 we have $\text{Anth}(b, c_1) = \text{Anth}(c_1, c_2)$ (namely $b/c_1 = c_1/c_2$, according to the Theaetetus' definition of proportion for magnitudes). Then $\text{Anth}(c_1, c_2) = \text{Anth}(b, c_1) = [2, \text{Anth}(c_1, c_2)]$. Hence, $\text{Anth}(a, b) = [1, 2, \text{Anth}(c_1, c_2)] = [1, 2, \text{Anth}(b, c_1)] = [1, 2, 2, \text{Anth}(c_1, c_2)] = \ldots = [1,\ \text{period}(2)]$. □

Notes This proof is not explicitly preserved in ancient sources but is the product of our interpretation of the *Meno*. Accepting this interpretation, this anthyphairetic proof is the most ancient proof preserved (and not the proof hinted in Aristotle's *Analytics Prior* 41a). The geometer who would conceive of such a proof would normally be Theaetetus, since Theaetetus was the first geometer to discover a theory of proportion of magnitudes based on equal anthyphairesis, the pre-Eudoxean theory mentioned by Aristotle in *Topics* 158b [8]. In the *Topics* passage, there is no explicit attribution, but there is something of a consensus among historians of Mathematics that the theory is due to Theaetetus.

The reconstruction of Theaetetus' theory of ratios of magnitudes has been obtained by Negrepontis and Protopapas [40]. Theaetetus' proof provides *the complete knowledge* of the diagonal with respect to the side of a square with one and a half anthyphairetic division steps plus the condition $bc_2 = c_1^2$, equivalently $b/c_1 = c_1/c_2$, an instance of Theaetetus' Logos Criterion (consisting in the *repetition/recollection* of Logos). Since the anthyphairesis of a to b is infinite, by Proposition X.2, a and b are incommensurable.

The computation depends on the representation of the first quotient c_1 as $a - b$ and the second quotient c_2 as equal to $3b-2a$. These representations use *the side and diameter numbers* q_n, p_n; thus $c_1 = p_1a - q_1b = a - b$, $c_2 = q_2b - p_2a = 3b - 2a$.

The anthyphairetic proof of the Pythagorean incommensurability extracted from the *Meno* is the oldest known to us from ancient sources, but it is certainly not the original Pythagorean proof, since it employs Theaetetus' theory of ratios of magnitudes.

13.5 Heuristic Discussion: The Restoration of Book II of the *Elements* to Its Original Pythagorean Form Is Expected to Reveal the Original Pythagorean Proof of Incommensurability

At this point we have tentatively rejected the arithmetical in favor of an anthyphairetic method, as the original Pythagorean proof of incommensurability, thanks to a combination of the accounts by the ancient authors Theon, Iamblichus, Proclus and our knowledge of modern continued fractions, according to which the Pythagorean side and diameter numbers coincide with the modern convergents of the square root of 2, and are thus closely connected with the infinite anthyphairesis of the diameter to the side of a square; in addition we shall see in Section 13.9 that, as explained by Proclus in his *Commentary to Plato's Republic* 2,27,24–28,4 [49], Proposition II.10 of the *Elements* appears to be the tool for the inductive proof of the fundamental Pell property of the side p_n and diameter q_n numbers, namely the property $q_n^2 = 2p_n^2 + (-1)^n$ for all natural numbers n. Thus, even though there is no reference to incommensurability in Book II, it appears that Book II has some connection with the infinite anthyphairesis, and hence with the incommensurability of the diameter to the side of a square.

Secondly, there are at least three modern anthyphairetic reconstructions of the proof of the incommensurability of the diameter to the side of a square, each a possible candidate for the original Pythagorean proof, (a) by Chrystal [17] adopted by Rademacher and Toeplitz [50], Toeplitz [58] and van der Waerden [61], (b) by Tennenbaum, about 1950, unpublished, made known by Conway [18] and by

Conway and Shipman [19] and (c) by Fowler [21]. These reconstructions could well have been realized in ancient times, because, as we shall see in Sect. 13.9, a common, not immediately obvious, thread in these reconstructions is their reliance *either* on the *Elegant Theorem* (if $a^2 = 2b^2$, then $(a + 2b)^2 = 2(a + b)^2$), as called by Proclus, in his *Commentary to the Republic* 2,27,24–28,4 [49], an immediate corollary to Proposition II.10 of the *Elements*, for the case of Fowler's [21] reconstruction, *or* on the corresponding statement we call *Subtractive Elegant Theorem* (if $a^2 = 2b^2$, then $(2b - a)^2 = 2(a - b)^2$), an immediate corollary to its twin Proposition II.9 of the *Elements*, for the case of Chrystal [17] or Tennenbaum. Knorr [33] had realized that Chrystal's proof is related to Proposition II.9. Thus again, Propositions II.9 & 10 of the *Elements* appear to be closely related to Pythagorean incommensurability.

On the other hand, several earlier historians of Greek Mathematics including Bretschneider [15], Hankel [25], Heath [28], vol. 1, pp. 354, 383–385, 386–388, have suggested that there are good reasons to believe that several fundamental Pythagorean discoveries, including the Pythagorean theorem and the Pythagorean Application of Areas, that are not part of Book II of the *Elements*, should nevertheless be included in it. Thus, in our quest for the original proof of the Pythagorean incommensurability, we will examine, in Sections 13.6–13.9, Book II of the *Elements* and its restoration/completion to its original Pythagorean form.

13.6 Book II of Euclid's *Elements*

Book II of Euclid's *Elements*, generally considered Pythagorean in origin, consists of just 14 Propositions. It is based on the Fifth Postulate, (a) in contrapositive form Proposition I.29 with consequences Propositions I.30, I.32–34, and the construction of the square of a given side Proposition I.46, and (b) in a direct form with consequences, the first part of the proof of the squaring of every rectilinear area Propositions I.42–45.

Definition II.2. of *Gnomon*
Proposition II.1. If n is a natural number, and $a_1, a_2, \ldots, a_n, b$ are lines, then $(a_1 + a_2 + \ldots + a_n)b = a_1 b + a_2 b + \ldots + a_n b$.
Proposition II.2. If a, b, c are lines and $a = b + c$, then $a^2 = ba + ca$.
Proposition II.3. If a, b, c are lines and $a = b - c$, then $a^2 = ba - ca$,
Proposition II.4. If a, b are lines, then $(a + b)^2 = a^2 +$ (*Gnomon* in the square $(a + b)^2$ about the square a^2) $= a^2 + b^2 + 2ab$.
Proposition II.5. If a, x are lines and $a/2 > x$, then $(a/2)^2 - (a/2 - x)^2 = x(a - x)$.
Proposition II.6. If a, x are lines, then $(a/2 + x)^2 = (a/2)^2 + x(a + x)$.
Proposition II.7. If a, b are lines, then $(a - b)^2 = a^2 + b^2 - 2ab$.
Proposition II.8. If a, b are lines, then $(a + 2b)^2 = a^2 + 4b(a + b)$.
Proposition II.9. If a, b are lines with $b < a < 2b$, then $a^2 + (2b - a)^2 = 2b^2 + 2(a - b)^2$.
Proposition II.10. If a, b are lines, then $(a + 2b)^2 + a^2 = 2(a + b)^2 + 2b^2$.
Proposition II.11. If a is a line, construct a line b, such that $a^2 = ab + b^2$.
Proposition II.12. If a, b, c are the sides of a triangle, with the sides a and b forming an obtuse angle, then $a^2 + b^2 < c^2$.

Proposition II.13. If a, b, c are the sides of a triangle, with the sides a and b forming an acute angle, then $a^2 + b^2 > c^2$.
Proposition II.14. (The second part of the squaring of every rectilinear area) If a, b are lines, construct a line m, such that $ab = m^2$.

Proposition II.8 is an easy consequence of Proposition II.4, and is not used anywhere in the Elements.

Question Q3 What is the role of Proposition II.8? Where is it used?

Propositions II.12 & 13 constitute the converse of the Pythagorean theorem II.4/5; their proof needs Propositions II.4,4/5, and II.7.

Question Q4 Why is the converse of the Pythagorean theorem proved so late in Book II, in particular after Proposition II.10?

We will attempt to gain essential understanding of the Pythagorean Mathematics and Philosophy by successive steps of restoration of Book II in its original, Pythagorean form.

13.7 Partial Restoration of Book II in Its Original Pythagorean Form: Proposition II.4/5=the Pythagorean Theorem, Proposition II.5/6=the Pythagorean Application of Areas in Defect, Proposition II.6/7=the Pythagorean Application of Areas in Excess

Several historians of Mathematics, including Heath [28] have suggested that three more Propositions must have been originally within Book II.

(1) *The Pythagorean theorem as Proposition II.4/5.*

Proclus, in his *Commentary to the First Book of Euclid's Elements* 426,11–12 [48], tells us that the proof of the Pythagorean theorem presented in Proposition I.47 of the *Elements*, relying on the "paradoxical" Propositions I.35–41, is Euclid's proof, and therefore, we may conclude that it is not the original Pythagorean one. It has been suggested a long time ago by Bretschneider [15], and Hankel [25] that the original proof of the Pythagorean theorem must have been in Book II, before II.9, where it is first used. Based on Plato's *Meno* 84d3–85b7, analyzed in Negrepontis and Farmaki [39] and Negrepontis [38], we have argued that the original position of the Pythagorean theorem is at II.4/5, receiving a proof in terms of the emblematic Proposition II.4. Thus

Proposition II.4/5 (Pythagorean theorem).

If a, b, c are the sides of a triangle, with the sides a and b forming a right angle, then $a^2 + b^2 = c^2$.

(2) *Pythagorean Application of Areas as Propositions II.5/6 in defect and II.6/7 in excess.*

Several historians of Mathematics, including Tannery [55], Zeuthen [66], Heath [28], van der Waerden [61] have suggested that the role and purpose of the two Propositions II.5 & 6, variations of the emblematic Proposition II.4 in the form of "completing a square", is to prove the two Application of areas in Defect and in Excess, respectively, that these Applications were originally in Book II, but that Euclid moved them to Book VI, where he proved more general versions (in defect VI.28, in excess VI.29). Thus, we add

Proposition II.5/6. Pythagorean Application of Areas in Defect.

For a and m lines, with $a/2 > m$, to construct a line x, such that $x(a - x) = m^2$.

Proposition II.6/7. Pythagorean Application of Areas in Excess.

For a and m lines, to construct a line x, such that $x(a + x) = m^2$.

13.8 The Controversy on Pythagorean Geometric Algebra

Tannery [55], Zeuthen [66], van der Waerden [61] and other historians of Greek Mathematics proposed the hypothesis that the two Propositions on the Application of Areas were really solutions of quadratic algebraic equations in the guise of geometry, and that Book II should be interpreted as *Pythagorean Geometric Algebra*.

This hypothesis was later disputed by some historians of Greek Mathematics, including Szabó [54], Unguru [59], Mueller [34], Saito [52].

Even though we believe that the arguments already given in support of the thesis on Geometric Algebra, including the original ones, the additional ones by van der Waerden [62, 63], Weil [65], Rashed [51], Blåsjö [13], Negrepontis and Farmaki [39], Negrepontis et al. [41] and Papadopoulos [43] are strong and valid, we nevertheless intend to provide in the present paper some novel arguments in support of Geometric Algebra, and therefore we will consider the additions of Propositions II.5/6 and II.6/7 at the present time as *tentative and not final*, subject to further evidence.

Question Q5 Do Propositions II.5/6, 6/7 really belong in Book II? Do Prepositions II,5/6, 6/7 indicate that the Pythagoreans were practicing Geometric Algebra? And if so, is there a Pythagorean use of Application of Areas II.5/6 or II.6/7?

13.9 The Proof by Mathematical Induction, Using Propositions II.10, that the Side p_n and Diameter q_n Numbers Satisfy the Pell Property $q_n^2 = 2p_n^2 + (-1)^n$ for Every Natural Number n

Proclus, in his *In Republic* [=*Commentary to Plato's Republic*] 2,28,17–29,1 [49], states that the Pythagoreans, after defining the side and diameter numbers, (1) first verified the first few cases, (2) then conjectured, and (3) finally proved that these numbers are solutions of *the Diophantine Pell equation*, namely that

$$q_{2n}^2 = 2p_{2n}^2 + (-1)^n \text{for every natural number} n$$

According to some modern scholars (including Freudenthal [24]; van der Waerden [61]), the Pythagoreans went on *to prove* the Pell property by an inductive argument using Proposition II.10 of the *Elements*, but more recently other modern scholars (including Unguru [60]; Acerbi [1]) have challenged the thesis that the Pythagoreans had obtained a proof by induction, or in fact any kind of proof of the Pell property, claiming that the only thing that Proclus' account shows is the verification of the Pell property for the first few cases.

Freudenthal and van der Waerden did not have convincing arguments that Proclus' account amounted to an inductive proof, and they did not have an argument that the Greeks were in possession of the modern Peano principle of Mathematical Induction. (The Wedberg [64] and Acerbi [1] claim that Plato's *Parmenides* 149a7-c3 rule "number of terms = contacts + 1" was proved by an appeal to a *bona fide* principle of Mathematical Induction, has been questioned by Negrepontis, [36] because this rule, if properly understood, applies to a finite recursion only).

In our work (Negrepontis and Farmaki [39], Chapter 8), we have presented some novel arguments showing that the Pythagoreans had indeed a step-by-step inductive proof, but without the principle of mathematical induction. We base our claim on two remarkable linguistic bonds, in Proclus' detailed account, that, to the best of our knowledge, have not been noted before.

The first linguistic bond is between	
the wording of *the heuristic arithmetical idea* of the Pythagoreans of (Proclus 2,25,9–12) that the sum of the squares $q_n^2 + q_{n+1}^2$, being equal exactly to $2p_n^2 + 2p_{n+1}^2$, makes both of them *true diameters*,	
	and the wording by Proclus (2,27,24–28,4) of the statement of *the geometrical Proposition II.10*
	An immediate consequence of the geometric Proposition II.10 is the geometric Elegant Theorem.
The second linguistic bond is between	
	the wording of the statement of *the geometric Elegant Theorem* (2,27,13–16),
and the wording of the first (2,28,17–21), the second (2,28,22–24) and the third (2,28,27–29,1) inductive step for the proof of *the arithmetical Pell property.*	

According to the first linguistic bond, the heuristic Pythagorean idea about the fact that two successive arithmetical diameters behave like true diameters was the inspiration for the proof of the geometric Proposition II.10, whose immediate consequence is the geometric Elegant theorem, which in effect serves as the geometric model for every step of the proof by induction of the arithmetical Pell property for the side and diameter numbers (for an argument that was initially inspired by our approach in [39], but followed a different reasoning to the same positive conclusion, see Baloglou and Thomaidis [11]).

Proof Outlined

The arithmetical form of II.10 is an immediate consequence of the geometric. From the recursive definition of the side and diameter numbers and Proposition II.10 arithmetized, it follows that

$$q_{n+1}^2 + q_n^2 = 2p_{n+1}^2 + 2p_n^2.$$

We now proceed by induction. The proposition holds for $n = 1$. Assume that it holds for n, namely that

$$qq_n^2 = 2p_n^2 + (-1)^n.$$

Thus $q_{n+1}^2 + (-1)^n = 2p_{n+1}^2$, namely

$$q_{n+1}^2 = 2p_{n+1}^2 + (-1)^{n+1}.$$

We claim that in its original Pythagorean form Book II included the following Definition and Propositions:

The Recursive Definition (at the position) II.9—of the side and diameter numbers:
$p_1 = q_1 = 1$,
$p_{n+1} = p_n + q_n, \quad q_{n+1} = 2p_n + q_n$ for every natural number n,
Proposition II.10/11.a (= the Elegant Theorem):
if $a^2 = 2b^2$, then $(a + 2b)^2 = 2(a + b)^2$, and
Proposition II.10/11.b (= the Pell property of the side and diameter numbers):
$q_n^2 = 2p_n^2 + (-1)^n$ for every natural number n.

Although this is not mentioned in some ancient source, the Pythagoreans would surely have no difficulty in deducing from Proposition II.9 the following

Proposition II.9/10 (=the Subtractive Elegant theorem):
if $d^2 = 2s^2$, then $(2s - d)^2 = 2(d - s)^2$.

Despite the relevance of our arguments, we will still consider at the present stage of our paper our claim that in fact Proclus describes an *inductive* proof of the Pell property by means of Proposition II.10, as *tentative* and not final.

Question Q6 Did the Pythagoreans know proof by mathematical induction, and did they prove the Pell property of the side and diameter numbers, or did they simply verify a few initial cases of the Pell property?

13.10 The Chrystal, Tennenbaum, Fowler Anthyphairetic Proofs of the Incommensurability of the Diameter to the Side of a Square, Their Reliance on Propositions II.9 & 10, and Their Rejection as Reconstructions of the Original Pythagorean Proof of Incommensurability

In this section we will consider the Chrystal, Tennenbaum, Fowler anthyphairetic proofs of the incommensurability of the diameter to the side of a square, and we will show that these, and some variations, are *consequences* of the two Elegant theorems II.9/10 and II.10/11.a, and thus consequences of Propositions II.9 and 10. These proofs are then rejected as reconstructions of the original Pythagorean proof of incommensurability because of their close ancient link with Propositions II.9 and 10, propositions that are connected with the side and diameter numbers, which, as we argued, *follow* the discovery of incommensurability.

13.10.1 Ancient and Modern Proofs of the Subtractive Elegant Theorem II.9/10; if $d^2 = 2s^2$, then $(2s - d)^2 = 2(d - s)^2$

Proof from II.9
Immediate.

Proof from the Theory of Gnomons of Book II
Assume $d^2 = 2s^2$. Then $s < d < 2s$, and we form the lines $d - s, 2s - d$. By Proposition II.7, $(2s - d)^2 = 4s^2 + d^2 - 4sd = 6s^2 - 4sd$. By Proposition II.7, $2(d - s)^2 = 2d^2 + 2s^2 - 4ds = 6s^2 - 4ds$. Hence, $(2s - d)^2 = 2(d - s)^2$.

Third Proof
First step. Since $d^2 = 2s^2$, we construct a square $ABCD$ with side $AB = s$ and diameter $BD = d$ (see Fig. 13.1).

Second step. We construct a point E on the diameter such that $BE = s$. Set $ED = s_1$. Then $d = s + s_1$, hence $s_1 = d - s$.

Third step. We construct EF perpendicular on diameter DB at point E, and set F the point of its intersection with AD.

Claim. $EF = s_1$

[By Proposition I.34, since DB is a diameter, the angle FDE is equal to half a right angle, and by construction the angle EDF is a right angle. By Proposition I.32, the angle DFE is half a right angle. Hence, by Proposition I.5, $DE = EF = s_1$.]

Claim. $AF = FE = s_1$.

[The triangles ABF, EBF are right angled, with hypotenuse and one side of the right angle equal]

Fourth step. We set $DF = d_1$. Then $DEFG$ is a square with side s_1. Thus $s_1 = d - s$, $d_1 = s - s_1 = s - (d - s) = 2s - d$. Thus, $s_1 = d - s$ and

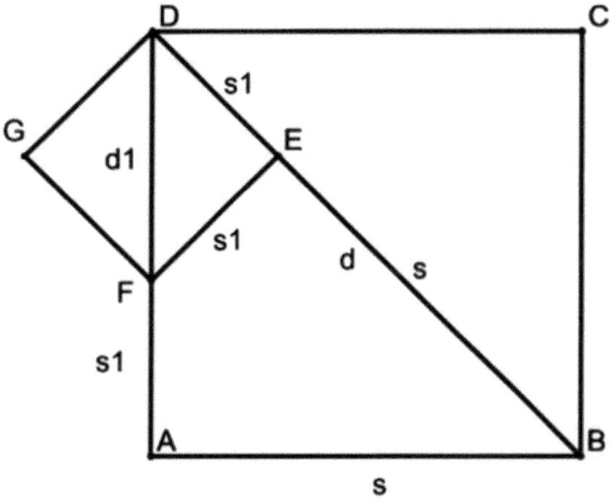

Fig. 13.1 Representation of the squares used in the third proof

$d_1 = 2s - d$ are the side and the diameter, respectively, of the square $DEFG$. This is the Subtractive Elegant Theorem.

Note Chrystal, vol. I, p. 270 [17]; Rademacher and Toeplitz, p. 16 [50]; Toeplitz, p. 4 [58]; van der Waerden, p. 127 [61]; Knorr, pp. 35–36; [32] Knorr [33] first realized the close connection of Chrystal's proof with Proposition II.9. We note more specifically that what is needed is just the proof of the Subtractive Elegant theorem.

Fourth Proof

Let s, d be the side and the diameter of a square, so that $d^2 = 2s^2$. The idea of this proof is to form the great square with side d, and to place inside it the two smaller squares with side s *below left and above right*. Since $d^2 = 2s^2$, as depicted in Fig. 13.2, it is clear that the intersection of the two smaller squares must be equal to the sum of the two remaining squares located *above left and below right* (see Fig. 13.3).

It is easy to see that this equality takes the form $(2s - d)^2 = 2(d - s)^2$. The Subtractive Elegant Theorem follows.

Note Stanley Tennenbaum, about 1950, unpublished, devised this argument for his beautiful proof of the incommensurability of the diameter to the side of a square. This proof became known by Conway [18]. We realized the close connection of

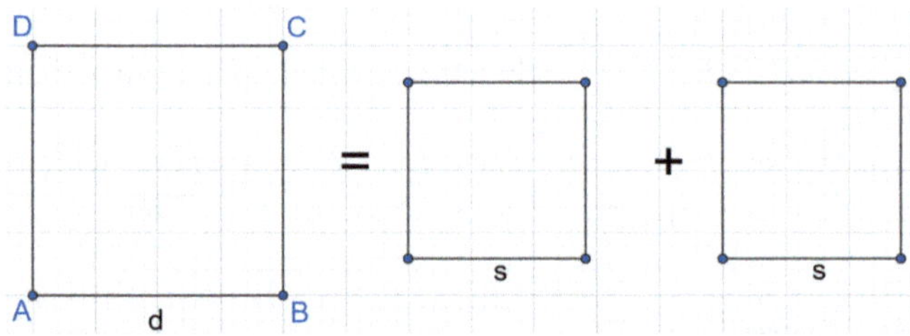

Fig. 13.2 The larger square's area is equal to the sum of the areas of the two smaller, equal squares

Fig. 13.3 The area of MFLI is equal to the sum of the areas of DKMG and LHBE

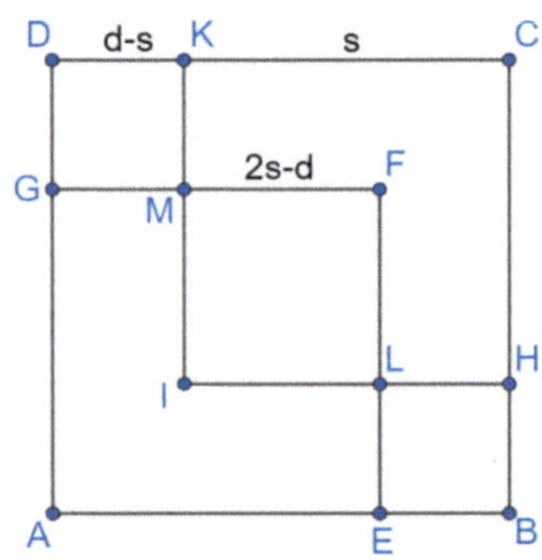

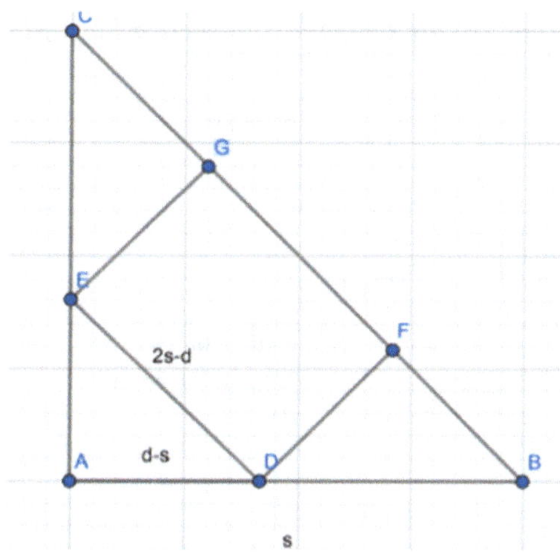

Fig. 13.4 Representation of the triangle used in the fifth proof

Tennenbaum's proof with Proposition II.9, in fact, it is really another proof of the Subtractive Elegant Theorem.

Fifth Proof
Let ABC be a right triangle with side $s = AB = AC$ and hypotenuse/diameter $d = BC$, as depicted in Fig. 13.4. We set $AD = AF = d - s$. We set $x = DE$, and *claim* $x = 2s - d$. Indeed,

$x^2 = 2(d - s)^2$ (by the isosceles Pythagorean theorem)
$= 2d^2 + 2s^2 - 4ds$ (Proposition II.7)
$= 3d^2 - 4ds$ (isosceles Pythagorean theorem)
$= d^2 + 4s^2 - 4ds$ (isosceles Pythagorean theorem)
$= (2s - d)^2$ (Proposition II.7)
Thus $x = 2s - d$, and hence
$2s - d$ is equal to the hypotenuse/diameter of the right triangle with sides $d - s$.

Note We realized that Fowler's argument, given below in 13.10.3, can be turned into still another proof of the Subtractive Elegant theorem.

13.10.2 The Anthyphairetic Proof of the Incommensurability of the Diameter to the Side of a Square from the Subtractive Elegant Theorem

Proposition *If* $d^2 = 2s^2$, *then* $Anth(d, s) = [1, 2, 2, 2, \ldots]$, *and thus the diameter and the side of a square are incommensurable to each other.*

***Proof** (Using Only the Subtractive Elegant Theorem)*

	$d^2 = 2s^2$		
		Def. $s_1 = d - s$	
		$d = s + s_1$	
			$d = s + s_1$
Def. $d_1 = s - s_1$			
$s = d_1 + s_1$			
	by the Subtractive Elegant Theorem $d_1^2 = 2s_1^2$		
		Def. $s_2 = d_1 - s_1$	
		$d_1 = s_1 + s_2$	
			$s = 2s_1 + s_2$
Def. $d_2 = s_1 - s_2$			
$s_1 = d_2 + s_2$			
	by the Subtractive Elegant Theorem $d_2^2 = 2s_2^2$		
		Def. $s_3 = d_2 - s_2$	
		$d_2 = s_2 + s_3$	
			$s_1 = 2s_2 + s_3$
	ad infinitum		

The general inductive step is as follows:

Inductive assumption $s_{n-1} = d_n + s_n$			
	Inductive assumption $d_n^2 = 2s_n^2$		
		Def. $s_{n+1} = d_n - s_n$	
		$d_n = s_n + s_{n+1}$	
			$s_{n-1} = 2s_n + s_{n+1}$
Def. $d_{n+1} = s_n - s_{n+1}$			
$s_n = d_{n+1} + s_{n+1}$			
	by the Subtractive Elegant Theorem $d_{n+1}^2 = 2s_{n+1}^2$		

If s_{n-1}, s_n, d_n have been defined so that $s_{n-1} = d_n + s_n$ and $d_n^2 = 2s_n^2$, then, setting $s_{n+1} = d_n - s_n$ and $d_{n+1} = 2s_n - d_n$, we obtain that $s_n = d_{n+1} + s_{n+1}$, and by the *Subtractive Elegant theorem*, that $d_{n+1}^2 = 2s_{n+1}^2$. We then obtain $\text{Anth}(d, s) = [1, 2, 2, \ldots]$. Incommensurability follows from Proposition X.2. □

Note. The arithmetical form of the Subtractive Elegant theorem implies an arithmetical proof of the incommensurability of the diameter to the side of a square

Proof (S. Tennenbaum's Original Proof, Using the Tennenbaum Proof of the Subtractive Elegant Theorem; T.M. Apostol's [4] Proof Using the Chrystal Proof of the Subtractive Elegant Theorem) Let s, d be the side and the diameter of a square, so that $d^2 = 2s^2$, and suppose that d, s are commensurable. Then there are least numbers m, n and a line segment c, such that $s = mc, d = nc$. We obtain the equation $n^2 = 2m^2$. By the arithmetical form of the Subtractive Elegant Theorem, $(2m - n)^2 = 2(n - m)^2$; and $0 < 2m - n < n$; $0 < n - m < m$; a contradiction, since we have found numbers that express the equality $d^2 = 2s^2$ and which are strictly smaller than the numbers m, n, respectively. □

13.10.3 Ancient and Modern Proofs of the Elegant Theorem II.10/11.a: if $d^2 = 2s^2$, then $(d + 2s)^2 = 2(d + s)^2$

Proof from II.10
Immediate, as Proclus notes.

Proof from the Theory of Gnomons of Book II
Assume $d^2 = 2s^2$. We form the lines $d+s, d+2s$. By Proposition II.4, $(d+2s)^2 = d^2 + 4s^2 + 4sd = 6s^2 + 4sd$. By Proposition II.7, $2(d + s)^2 = 2d^2 + 2s^2 + 4ds = 6s^2 + 4ds$. Hence, $(d + 2s)^2 = 2(d + s)^2$.

Third Proof ***(Fowler [21])***
Let ABC be a right-angle isosceles triangle with $AB = AC = s$, $BC = d$ (see Fig. 13.5). By the isosceles Pythagorean theorem $d^2 = 2s^2$. We construct points D, E, such that $A*B*D$ in the same line and $BD = d$ and $A*D*E$ in the same line and $DE = d$, and we construct the line DE. Then ADE is an isosceles right triangle

Fig. 13.5 Representation of the triangle used in the proof by Fowler

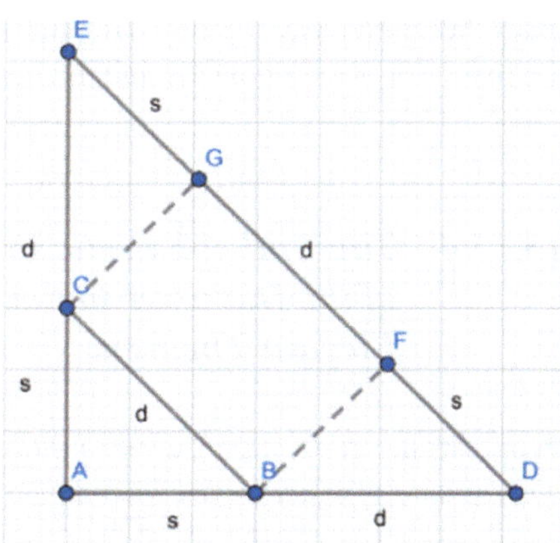

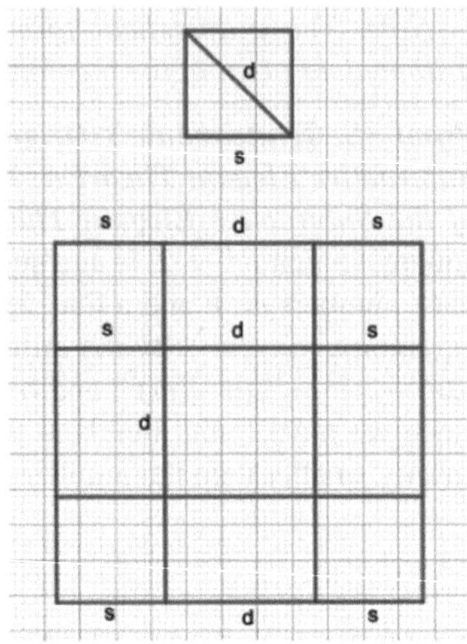

Fig. 13.6 Representation of the squares and rectangles of the fourth proof

with sides $s + d$. We construct lines BF and CG perpendicular to the line DE through the points B, C, respectively. It is clear that $DF = s$, $EG = s$ and $FG = d$. Hence, $ED = 2s + d$. By the isosceles Pythagorean theorem $(d + 2s)^2 = 2(s + d)^2$.

Fourth Proof *(Inverse Tennenbaum)*
Let a square have side s and diameter d. We construct the square with side $s + d + s$; this square consists of a square d^2, four squares s^2 and four rectangles sd (see Fig. 13.6). Thus, $(s + d + s)^2 =$ (by Proposition II.8) $d^2 + 4s^2 + 4ds =$ (since $d^2 = 2s^2$)
$2d^2 + 2s^2 + 4ds =$
$2(d^2 + s^2 + 2ds) =$ (by Proposition II.4)
$2(s + d)^2$.

Fifth Proof *(Inverse Chrystal)*
Let a right triangle have right side s and hypotenuse/diameter d, as depicted in Fig. 13.7. Extend the side to a line $s + d + s$, and extend the hypotenuse d to a line $d + s$. We then prove, by reversing the steps in the Chrystal proof (proof 9.1.3), that the line $s + d$ is the right side and the line $s + d + s$ the hypotenuse of a right triangle.

13.10.4 The Anthyphairetic Proof of the Incommensurability of the Diameter to the Side of a Square from the Elegant Theorem

Proposition *If $d^2 = 2s^2$, then $Anth(d, s) = [1, 2, 2, 2, \ldots]$, and thus the diameter and the side of a square are incommensurable to each other.*

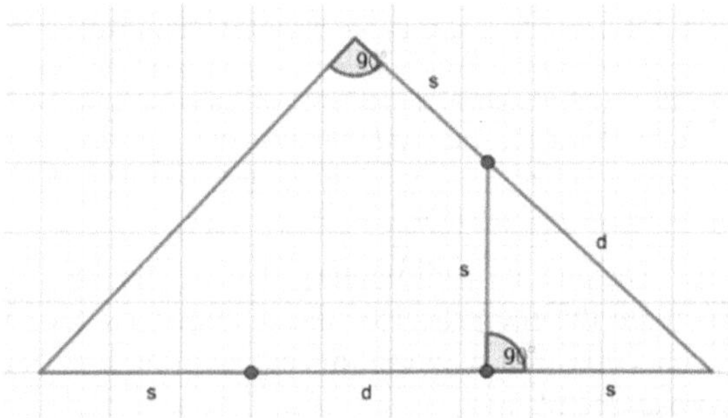

Fig. 13.7 Representation of the triangle used in the fifth proof

Proof (Using *Only* the Elegant Theorem) If $d^2 = 2s^2$, namely, if d is the diameter and s is the side of a square, then $(d+2s)^2 = 2(d+s)^2$, namely, $d+2s$ is the diameter and $d + s$ is the side of a square; It is seen immediately that $d(d + s) = s(d + 2s)$. Then by Proposition 13.4.1, we have $\text{Anth}(d, s) = \text{Anth}(d + 2s, d + s)$. On the one hand $d > s$, and $d = s + (d - s)$, with $d - s < s$ (since $d^2 = 2s^2 < 4s^2$, hence $d < 2s$), Hence $\text{Anth}(d, s) = [1, \text{Anth}(s, d - s)]$.

On the other hand, it is clear that $d + 2s = (d + s) + s$ with $s < d + s$, and $d+s = 2s+(d-s)$ with $d-s < s$. Hence $\text{Anth}(d+2s, d+s) = [1, 2, \text{Anth}(s, d-s)]$. Thus $[1, \text{Anth}(s, d - s)] = [1, 2, \text{Anth}(s, d - s)]$. Hence $\text{Anth}(s, d - s) = [2, \text{Anth}(s, d-s)]$. By induction we have $\text{Anth}(s, d-s) = [2, 2, 2, \ldots]$, and finally $\text{Anth}(d, s) = [1, 2, 2, 2, \ldots]$. The incommensurability of the diameter to the side follows, by Proposition X.2 of the *Elements*, from the fact that the anthyphairesis of d to s is infinite. □

Note Although, as we have seen in Sect. 13.8, for Proclus the Elegant Theorem has the role of a geometrical model for the arithmetical inductive step of the proof of the Pell property of the side and diameter numbers, it is nevertheless true that the Elegant Theorem can provide a very simple proof of the computation of the infinite anthyphairesis of the diameter to the side of a square. It is not known if the ancients had ever used the Elegant Theorem for this proof. The modern proof was first discovered by D. H. Fowler.

> "But, just as Socrates unblocks that impasse [in the *Meno*] by conjuring a clever figure out of thin air, so I here draw Figure [1]:

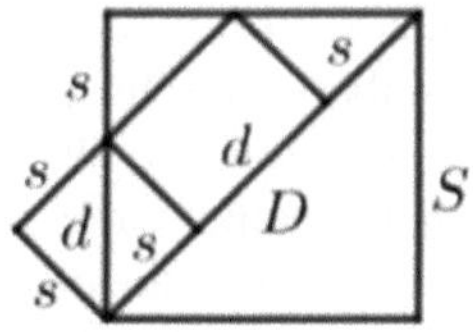

> Starting from the small diagonally placed square in the left-hand corner, with side s and diagonal d, we construct a larger square whose side S is $s + d$, and check that its diagonal D is $2s + d$.... With one eye on Figure [1], and with the insight that the ratios are unaffected by the scale or orientation of the figures involved, we take up our problem again, and see

that the ratio (big side to big diagonal minus side) is the same as the ratio (little side plus diagonal to little side); and we can now evaluate this as twice, followed by the ratio (little side to little diagonal minus side) which is -scale and orientation aside - what we just started from. Hence *the ratio (side to diagonal minus side) is twice, twice, twice, twice, continuing thus indefinitely*, and so the ratio (diagonal to side of a square) is once, twice, twice, twice, ...' [D. H. Fowler, p. 231 [21]]

Note Despite its rudimentary nature, the idea of using the Elegant Theorem to calculate the reciprocal of the diameter by the side is not self-evident. Although Proclus does not mention this calculation, it is quite possible that the Pythagoreans were aware of it.

However, this calculation clearly *follows* the discovery of the side-diameter numbers, which demonstrates the crucial relation of the pairs (a, b) and $(a + 2b, a + b)$ between "diameter-side" pairs, and this arithmetic recursive relationship is imitated in the geometrical Elegant Theorem.

So the calculation of the anthyphairesis via the Elegant Theorem *follows* the discovery of the diameter-to-side incommensurability, and it must be rejected as the reconstruction of the original Pythagorean calculation of the diameter-to-side anthyphairesis.

13.11 The Structure of the Partially Restored Book II of the *Elements* Suggests the Rejection of the Anthyphairetic Proofs Based on Propositions II.9 and 10, and Leads Us to the Expectation of a Reconstruction of the Original Pythagorean Proof of the Incommensurability of the Diameter to the Side of a Square at the Position II.8/9.a of Book II

But now this realization has a fascinating consequence in our quest for the discovery of the original Pythagorean proof of incommensurability. We note that Propositions II.9 and 10 deal with the side and diameter numbers, in fact with the proof of the Pell property of these numbers. But according to common sense, and, in fact, according to the explicit statement by Proclus, the proof of the incommensurability of the diameter to the side of a square must *precede* the introduction, definition and study of the side and diameter numbers, and thus, in the complete restored form of Book II, we expect that the incommensurability proof will *precede* the introduction of the side and diameter numbers, namely the incommensurability proof will *precede* Proposition II.9. We thus expect that the proof of incommensurability of the diameter to the side of a square will be *placed at the position II.8/9.a*. Schematically:

We are led to the position II.8/9.a for the original position and proof of the Pythagorean incommensurability by looking *backwards* from Propositions II.9 and 10. But this scheme will have a chance to work, only if the tools developed in Book

Book II partially restored suggests the position and the original Pythagorean method of proof of the incommensurability of the diameter to the side of a square

II.1, 2, 3		II.9— def. of side & diameter numbers
II.4		II.9
II.4/5 Pythagorean theorem		II.9/10 Subtractive Elegant Theorem
		II.10
II.5		II.10/11.a *Elegant theorem*
II.5/6 Application of Areas in Defect		II.10/11.b Pell property
II.6		*anthyphairetic definition of side & diameter numbers (Proclus,* Commentary to Republic *2,27,2–10)* II.10/11.c Anth(q_n, p_n)= $[1, 2, 2, \ldots, 2$ ($n-1$ times)]
II.6/7 Application of Areas in Excess		II.11 Mean and extreme ratio
II.7		II.12 & II.13 Converse Pythagorean theorem
II.8 $(a+2b)^2 = a^2 + 4b(a+b)$		II.14 Mean proportional
	II.8/9.a *Original Pythagorean proof of incommensurability before the third column and, hopefully, with the tools from the first column*	

II *before* II.8/9.a, namely the Propositions *up to and including Proposition II.8*, will serve as the tools for this incommensurability proof.

13.12 The Reconstructed Original Pythagorean Proof of Incommensurability at the Position II.8/9.a of Book II of the *Elements*, with the Use of the Application of Areas in Excess

13.12.1 Proposition II.8/9.a and Its Pythagorean Proof

Proposition 13.6 (II.8/9.)) *If a is the diameter and b is the side of a square then $a^2 = 2b^2$, $Anth(a, b) = [1, 2, 2, 2, \ldots]$, and a, b are incommensurable.*

Proof The proof is in four steps.

Step 1. Since $a^2 = 2b^2$, we have that $b < a < 2b$, we get

$$a = b + c_1, c_1 < b.$$

Thus Anth$(a, b) = [1,$ Anth$(b, c_1)]$.

Step 2. We proceed in the anthyphairetic substitution $b + c_1$ for a in the original equality, and get

$$(b + c_1)^2 = 2b^2.$$

By Propositions II.4 and II.1, we find $a^2 = (b + c_1)^2 = b^2 + c_1^2 + 2bc_1 = 2b^2$, $b^2 = 2bc_1 + c_1^2$ (see Fig. 13.8).

Then $b(b - 2c_1) = c_1^2$. We set

$$x = b - 2c_1,$$

hence $b = 2c_1 + x$. Then

$$x(2c_1 + x) = c_1^2,$$

a Pythagorean Application of Areas in Excess II.6/7
(in fact the very first use of II.6/7),
with $a = 2c_1, m = c_1$. Thus, the substitution of the first anthyphairetic relation into the original equality results in a Pythagorean application of areas in excess. *Proposition II.6/7*, the solution of such an application of areas, is the ideal tool for finding, computing the second anthyphairetic quotient.

We set y for the hypotenuse of the right triangle with right sides $m = c_1$ and $a/2 = c_1$, as depicted in the above Fig. 13.9. Then $y^2 = c_1^2 + c_1^2 = 2c_1^2$, hence $y = c_1 + c_2$, for some c_2, with $c_2 < c_1$. By *Proposition II.6/7*, the solution x is equal to $x = y - c_1$, thus $x = y - c_1 = (c_1 + c_2) - c_1 = c_2 < c_1$, hence

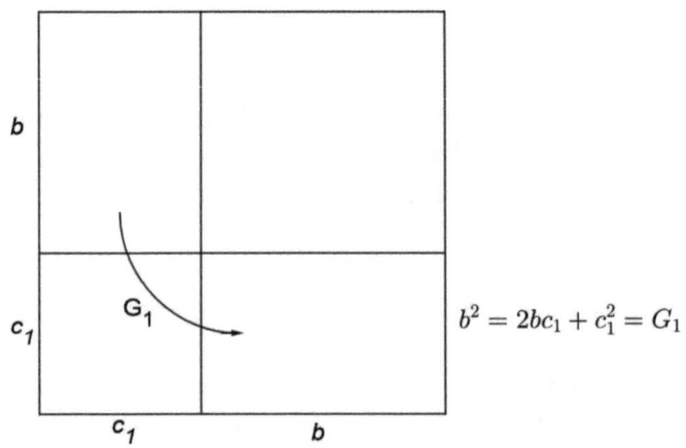

Fig. 13.8 The original square is divided into two squares of different size and two equal rectangles

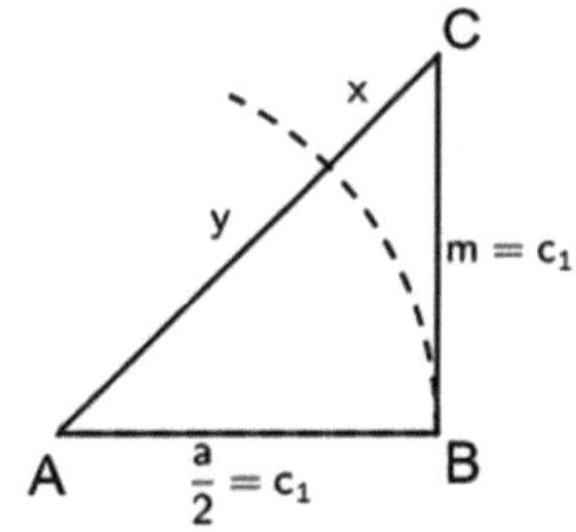

Fig. 13.9 The triangle used for the computation of the second anthyphairetic quotient

$b = x + 2c_1 = 2c_1 + c_2, c_2 < c_1$, $b = 2c_1 + c_2, c_2 < c_1$. Thus, the second anthyphairetic quotient is 2, and we have

$$\text{Anth}(b, c_1) = [2, \text{Anth}(c_1, c_2)], \text{ and}$$

$$\text{Anth}(a, b) = [1, \text{Anth}(b, c_1)] = [1, 2, \text{Anth}(c_1, c_2)].$$

We note that this use of Application of Areas is purely *"high-school" algebraic*, in the sense meant by van der Waerden. Essentially, we solve a quadratic equation, and thus find its integer part.

Step 3. Now we proceed in exactly the same way as in the second step. We substitute b by its anthyphairetic equal $2c_1 + c_2$ in the application of areas $(2c_1 + c_2)^2 = 2(2c_1 + c_2)c_1 + c_1^2$ (see Fig. 13.10).

By Proposition II.8 (first and only use of II.8!), we find $c_2^2 + 4c_1^2 + 4c_1c_2 = 4c_1^2 + 2c_1c_2 + c_1^2$, then $c_2^2 = c_1^2 - 2c_1c_2 = c_1(c_1 - 2c_2)$, and $c_2^2 = c_1(c_1 - 2c_2)$. We set $x = c_1 - 2c_2$, then $c_1 = x + 2c_2$, then $x(x + 2c_2) = c_2^2$. We observe first that with the anthyphairetic substitution, we have come up again with *an application of areas in excess II.6/7*, but secondly that this application of areas $x(x + 2c_2) = c_2^2$ is the same as the previous one: $x(x + 2c_1) = c_1^2$.

Step 4. For the completion of the proof we will make use of some propositions making heavy use of the elementary Proposition I.44 of the *Elements* plus a rudimentary induction. We note that Proposition I.44 is in the second part of Book I, which is, as we have already argued, preparatory for Book II, culminating, according to our interpretation with the proof of incommensurability at the exact position II.8/9. Proposition I.44 then finds its use in the proof of Propositions 5, 6 and 7. □

13.12.2 Proposition: The Preservation of Application of Areas (In Excess)/The Preservation of Gnomons

If $Aa^2 = Bab + Cb^2$, and $Ac^2 = Bcd + Cd^2$, then $ad = bc$.

Lemma *If* $Aa^2 = Bab + Cb^2$ *and* $Aa^2 = Bae + Ce^2$, *then* $b = e$.

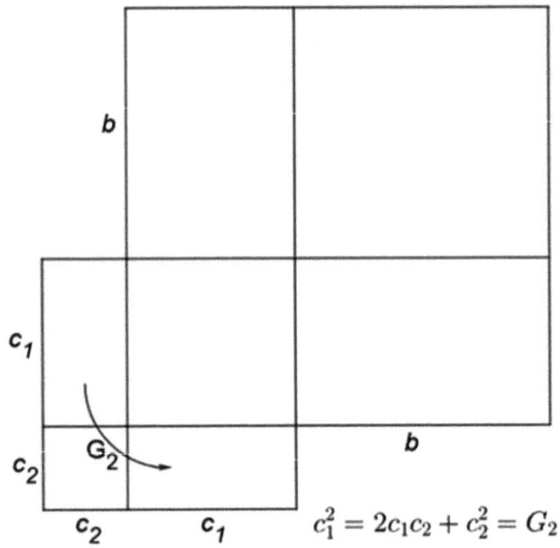

Fig. 13.10 Representation of the second gnomon G_2 of the process

Proof of Lemma By hypothesis $Bab + Cb^2 = Bae + Ce^2$. Suppose b unequal to e, say $b > e$. Then $Bab > Bae$ and $Cb^2 > Ce^2$, hence $Bab + Cb^2 > Bae + Ce^2$, a contradiction. □

Proof of Proposition 13.12.2 By Proposition I.44, there is a line e such that $ae = bc$. From the assumption,

$Aa^2c = Babc + Cb^2c$, whence
$Aa^2c = Baae + Cbae$, whence
$Aac = Bae + Cbe$, whence
$Aacc = Baec + Cbec$, whence
$Ac^2a = Bcea + Ce^2a$, whence
$Ac^2 = Bce + Ce^2$.
By Lemma, $d = e$.
Hence $ad = bc$. □

13.12.3 *Proposition: If* $Aa^2 = Bab + Cb^2$ *and* $Ac^2 = Bcd + Cd^2$, *then* $Anth(a, b) = Anth(c, d)$

Proof of Proposition 13.12.3 If $Aa^2 = Bab + Cb^2$, and $Ac^2 = Bcd + Cd^2$, then $ad = bc$ (Proposition 13.12.2); and,
if $ad = bc$, then $\mathrm{Anth}(a, b) = \mathrm{Anth}(c, d)$ (Proposition 13.4.1). □

13.12.4 The Two Pythagorean Uses of the Pythagorean Application of Areas in Excess II.6/7 in the Original Pythagorean Incommensurability Proof: (a) Geometric Algebra for Steps 2 and 3, and (b) Preservation of Gnomons for Step 4

The *Pythagorean method of the application of areas* has been proven crucial in this reconstruction. The application of areas is used in both stages of the proof, first, in order to construct a solution of the application of areas that is formed at every stage of the division and use this construction/solution in order to calculate the quotient of the anthyphairetic division step, and secondly, in order to note, when at some stage of the anthyphairesis we come to an application of areas having exactly the same form as in a previous stage, *the preservation of areas/Gnomons*, that the anthyphairesis will be repeated from there on ad infinitum.

The first use, in our opinion, resolves the long conflict between the supporters and opponents of *Geometric Algebra* in favor of the supporters. The Pythagoreans/Hippasus wanted *to compute the anthyphairesis* of the diameter to the side of a square so as to prove it is infinite. They needed to compute the *anthyphairetic quotient* for each one of the infinite steps. They succeeded in doing this by solving the quadratic equation $x^2 = 2xy + y^2$. In modern terms setting $\omega = x/y$, $\omega^2 - 2\omega - 1 = 0$, we find $\omega = (2 + \sqrt{8})/2$, hence the integer part of ω is equal to 2, and this is precisely the anthyphairetic quotient. The Pythagoreans are thus doing algebra precisely in the sense of van der Waerden, p. 199 [62]:

> When I speak of Babylonian or Greek or Arab algebra, I mean algebra in the sense of AL-KHWARIZMI, or in the sense of CARDANO'S "Ars magna", or in the sense of our school algebra. Algebra, then, is: the art of handling algebraic expressions like $(a + b)^2$ and of solving equations like $x^2 + ax = b$.

High School Algebra, indeed:

Starting with the Application of Areas in excess $b^2 = 2bc_1 + c_1^2$, transforming it to modern notation $(b/c_1)^2 = 2(b/c_1) + 1$, and setting $b/c_1 = \omega$, we obtain the quadratic algebraic equation $\omega^2 - 2\omega - 1 = 0$. Solving it we find $\omega = (2 + \sqrt{2^2 + 4 \cdot 1 \cdot 1})/2 = (2 + \sqrt{8})/2 = 1 + \sqrt{2}$, thus $2 < \omega < 3$. Reverting back to ancient notation $2/1 < b/c_1 = \omega < 3/1$, namely $2c_1 < b < 3c_1$, namely $b = 2c_1 + c_2$, $c_2 < c_1$. This is precisely the second anthyphairetic step.

Our reconstruction of the original Pythagorean proof of the incommensurability of the diameter to the side of a square with the use of the Pythagorean Application of Areas in Excess II.6/7, a use that is clearly algebraic in the van der Waerden sense of "high school algebra" provides the only use of the method of Application of Areas by the Pythagoreans and, in our opinion, settles the question of *Geometric Algebra*, deciding in favor of the thesis of those historians of Mathematics, such as Tannery, Zeuthen, van der Waerden, who originally proposed it and had the right intuition, but, as far as we know, never explained the use that the Pythagoreans had for the method they discovered.

The second Pythagorean use of the Pythagorean Application of Areas is related to the preservation of the Gnomons, namely that the preservation of the Gnomons, the preservation of the Application of Areas in excess, implies the preservation of the anthyphairetic quotient, and thus constitutes a finitization of the infinite anthyphairesis.

13.13 The Anthyphairetic Nature of the Two Pythagorean Philosophic Principles Infinite and Finite at the Position II.8/9.b Provides a Confirmation of Our Reconstruction of the Original Pythagorean Proof of Incommensurability in Sect. 13.12

We will briefly examine the Pythagorean philosophy as presented by Philolaus, Plato and Aristotle. The most impressive element of the Pythagorean philosophy is that it posits as principles two abstract concepts, the Infinite and the Finite, concepts which derive, according to Aristotle's categorical assurance, from their dealing with "lessons" (Mathematics). We should thus be able to trace, within the Pythagorean mathematical discoveries, the origin of these two principles.

Question 7 What is the meaning of the two Pythagorean principles Infinite and Finite?

Indeed, we will demonstrate that the Pythagorean philosophy is a faithful rendering of the fundamental geometric reconstruction of the Pythagorean anthyphairetic proof of the incommensurability of the diameter to the side of a square, presented in Sect. 13.12. Specifically, the two Pythagorean principles Infinite/Apeiron and Finite/Peras correspond to the infinite anthyphairesis of the diameter by the side of a square, and its Finitization by the fact that at each stage of the infinite process the shape and form are preserved, with the subtraction of a Gnomon, in the sense that the anthyphairetic dyad of that stage satisfies the same application of areas in excess.

13.13.1 According to Aristotle, the Two Philosophic Pythagorean Principles, Infinite and Finite, Have Mathematical Origin

(1) Aristotle is clear: *the Pythagorean philosophic principles were the principles that were formed by their pre-occupation with Mathematics.*

> Among these thinkers and before them,
> the so-called Pythagoreans were the first to latch on to mathematics.
> They both advanced these inquiries and, having been brought up in mathematics,
> thought that its starting-points were the starting-points of all beings.

Aristotle, *Metaphysics* 985b23–26 [10]

(2) The two Pythagorean principles are *the Finite and the Infinite*,

> The Pythagoreans have spoken of the principles as two ... *the Finite* and *the Infinite*.
> Aristotle, *Metaphysics* 987a13–16 [10]

More carefully Philolaus in his Fragment 1 refers to the two Pythagorean principles as the Infinite and the *Finitizing* (περαινόντων).

Plato in the *Philebus* is crediting some almost mythical Prometheus, who is certainly Pythagoras.

> *Socrates* A gift of gods to men, as I believe, was tossed down from some divine source *through the agency of a Prometheus* together with a gleaming fire; and the ancients, who were better than we and lived nearer the gods, handed down the tradition that things which are ever said to exist are sprung from one and many and have inherent in them *the finite and the infinite*.
> Plato, *Philebus* 16c5–10 [45]

We then expect to find in the Pythagorean Mathematics the source of these two fundamental Pythagorean philosophical principles.

13.13.2 The Pythagoreans Gave the Symbolic Names Even/Artion for the Principle of the Infinite/Apeiron, and Odd/Peritton for the Principle of the Finite/Peras

The Pythagoreans liked to express themselves symbolically; thus the Finite was called Odd and the Infinite Even. The linguistic similarity of Finite (Peras) with Odd (Peritton) and of Infinite (Apeiron) with Even (Artion) is lost in English.

> the odd and the even,
> and that of these the odd is the Finite and the even the Infinite
> Aristotle, *Metaphysics* 986a18–19 [10]

Our task is to obtain from the symbolic descriptions the real content of the two Pythagorean principles.

13.13.3 Aristotle and Simplicius Clarify the Symbolic Description of the Pythagorean Principle of the Finite as Odd: An Odd Number of Units in the Figure of an Arithmetical Gnomon Surrounding a Square and thus Forming a Larger Square, thus Preserving the Square Figure; the Preservation of the Square Arithmetical Figure During This Process Ad Infinitum Is Precisely the Symbolic Finitization

But there are things that increase *without altering*,
as the square
by surrounding it with *a gnomon*
is increased *but is not thereby altered*;
Aristotle, *Categories* 15a30–31 [5]

The Odd number possesses a finitizing power, in the sense that an Odd number in the form of an arithmetical Gnomon, by surrounding and comprehending a square, increases the square, but preserves the form of the square. Simplicius explains in detail, in his *Commentary to Aristotle's Physics* 456,16–457,8 [53] and we paraphrase:

Why the principle of the *Finite* is symbolized by *the Odd*?
Every *Odd* number $2n + 1$ can be represented as an arithmetical *Gnomon* G_n consisting of n horizontal units + 1 corner unit + n vertical units.
And what is the Infinite we may ask that is finitized by this Gnomon and in what sense?
The Infinite is the infinite collection of all squares.
The connection of "Gnomon" and "Peras-Peritton" is understood always on a symbolic level, if we note that the arithmetical Gnomon G_n with which we go from a small arithmetical square T_n to the immediately larger square T_{n+1} is always an odd number and has the shape of the Gnomon.
That is, for each arithmetical square
$T_1 = 1 \times 1, T_n = n \times n$
the Gnomon/Odd
$G_n = T_{n+1} - T_n = 2n + 1$
"comprehends" (εναπολαμβάνει) the square by "surrounding" (περιθέσεως περί) it and thus generates the immediately greatest square
$T_{n+1} = (n + 1) \times (n + 1)$
in every step of an infinite process.

Thus the symbolic name Odd for the principle of the Finite is explained by the fact that it is a Gnomon, and the finitizing action of the Gnomon is clearly its comprehension and surrounding of a square, by which it has the power to preserve the shape of the square throughout infinite process. The new square is greater than the old square, but it has not changed in the sense that it retains the same shape and form (as in Fig. 13.11).

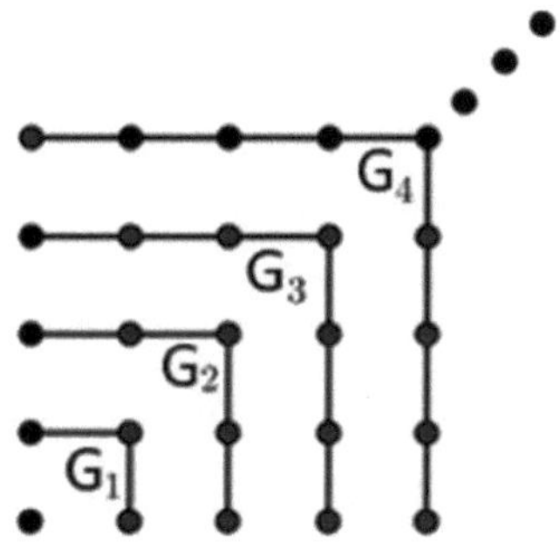

Fig. 13.11 The sequence of arithmetical Gnomons

The Odd/Gnomon G_n surrounding the square T_n increases to the square T_{n+1} but does not alter its shape, since after the circumscription of the square with the Odd/Gnomon, the resulting figure is again a square, exactly the same in form and shape as the original smaller square.

13.13.4 The Symbolic Description of the Infinite as Even Implies that the Real Pythagorean Principle of the Infinite Is a Principle of Binary Division Ad Infinitum, and in Fact, as Simplicius and John Philoponus Make Clear, a Principle of Binary Division Ad Infinitum of Magnitudes into Equal or Unequal Parts

ἅμα γαρ [οι Πυθαγόρειοι] ουσίαν ποιούσι το ἄπειρον και μερίζουσιν.
With the same breath they [the Pythagoreans] treat the infinite as substance, and divide it into parts.

According to Aristotle, *Physics* 204a33–34 [6] the Pythagorean principle of the Infinite is in fact a Principle of division ad infinitum. Simplicius, *Commentary to Aristotle's Physics* 455,20–23 [53] writes:

These [the Pythagoreans] said that the infinite was the even number,
because, according to their expositors, everything even can be divided into equals,
and what is divided into equal parts is infinite through bisection.
For division into equal halves proceeds *ad infinitum*.

But, shortly after, Simplicius specifies that the principle of the Infinite is a principle of binary division of magnitudes, not of numbers, and in fact in unequal parts.

So in this way the expositors ascribe the Infinite to the Even
through the division into equal parts,
and it is clear that they take the division *ad infinitum*
not as being of numbers but of *magnitudes* (οὐχ ἐπ' ἀριθμῶν ἀλλ' ἐπὶ μεγεθῶν).
For numbers, including the even, do not all divide into equals for long,
and those that are divided reach the one and halt the division,

while in the case of *magnitudes* what prevents *the remainder* (*τὸ καταλειπόμενον*, 455,29) from being divided *ad infinitum, even if the division is not into equal parts*? 455, 24–30 [53]

Finally, Simplicius states that in fact Aristotle does not have in mind division into equal parts, but rather into unequal parts.

In general *Aristotle does not seem to ascribe being Infinite*
to the division into equal parts.455,33–35

The similar comments by John Philoponus on the Infinite

A similar comment to Aristotle's *Physics* 202b30–203a16 is found in John Philoponus' *Commentary on Aristotle's Physics*, 386.14–395.7, especially in 388.24–389.20.
According to this Commentary, as the Pythagoreans presented their teachings "symbolically" ("συμβολικώς"), "from the numbers", "εκ των αριθμών",
so also the principle "the Infinite" was described "symbolically", as "Even",
something that already becomes clear from what Aristotle says.
For, Even "pertains to division" ("προς της διαιρέσεως εστί") and
"it is the cause of infinite division", as the dichotomy
"it is the cause of the division of the magnitudes ad infinitum"
and the Infinite is
"the cause" (*η αιτία*)
"of disunity, indeterminacy, division ad infinitum, and multitude"
(της διαστάσεως και αοριστίας και της επ άπειρον τομής και πλήθους),
"of the magnitudes and of all the entities in nature"
(*τοις μεγέθεσι* και πάσι τοις φυσικοίς).

This brings the Pythagorean principle of the Infinite close to a philosophical form of infinite anthyphairesis.

13.13.5 The Symbolic Arithmetic Gnomons Refer to the Real Geometric Gnomon of Book II of the Elements, and thus the Two Pythagorean Principles Arose from the Mathematics in Book II

Simplicius, *Commentary to Aristotle's Physics*

The Pythagoreans called *the odd numbers gnomons*
because when placed around squares they preserve the same shape,
like the gnomons in geometry. For they call the two complements together with one of the parallelograms around the same diagonal *a gnomon*, which, when added to the other of the parallelograms around the same diagonal *makes the whole similar to that to which it was added*.
So *the odd numbers* are also called *gnomons*
because, when added around the already existing squares,
they always preserve the square shape. 457,1–8 [53]

The arithmetical Gnomons that symbolize the Pythagorean principle of the Infinite is correlated with the geometric Gnomons of Book II of the *Elements*.

13.13.6 The Geometric Gnomons in the Real Principle of the Infinite Must Be Not Increasing, as in the Symbolic Description, but Decreasing, as Hinted by Aristotle and Specified Explicitly by Proclus

Since the *actual* Pythagorean Infinite refers to binary infinite *division* of magnitudes (and not to the symbolic incremental infinite sequence of numbers), what would the *actual* Pythagorean Finite/Peras be? The finitization of the symbolic incremental infinite of the sequence of numbers is carried out with Odd arithmetical Gnomons which preserve the shape unchanged throughout the infinite process. Since the Pythagorean principle of the Infinite is *symbolically* about arithmetical Gnomons increasing by addition ad infinitum, *but actually* about *geometric Gnomons* decreasing by division ad infinitum; and the principle of the Finite acts upon the Infinite by preserving, not the arithmetic, *but the geometric* Gnomons,

> it follows that the Gnomons that preserve the form in the real Pythagorean Infinite must be not arithmetical, but geometric, like the Gnomons of Book II of the *Elements*;

and, further, since the Infinite is a dyadic division ad infinitum into decreasing magnitudes, it also follows that

> *the geometrical Gnomons that preserve the form in the actual Pythagorean Infinite must be not increasing but decreasing.*

Decreasing Gnomons preserving the form are indeed foreseen in a passage in Aristotle's *Categories*:

> Thus alteration would be distinct from the other changes.
> For if it were the same, a thing altering would, as such, have to be
> *increasing too or decreasing*, or one of the other changes would have to follow;
> but this is not necessary.
> Equally, a thing increasing - or undergoing some other change - would have to be altering.
> But there are things that *increase without altering*,
> *as a square is increased by the addition of a gnomon but is not thereby altered*;
> similarly, too, with other such cases.
> Hence the changes are distinct from one another.
> Aristotle, *Categories* 15, 25–33 [5]

Proclus' passage, in his *Commentary to the First Book of Euclid's Elements* 59,15–61,17, refers explicitly to the presence of an infinite sequence of descending Gnomons, a feature characteristic of incommensurability:

> The statement that every ratio is commensurable belongs to arithmetic only
> and not to geometry, for geometry contains incommensurable ratios.

> Likewise, the principle that there is a least for the division of square gnomons is peculiar to arithmetic: in geometry there is no least gnomon. 60, 7–12 [48]

Here Proclus introduces a criterion of incommensurability, not simply by means of anthyphairetic infinity, but, more specifically, by the presence of an infinite decreasing sequence of Gnomons.

13.13.7 Anthyphairetic Interpretation of the Two Pythagorean Principles Infinite and Finite/Finitizing

According to Sections 13.13,1-6, behind the Pythagorean symbolism, *the real Pythagorean principle of the Infinite* is the decrease by binary division ad infinitum of magnitudes, in fact of squares and surrounding Gnomons, and *the real Pythagorean principle of the Finite* is the preservation, despite the ad infinitum decrease of magnitudes, of the same form and shape of the squares by the surrounding Gnomons, and in this sense the finitization of the Infinite.

At this point we cannot fail to recognize that a mixture of Infinity and Finiteness, with characteristics in striking agreement with what is required, appears in one and only one case in Pythagorean geometry: in our reconstruction of the proof of diameter-to-side square incommensurability, in Sect. 13.12, where the Infinite is the infinite process of anthyphairesis of the diameter by the side of a square, and the Finite is the preservation of the shape and form of the application of areas in excess/Gnomon, at each step.

The Finitization of the Infinite in the *anthyphairesis* of the diameter to the side of a square is analogous to the symbolic-arithmetic one, in the following sense:

In the symbolic description with the addition of the Gnomon, the shape and form of the square is increased but preserved indefinitely, in the infinite multiplication of a diameter by a side with the removal of the Gnomon, the shape and form of this equation of the parabola of sections in hyperbola is preserved indefinitely.

We conclude that the Pythagoreans did receive, as Aristotle expressly informs us, the two principles Apeiron and Peras from their engagement with the lessons, and specifically from their most fundamental discovery, that is, the anthyphairetic proof of the incommensurability of the diameter and side of a square.

13.13.8 According to Philolaus, Simplicius, Hesychius, Eustratius, Anonymon Scholion to Elements, the Gnomon Possesses the Power to Provide Knowledge

An additional confirmation of our interpretation is provided by the numerous ancient comments on the power of the Gnomon to provide knowledge.

Preservation of Gnomons/Finite	Anthyphairetic Division/Infinite
$a^2 = 2b^2$	
	$a = b + c_1$
$b^2 = c_1(2b + c_1) = G_1$	
	$b = 2c_1 + c_2$
$c_1^2 = c_2(2c_1 + c_2) = G_2$	
	$c_1 = 2c_2 + c_3$
$c_2^2 = c_3(2c_2 + c_3) = G_3$	
	$c_2 = 2c_3 + c_4$
...	
$c_{2n-2}^2 = G_{2n-1}$	
	$c_{2n-2} = 2c_{2n-1} + c_{2n}$
$c_{2n-1}^2 = G_{2n}$	
	$c_{2n-1} = 2c_{2n} + c_{2n+1}$
...	

According to Philolaus' (possibly not genuine) *Fragment* 11 the gnomon has cognitive power.

πάντα γνωστὰ καὶ ποτάγορα ἀλλάλοις κατὰ γνώμονος φύσιν ἀπεργάζεται
renders all things
makes [all things] known and agreeable with one another according to the nature of the gnomon
[[translation by Huffman, 1983 [29], with modifications by the authors]]

In the passage of the geometric number in *Republic* 546b3-c7, which has a close relationship with Philolaus' *Fragment* 11, the position of the guide is taken by Logos, which also has cognitive power:

πάντα προσήγορα καὶ ῥητὰ πρὸς ἄλληλα ἀπέφηναν· *Republic* 546b7-c1
render all things conversable and rational with one another

Simplicius, *Commentaries on Categories* 430.5–431.5, states that the geometric gnomon, consisting, according to Definition II.2 of Euclid's *Elements*, of the two complements with one of the parallelograms

γνωματεύει και κρίνει
makes known and judges the remaining rectangle.

Hesychius, *Lexicon* Γ 748, [3] p. 515 and Eustratios, *In Nicomachean Ethics* 23,15–30 [57] associate Gnomon with knowledge.

Anonymous Commentary II.11 *on Euclid's Elements*, which is a comment on Definition II.2 of the geometric Gnomon, justifies the name of the Gnomon by saying that

απ' αυτού. . . το όλον γνωρίζεται
by Gnomon...the whole becomes known

just as the astronomical Gnomon has as his sole mission

> το τας ενεστώτας ώρας ποιείν γνωρίμους
> the Gnomon is that which makes known the present hours

The cognitive nature and power of the Pythagorean principle is in complete agreement with the role of the geometric principle of the preservation of Gnomons in the reconstruction of incommensurability given in Sect. 13.12, where the acquisition of knowledge of the infinite anthyphairetic expansion of the diameter to the side of a square is clearly achieved because of the finitizing power of Gnomon. So the cognitive nature and power of the gnomon in Pythagorean philosophy is the transference of the corresponding cognitive nature and power of the gnomon in Pythagorean geometry, and especially in the Pythagorean proof of the incommensurability of diameter to side of a square. This association of the proof method, i.e. the anthyphairetic method, with the acquisition of complete knowledge of diameter to side is an additional substantial argument in favor of the anthyphairetic reconstruction, as the traditional arithmetic incommensurability method of proof is far from providing this complete knowledge.

13.14 The Two Philosophical Principles Infinite and Finite Reappear in Zeno's *Fragment* B3, with Substantially the Same Anthyphairetic Meaning, Establishing an Early Dating of the Pythagorean Proof of Incommensurability and a Further Confirmation of Their Method of Proof

Question 8 Dating of the discovery of the Pythagorean incommensurability

Democritus, born about 470 BC, wrote 'on irrational lines and solids;' and as Hardy & Wright, p. 42 [26] have observed

> 'it is difficult to resist the conclusion that the irrationality of the diameter to the side of a square was discovered before Democritus' time.'

But the relation of the Pythagorean discovery of incommensurability with Zeno of Elea is even stronger and older. We will be content here to consider

> Καὶ τί δεῖ πολλὰ λέγειν, ὅτε καὶ ἐν αὐτῷ φέρεται τῷ τοῦ Ζήνωνος συγγράμματι; πάλιν γὰρ δεικνύς, ὅτι
> εἰ πολλά ἐστι, τὰ αὐτὰ πεπερασμένα ἐστὶ καὶ ἄπειρα,
> γράφει ταῦτα κατὰ λέξιν ὁ Ζήνων·
>
> [Zeno's *Fragment* B3]
> "εἰ πολλά ἐστιν,
> ἀνάγκη τοσαῦτα εἶναι ὅσα ἐστὶ καὶ οὔτε πλείονα αὐτῶν οὔτε ἐλάττονα.
> εἰ δὲ τοσαῦτά ἐστιν ὅσα ἐστί, πεπερασμένα ἂν εἴη.

εἰ πολλά ἐστιν, ἄπειρα τὰ ὄντα ἐστίν.
ἀεὶ γὰρ ἕτερα μεταξὺ τῶν ὄντων ἐστί, καὶ πάλιν ἐκείνων ἕτερα μεταξύ.
καὶ οὕτως ἄπειρα τὰ ὄντα ἐστί."
Simplicius, *Commentary to Aristotle's Physics* 140,27–33 [53]

And why should I say any more, for it also exists in the treatise of Zeno?
For again, showing that if there are many,
the same things will be Finite and Infinite,

Zeno writes thus verbatim:

"If there are many,
necessarily they are as many as they are, and neither more of them nor fewer.
But if they are as many as they are,
they would be Finite.

If they are many,
existing things will be Infinite.
For there are always other things generated by existing things,
and again other things generated by them.
And in this way existing things will be infinite."

According to the analysis of Zeno's arguments and paradoxes, undertaken by Negrepontis [37]

(a) Zeno's arguments and paradoxes, including Zeno's *Fragment* B3, aim to establish, by a contradiction, *not* the impossibility of motion or the impossibility of multiplicity, as is generally believed, *but* that the true [Plato's intelligible] Beings and the Beings of Opinion [Plato's sensibles] are separate and different entities;
(b) the true Beings satisfy the almost contradictory property *Infinite and Finite*, namely of having at the same time an *infinite multitude* of parts, by the principle of Otherness, and being *One*, in the self-similar sense, according to which each of these infinitely many parts has *as many* (*tosa*) parts *as* (*hosa*) *of* any other part, namely *the same finite number* of parts, while the sensible entities do not satisfy the *Infinite* and *Finite* property;
(c) the contradiction results from the assumption that the true Beings coincide with the sensibles, called the many;
(d) the One of the second hypothesis in the Platonic dialogue *Parmenides*,
(d1) is the philosophical analogue of *periodic anthyphairesis*,
–of *infinite anthyphairesis*, by means of the generation ad infinitum of parts, by the principle of Otherness (*Parmenides* 143b1–8), exactly as with the generation of the Infinite in Zeno's *Fragment* B3, and
–*of circularity and periodicity of the anthyphairesis*, resulting in each of these infinitely many parts possesses *as many* (*tosa*) parts *as* (*hosa*) *of* any other part, namely *the same finite number* of parts (*Parmenides* 144d4-e3), exactly as with the self-similar One and the Finite in Zeno's *Fragment* B3,
(d2) is a paradigmatical Platonic intelligible Being, and
(d3) coincides with Zeno's true Being.

Thus *Zeno's true Being is also the philosophic analogue of periodic anthyphairesis*, something that can only mean that Zeno's true Being is a philosophic imitation of the Pythagorean discovery of the infinite periodic anthyphairesis of the diameter to the side of a square. Since the only way that Zeno would have as principles for his true Beings the anthyphairetic Infinite and the Finite is by imitating the Pythagoreans, it follows that the Pythagorean proof of incommensurability has occurred at an early date, namely before Zeno formed his paradoxes and arguments.

For details on our anthyphairetic interpretation of Zeno's paradoxes and arguments the reader is referred to Negrepontis [37]. Our analysis of the heavy reliance of Zeno's philosophical arguments and paradoxes on Pythagorean incommensurability undermines the basis of Burkert's thesis ([16], p. 456), summed up as

> Certain as it is that Zeno's arguments are relevant to mathematics and even contributed in a certain way to the development of Greek mathematics, still, from the historical point of view ontology is prior to mathematics.

In fact, Zenonian ontology is shown to be posterior to the Pythagorean mathematics of incommensurability.

13.15 The Second Confirmation of Our Reconstruction: Our Interpretation of the Pythagorean Definition of the Three Kinds of Angles at the Position II.13/14.a and of the Postulate 4 of the *Elements* at the Point II.13/14.b, in Terms of the Two Pythagorean Principles Infinite and Finite

In the long fascinating passage Proclus, *In Euclidem* 131.13–134.7 [48] describes how the Pythagoreans maintained that the cause of the right angle is the Pythagorean principle of the Finite, and the cause of the acute and obtuse angles is the Pythagorean principle of the Infinite. We will outline our interpretation of the Pythagorean approach to the three kinds of angles, not by simply defining them, but by appealing to the two Pythagorean principles.

13.15.1 *The Cause of the Obtuse and Acute Angles on the One Hand Is Said by Proclus to Be the Pythagorean Principle of the Infinite, and on the Other Hand by Our Interpretation Is the Infinite Anthyphairesis of the Diameter to the Side of a Square, Suggesting that the Pythagorean Principle of the Infinite Coincides with the Infinite Anthyphairesis of the Diameter to the Side of a Square*

We will argue that Propositions II.12, 13 for isosceles triangles express the inequalities characterizing the acute and obtuse angles, and the isosceles Pythagorean theorem II.4/5 expresses the equality characterizing the right angle.

The Proclus passage describes how the acute and obtuse angles are being caused by the Pythagorean principle of the Infinite. The principle of the Infinite appears in various expressions, such as *apeiron, aperantos, aoriston* (indefinite), *mallon-kai-hetton* (more-and-less), *meizon-kai-ellason* (greater-and-smaller), *ametria* (incommensurability), infinite motion and change, *ep'apeiron proodos* (infinite progress). These expressions clearly point to an anthyphairetic infinite, e.g. the more-and-less or the greater-and-smaller are the words used by Plato in the *Philebus* 23b-25e to express the philosophical analogue of the anthyphairetic infinite (cf. Negrepontis [35]).

1. The Pythagorean principle of the Infinite has been interpreted, in Sect. 13.12, as the infinite anthyphairesis of the diameter d to the side s of the square.
2. The infinite anthyphairesis of the diameter d to the side s of a square is equivalently described by the infinite double sequence of the side p_n and diameter q_n numbers, $n = 1, 2, \ldots$ (since by Proposition II.10/11,
   $\text{Anth}(d, s) = [1, 2, 2, 2, \ldots]$, and
   $\text{Anth}(q_n, p_n) = [1, 2, 2, 2, \ldots, 2 \text{ (n-1 times) for } n = 1, 2, \ldots]$.
3. The sequence of the side and diameter numbers satisfies the Pell property $q_n^2 = 2p_n^2 + (-1)^n$ (Proposition II.10/11.b).
4. The angle ω_n, given by

   Definition. We let ω_n be the angle contained by the equal sides p_n, p_n, of the isosceles triangle with sides p_n, p_n, q_n.

   is determined by the side and diameter numbers, and thus by the anthyphairesis of the diameter to the side of a square. The Pell property of the side and diameter numbers, together with Propositions II.12, 13 applied to the isosceles triangle with sides p_n, p_n, q_n, show the following.

Proposition 13.7 *The angle ω_{2n-1} is acute, and the angle ω_{2n} is obtuse for every $n = 1, 2, \ldots$ (as shown for the first few cases in Fig. 13.12).*

Proof By *the Pell property*, $q_{2n-1}^2 = 2p_{2n-1}^2 - 1$, hence $q_{2n-1}^2 < 2p_{2n-1}^2$, hence, by Propositions II.4/5, II.12, and II.13, the angle ω_{2n-1} contained by the equal sides

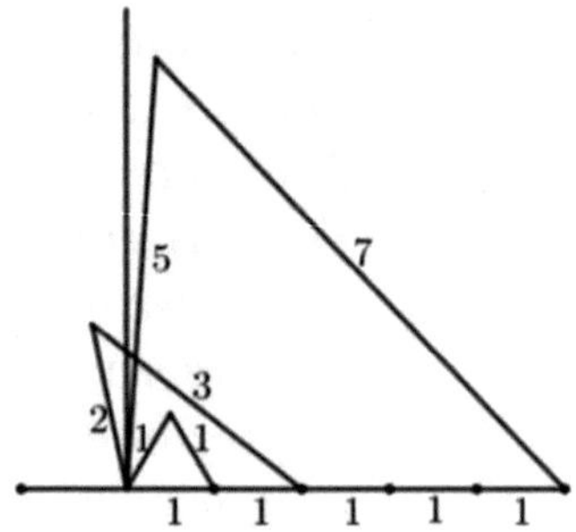

Fig. 13.12 Representation of the first three isosceles triangles generated by the side and diameter numbers, approximating the limit right triangle

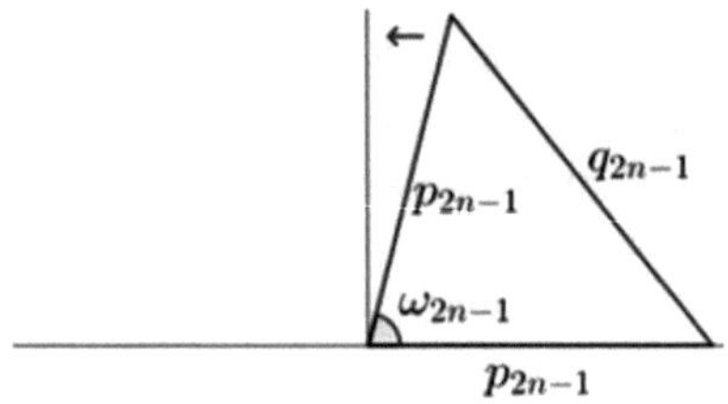

Fig. 13.13 General isosceles triangle generated by side and diameter numbers of odd index, approximating the right triangle from below with an acute angle

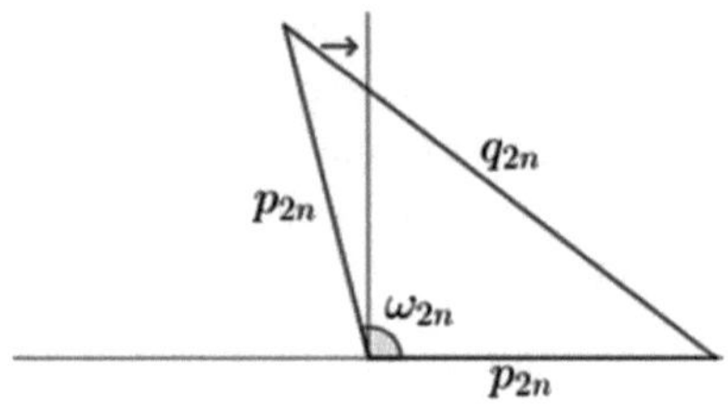

Fig. 13.14 General isosceles triangle generated by side and diameter numbers of even index, approximating the right triangle from above with an obtuse angle

p_{2n-1}, p_{2n-1} of the isosceles triangle with sides $p_{2n-1}, p_{2n-1}, q_{2n-1}$ is *acute* (see Fig. 13.13); and by *the Pell property*, $q_{2n}^2 = 2p_{2n}^2 + 1$, hence, $q_{2n}^2 > 2p_{2n}^2$, hence, by Propositions II.4/5, II.12, and II.13, the angle ω_{2n} contained by the equal sides p_{2n}, p_{2n} in the isosceles triangle with sides p_{2n}, p_{2n}, q_{2n} is *obtuse* (see Fig. 13.14). □

5. Both the increasing sequence of acute angles ω_{2n-1} and the decreasing sequence of obtuse angles ω_{2n} converge to the right angle as n goes to infinity.

q_1, p_1										
										q_2, p_2
	q_3, p_3									
									q_4, p_4	
		q_5, p_5								
								q_6, p_6		
			q_7, p_7							
							q_8, p_8			
				...						
						...				
					$d^2 = 2s^2$					

Indeed, the diameter q_n and the side p_n of the isosceles triangle tend to infinity, and informally we have

$$q_n^2/p_n^2 = 2/1 - (1/p_n^2),$$

and thus the sequence of angles (ω_n) approaches and converges to the right angle. Hence, the terms of the sequence (ω_{2n-1}) of acute angles eventually exceed every acute angle, and the terms of the sequence (ω_{2n}) of obtuse angles eventually become smaller than every obtuse angle.

6. This allows for a definition of the two kinds of angles, the obtuse and the acute angle, ultimately in terms of the infinite anthyphairesis of the diameter to the side of a square, as follows:

   Pythagorean definition of the acute and the obtuse angle.
   An angle ω is *acute* if there is an *odd* number $2n - 1$, such that $\omega < \omega_{2n-1}$, and
   an angle ω is *obtuse* if there is an *even* number $2n$, such that $\omega > \omega_{2n}$.

   Note. At this point it is explained why the two Propositions II.12, 13, although proved by means of the Pythagorean theorem and the Gnomonic Propositions II.4 and II.7, appear in Book II *later than expected*, in fact after the two Propositions II.9 and 10, intimately connected with the Pell property of the side and diameter numbers.

   The reason is that Propositions II.12 and 13 are first needed, together with *the Pell property* for the side and diameter numbers (Proposition II.10/11.c), to derive the fact that the Pythagoreans were producing the two of the three kinds of angles, the obtuse and the acute angle, from the Pythagorean principle of the Infinite.

   The position of Propositions II.12 and 13 suggests that we apply these two propositions to the side and diameter numbers in order to obtain acute and obtuse angles.
7. Thus Proclus' claim that *the Pythagorean principle of the Infinite* is the cause of the acute and the obtuse angles is confirmed if we interpret the Pythagorean principle of the Infinite as the infinite anthyphairesis of the diameter to the side of a square.
8. And this in turn confirms our interpretation, in Sect. 13.12, that *the Pythagorean principle of the Infinite* coincides with the infinite anthyphairesis of the diameter to the side of a square.

13.15.2 The Cause of the Right Angle on the One Hand Is Said by Proclus to Be the Pythagorean Principle Finite, and on the Other Hand by Our Interpretation Is the Preservation of the Gnomons in the Anthyphairesis of the Diameter to the Side of a Square, Suggesting that the Pythagorean Finite Coincides with the Preservation of the Gnomons in the Anthyphairesis of the Diameter to the Side of a Square

Proclus claims that

the right angle is produced from the principle of the Finite.

The principle of the Finite is described by the terms *peras* (finite), *horos*, *horismos* (limit).

1. The Pythagorean principle of the Finite has been interpreted, in Sect. 13.13, as the preservation of the Gnomons in the anthyphairesis of the diameter to the side of a square; in the notation of Sect. 13.12,

$$b^2 = 2bc_1 + c_1^2 = G_1,$$

$$c_1^2 = 2c_1c_2 + c_2^2 = G_2.$$

2. By Proposition 3, $bc_2 = c_1^2$, and since $c_1 = a-b$, $c_2 = b-2c_1 = b-2(a-b) = 3b-2a$, we have, by II.7, $b(3b-2a) = (a-b)^2 = a^2+b^2-2ab$,
$3b^2 - 2ab = a^2 + b^2 - 2ab$,
$2b^2 = a^2$.
Thus the preservation of the Gnomons results in the equality $a^2 = 2b^2$, which, by the Pythagorean theorem II.4/5 and its converse II.12 & 13, is equivalent to the right angle.
3. Thus Proclus' claim that *the Pythagorean principle of the Finite* is the cause of the right angle is confirmed if we interpret the Pythagorean principle of the Finite as the preservation of the Gnomons in the anthyphairesis of the diameter to the side of a square.
4. And this in turn confirms our interpretation, in Sect. 13.12, that *the Pythagorean principle of the Finite* coincides with the preservation of the Gnomons in the anthyphairesis of the diameter to the side of a square.

Note Morrow, p. 104 [48], writes in his Note 93 to Proclus' 129.17:

> A cryptic reference to the distinction between right angles and obtuse or acute angles. See 131.13–134.7.

This remark shows that he has not understood the connection of the Pythagorean treatment of the three kinds of angles with the anthyphairetic proof of incommensurability of diameter to the side of a square, and the resulting side and diameter numbers and their Pell property. To the best of our knowledge our interpretation is new.

13.15.3 The Meaning that the Pythagoreans Assigned to the Fourth Postulate in Euclid's Elements: All Right Angles Are Equal According to Proclus, In Euclidem 191,5–15

There is something anomalous with the status of Postulate 4. Proclus, *In Euclidem* 188,11–189,12 [48] gives a quite simple proof of Postulate 4; thus, in Euclid's scheme, Postulate 4 is not really a Postulate, but a Proposition with a proof. Proclus' passage describing the establishment of Postulate 4 from the Pythagorean and anthyphairetic principle of the Finite, strikes us as a Pythagorean form of Thales theorem. We describe our interpretation in Section 13.15.5, below.

13.15.4 Proclus, In Euclidem 191,5–15

[1] Φανερὸν δὲ καὶ ἐκ τοῦδε τοῦ αἰτήματος, ὅτι
ἡ ὀρθότης τῆς γωνίας
τῇ ἰσότητι
συγγενής ἐστιν,
ὥσπερ ἡ ὀξύτης καὶ ἀμβλύτης
τῇ ἀνισότητι. 191,5–7 [48]

This postulate also shows that
rightness of angles
is akin to equality,
as acuteness and obtuseness
are akin to inequality.

[2] καὶ γάρ ἐστιν·
ἡ μὲν ὀρθότης αὐτῇ
τῇ ἰσότητι
σύστοιχος
"ἀμφότεραι γὰρ ὑπὸ τὸ πέρας,
ὥσπερ δὴ καὶ ἡ ὁμοιότης"
ἡ δὲ ὀξύτης καὶ ἀμβλύτης
τῇ ἀνισότητι,
καθάπερ καὶ ἡ ἀνομοιότης·
ἀπειρίας γὰρ ἔκγονοι πᾶσαι.
διὸ καὶ
οἱ μὲν τὸ ποσὸν ὁρῶντες τῶν γωνιῶν
τὴν ὀρθὴν ἴσην τῇ ὀρθῇ λέγουσιν,
οἱ δὲ τὸ ποιὸν ὁμοίαν.
ὅπερ γάρ ἐστιν ἐν ποσοῖς ἡ ἰσότης
τοῦτο ἐν τοῖς ποιοῖς ἡ ὁμοιότης.

191,7–15 [48]
Rightness is in the same column with equality,
for both of them belong under the Finite,
as does likeness also.

But acuteness and obtuseness
are akin to inequality,
as is also unlikeness;
for all of them are the offspring of the Unlimited,

This is why
those who look at the quantity of angles
say that a right angle is equal to a right angle,
while others, looking at the quality, say it is similar.
For similarity has the same position among qualities
that equality has among quantities.

13.15.5 Our Interpretation of Proclus' Passage on Postulate 4

The first difference of the right angle from the class of acute and obtuse angles is the difference between *equality and inequality*, and this difference is a consequence of the fact that the cause of the right angle is the principle of Finite, while the cause of the acute and obtuse angles is the principle of the Infinite (in Proclus, *In Euclidem* [48] 131, 9–132, 17, and especially in 191, 5–15).

Postulate 4 All the right angles are equal.

Proof (From the Pythagorean Principle of the Finite) Let ω_1, ω_2 be two right angles. We construct two isosceles right triangles, one with right angle ω_1 and with hypotenuse a and sides b, the other with right angle ω_2 and with hypotenuse c and sides d. For example, assume that $a^2 = 2b^2$, and set $c = a + 2b, d = a + b$, as in the Elegant Theorem. Then we have two triangles, as depicted below in Fig. 13.15.

By the Pythagorean theorem II.4/5, $a^2 = 2b^2$, $c^2 = 2d^2$.

Then by Proposition 13.12.2, $ad = bc$.

By Proposition 13.12.3, $\text{Anth}(a, b) = \text{Anth}(c, d) = [1, \text{period}(2)]$. □

We thus see that indeed the equality of the infinite anthyphairesis follows from the Pythagorean principle of the Finite, namely the preservation of Gnomons/Application of Areas in Excess from the second to the third step of the

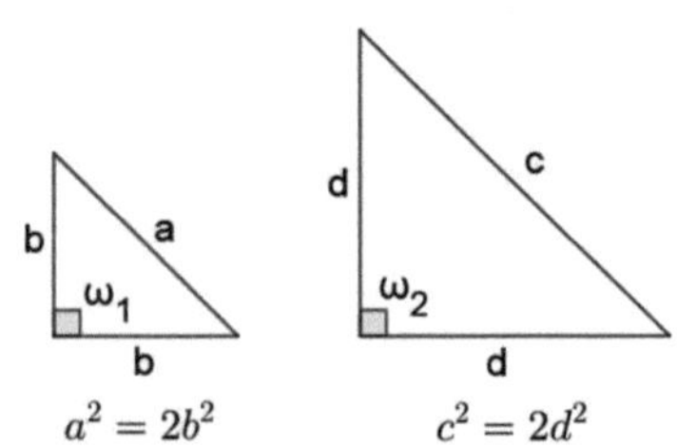

Fig. 13.15 Postulate 4

anthyphairesis of the diameter to the side of a square. Thus, by *equality of all right angles* (*Postulate 4*) the Pythagoreans mean that in *every isosceles right triangle* the anthyphairesis of the hypotenuse to the side is *equal* to $[1, \text{period}(2)]$. The Postulate is thus seen as a form of Thales' Theorem without ratios of magnitudes.

13.16 Our Answers to the Questions About Pythagorean Mathematics and Philosophy

At this point we feel justified to declare that our reconstruction of the original Pythagorean proof of incommensurability, given in Sect. 13.12, should no longer be regarded as tentative, but reasonably firmly established.

Q1. *What is the method of the original proof of incommensurability by the Pythagoreans?*

Answer to Q1. It is seen that Book II is divided into two parts, one before the Pythagorean incommensurability and the other after, and suggests an exact position for the Pythagorean Proposition of proof of incommensurability, at point II.8/9.a, right in the middle of these two parts, together with the tools that are to be used. This suggestion is similar to the earlier suggestion by which we were led to the reconstruction of the original Pythagorean proof of the Pythagorean theorem at Proposition II.4/5, by Plato's *Meno* 84d3–85b7. Proclus' passage on the three kinds of angles is of great importance and usefulness. It establishes that the Pythagorean proof of incommensurability is anthyphairetic and uses Application of Areas and Geometric Algebra.

Q2. *How did the Pythagoreans conceive of the side and diameter numbers?*

Answer to Q2. The Pythagoreans defined them first as the unique relatively prime numbers whose anthyphairesis is equal to a finite initial segment of the infinite anthyphairesis of the diameter to the side of a square.

Q3. *What is the role of Proposition II.8 in Book II? Where is it used?*

Answer to Q3. Proposition II.8 is used in our reconstruction of the Pythagorean incommensurability proof.

Q4. *Why is the converse of the Pythagorean theorem so late in Book II?*

Answer to Q4. The converse of the Pythagorean theorem is delayed because it is proved exactly where it is needed and used, namely for the proof of II.13/14.a, the Pythagorean approach to the three kinds of angles.

Q5. *Do Propositions II.5/6, 6/7 really belong in Book II? Do Propositions II,5/6, 6/7 indicate that the Pythagoreans were practicing Geometric Algebra? And if so, is there a Pythagorean use of Application of Areas II.5/6 or II.6/7?*

Answer to Q5. *Geometric Algebra* and *Application of Areas* are used in an algebraic manner in the original Pythagorean proof of incommensurability, specifically for the computation of the second and third anthyphairetic quotients. This appears to be the only case of use of the Application of Areas in Excessby the Pythagoreans. Our reconstruction of the original Pythagorean proof is confirmed (a) by our anthyphairetic interpretation of the two Pythagorean principles, Infinite and Finite/Finitizing, and (b) by the definition of the three kinds of angles and the derivation of Postulate 4 by the Pythagoreans, in terms of the Pythagorean principles Infinite and Finite. Raising the tentative nature of our reconstruction of the Pythagorean incommensurability implies that we raise as well the tentative nature of the Pythagorean use of Geometric Algebra.

Q6. *Did the Pythagoreans know proof by mathematical induction, and did they prove the Pell property of the side and diameter numbers, or did they simply verify a few initial cases of the Pell property?*

Answer to Q6. We have argued earlier on the basis of the accounts by Theon, Iamblichus, but mainly by Proclus' *Commentary to the Republic*, where we have noted two striking linguistic similarities that suggest an inductive proof of the Pell property by means of Proposition II.10. Proposition II.13/14.a provides a strong independent argument that the Pythagoreans had indeed a proof of the Pell property of the side and diameter numbers, since now the Pythagoreans actually ***use*** the Pell property in order to define the three kinds of angles from the higher principles of Infinite and Finite. *It is safe to assume that the Pythagoreans would not use the Pell property if they had not proved it rigorously*. Hence the very presence of this passage shows that, despite the claims of Mueller [34], Unguru [60] and Fowler [22], the Pythagoreans not only had knowledge of induction, but in fact employed induction in order to prove the Pell property of the side and diameter numbers. In fact we may regard symbolically the Pythagorean proof of the Pell property as *the birth of mathematical induction*.

Q7. *What is the meaning of the two Pythagorean principles Infinite and Finite?*

Answer to Q7. Propositions II.13/14.a,b confirm clearly that the two Pythagorean principles, the Infinite and the Finite are to be interpreted as infinite anthyphairetic division, and "preservation" of Gnomons/Application of Areas, respectively. That Pythagorean infinite is anthyphairetic can also be seen from the Philolaus *Fragment* 6, where the Infinite can hardly mean anything else besides the Infinity of the musical anthyphairesis (Section 13.13.5).

Q8. *Dating of the discovery of the Pythagorean incommensurability.*

Answer to Q8. Both the association of the Pythagorean discovery with Hippasus and more convincingly our anthyphairetic interpretation of Zeno's arguments and paradoxes, in particular of Zeno's *Fragment* B3, show that Zeno's philosophy was shaped in imitation of the Pythagorean anthyphairetic proof of the incommensurability of the diameter to the side of a square, and thus the Pythagorean discovery is prior to Zeno's time.

Q9. *Why did Euclid not retain the original Pythagorean proof of incommensurability in Book II of his* Elements*?*

Answer to Q9. It should be clear that Euclid made many changes in Book II. Thus the Pythagorean theorem was moved from II.4/5 to I.47, its converse from II.12 & 13 to I.48, the Application of Areas in Defect II.5/6 and in Excess II.6/7, to the more general Propositions VI.28 & 29, respectively, but II.11 was retained since it was needed before Book VI, for the construction of the regular pentagon in Book IV. In line with these changes, the Pythagorean proof of incommensurability of the diameter to the side of a square, which as we have shown belongs in position II.8/9, was omitted due to the presence of a more general theorem in the problematic Proposition X.9.

As a final conclusion, our approach unifies the restored/completed Book II of the *Elements*, which in its Euclidean form appears as a collection of mostly unrelated Propositions: it emerges that the principal original purpose of Book II is *to develop the tools and provide the proof for the incommensurability of the diameter to the side of a square* (and also for the incommensurability of the mean and extreme ratio following from Proposition II.11), and *to provide the basic properties of the side and diameter numbers*.

Note Unless otherwise indicated, the translations of the ancient Greek passages in the paper follow the translations given in this list, with modifications by the authors.

Acknowledgements We would like to thank Dimitrios Protopapas for several essential suggestions and Ayberk Zeytin for help on the technical aspect of the paper.

References

1. F. Acerbi, Plato: Parmenides 149a7-c3. A proof by complete induction? Arch. History Exact Sci. **55**, 57–76 (2000)
2. Alexander of Aphrodisias, *On Aristotle Prior Analytics 1.23–31*, Translated by Ian Mueller (Bloomsbury, 2006)
3. H. Alexandrinus, *Hesychii Alexandrini Lexicon*, ed. by P.A. Hansen, I.C. Cunningham. Founded by: K. Latte, vol. 1A, (De Gruyter, 2018)
4. T.M. Apostol, Irrationality of the square root of two – a geometric proof. Am. Math. Mon. **107**, 841–842 (2000)
5. Aristotle, *The Categories*, translated by H. P. Cook, William Heinemann (Harvard University Press, Cambridge, 1960)
6. Aristotle, *Physics*, translated by R.P. Hardie and R.K. Gaye, in *The Complete Works of Aristotle. The Revised Oxford Translation*, ed. by J. Barnes, vol. 1, Princeton (1984)
7. Aristotle, *Prior Analytics*, translated, with Introduction, Notes, and Commentary by R. Smith (Hackett Publishing Company Indianapolis I Cambridge, 1989)
8. Aristotle, *Topics Books I and VIII, with Excerpts from Related Texts.* Translated with a Commentary by R. Smith (Clarendon Press, Oxford, 1997)
9. Aristotle, *Nicomachean Ethics*, translated and edited by R. Crisp (Cambridge University Press, 2004)

10. Aristotle, *Metaphysics*, translated with Introduction and Notes by C. D. C. Reeve (Hackett Publishing Company, Indianapolis, 2016)
11. G. Baloglou, Y. Thomaidis, *From Plato's Rational Diameter to Proclus' Elegant Theorem*, arXiv:2011.07335v4 [math.HO] (2022)
12. O. Becker, Die Lehre vom Geraden und Ungeraden im neunten Buch der Euklidischen Elemente, Quellen und Studien zur Geschichte der Mathematik. Astron. Phys. **3**, 533–553 (1936)
13. V. Blåsjö, In defence of geometrical algebra. Arch. History Exact Sci. **70**, 325–359 (2016)
14. Boethius, *De Institutione Musica [Fundamentals of music]*, translated, with introduction and notes by C. M. Bower, ed. by C. V. Palisca (Yale University Press, New Haven, 1989)
15. C. A. Bretschneider, 1870. *Die Geometrie und die Geometer vor Euklides*, Leipzig.
16. W. Burkert, *Lore and Science in Ancient Pythagoreanism* (Harvard University Press, Cambridge, 1972) (original German version: Weisheit und Wissenschaft: Studien zu Pythagoras, Philolaos und Platon, Carl, Nurnberg, 1962)
17. G. Chrystal, *Textbook of Algebra* (A. and C. Black, Edinburgh, 1889)
18. J.H. Conway, *The Power of Mathematics*, in volume *Power*, ed. by A.F. Blackwell, D. MacKay (Cambridge University Press, 2006), pp. 36–50
19. J.H. Conway, J. Shipman, Extreme proofs I: The irrationality of $\sqrt{2}$. Math. Intell. **35**, 2–7 (2013)
20. D.H. Fowler, Book II of Euclid's elements and a pre-Eudoxan theory of ratio part 2: sides and diameters. Arch. History Exact Sci. **26**, 193–209 (1982)
21. D.H. Fowler, The story of the discovery of incommensurability, revisited, in *Trends in the Historiography of Science*, ed. by K. Gavroglu, J. Christianidis, E. Nicolaidis, Boston Studies in the Philosophy of Science, vol. 151 (Springer, 1994), pp. 221–235
22. D.H. Fowler, Could the Greeks have used mathematical induction? Did they use it? Physis **31**, 253–265 (1994)
23. D.H. Fowler, *The Mathematics of Plato's Academy, a New Reconstruction*, 2nd edn. (Clarendon Press, Oxford, 1999). (First edition, 1987)
24. H. Freudenthal, 1953. *Zur Geschichte der vollständingen Induktion*, In *Archives Internationales d' Histoire des Sciences*, 22, pp. 17–37.
25. H. Hankel, 1874. *Zur Geschichte der Mathematik im Altertum und Mittelalter*, Leipzig, 95–103.
26. G.H. Hardy, E.M. Wright, *An Introduction to the Theory of Numbers* (Clarendon Press, Oxford, 1938)
27. T.H. Heath, *A History of Greek Mathematics, volume I, From Thales to Euclid*, Oxford (1921)
28. T.L. Heath, *The Thirteen Books of Euclid's Elements* (Cambridge University Press, Cambridge, 1926). (First edition 1908, Second Edition 1926, Reprinted by Dover Publications, New York, 1956)
29. C.A. Huffman, *Philolaus of Croton, Pythagorean and Presocratic* (Cambridge University Press, 1993)
30. Iamblichus, *On the Pythagorean Life*, translated with Notes and Introduction by G. Clark (Liverpool University Press, 1989)
31. Jamblique, *In Nicomachi arithmeticam*, Edition and translation by N. Vinel (Fabrizio Serra Editore, Pisa, 2014)
32. W.R. Knorr, *The Evolution of Euclidean Elements: A Study of the Theory of Incommensurable Magnitudes and Its Significance for Early Greek Geometry* (Reidel, Dordrecht 1975)
33. W.R. Knorr, 'Rational diameters' and the discovery of incommensurability. Am. Math. Mon. **105**, 421–429 (1998)
34. I. Mueller, *Philosophy of Mathematics and Deductive Structure in Euclid's Elements* (MIT Press, Cambridge, 1981)
35. S. Negrepontis, The anthyphairetic nature of the Platonic principles of Infinite and Finite, in *Proceedings of the 4th Mediterranean Conference on Mathematics Education, Palermo, 28–30 January 2005, volume I*, ed. by A. Gagatsis et al. (2005), pp. 1–26

36. S. Negrepontis, The periodic Anthyphairetic nature of the one of the second hypothesis in Plato's Parmenides, in *Proceedings of the Conference Mathématiques et musique des Grecs à Euler, Strasbourg, 10–11 September, 2015*, ed. by X. Hascher, A. Papadopoulos (Hermann, Paris, 2015)
37. S. Negrepontis, Zeno's arguments and paradoxes are not against motion and multiplicity but for the separation of true Beings from sensibles, in *Essays in Geometry. Dedicated to Norbert A'Campo*, ed. by A. Papadopoulos (European Mathematical Society Press, 2023), pp. 949–998
38. S. Negrepontis, The anthyphairetic interpretation of knowledge as recollection in Plato's Meno 80d-86e and 97a-98b, in *Handbook of the History and Philosophy of Mathematical Practice*, ed. by Bh. Sriraman (Springer, 2024), pp. 471–519
39. S. Negrepontis, V. Farmaki, *History of Greek Mathematics, Volume I* (Ekkremes, Athens, 2019) [in Greek]
40. S. Negrepontis, D. Protopapas, The reconstruction of Theaetetus' theory of ratios of magnitudes, in *Essays in Topology. Dedicated to Valentin Poénaru*, ed. by L. Funar, A. Papadopoulos (European Mathematical Society, 2024)
41. S. Negrepontis, V. Farmaki, D. Kalisperi, Pre-Eudoxian geometric algebra, in *Ganita Bharati, Bulletin of the Indian Society for History of Mathematics*, vol. 44(2) (2022)
42. R. Netz, *A New History of Greek Mathematics* (Cambridge University Press, 2022)
43. A. Papadopoulos, *Zbl 07727353, Zentralblatt Review of L. Corry, British Versions of Book II of Euclid's Elements: Geometry, Arithmetic, Algebra (1550–1750)*, Springer Briefs in History of Science and Technology (Springer, Cham, 2022)
44. Pappus, *Commentary on Book X of Euclid's Elements*, Arabic text and translation by W. Thomson; Introductory Remarks, Notes, and a Glossary of technical terms by G. Junge and W. Thomson, Cambridge (1930)
45. Plato, *Parmenides, Theaetetus, Sophist, Politicus, Philebus, Epinomis*, translated by H. N. Fowler (Harvard University Press/W. Heinemann, Cambridge/London, 1921)
46. Plato, *Republic*, translated by P. Shorey (Harvard University Press/W. Heinemann, Cambridge/London, 1969)
47. Plutarch, *Lycurgus and Numa*, translated by B. Perrin, William Heinemann (Harvard University Press, Cambridge, 1967)
48. Proclus, *A Commentary on the First Book of Euclid's Elements*, translated with Introduction and Notes by G. R. Morrow (Princeton University Press, Princeton, 1970)
49. Proclus, *Commentary on Plato's Republic, Volume II, Essays 7–15*, translated with an introduction and notes by D. Baltzly, J.F. Finamore, G. Miles (Cambridge University Press, 2022)
50. H. Rademacher, O. Toeplitz, *Von Zahlen und Figuren Proben Mathematischen Denkens fur Liebhaber der Mathematik Ausgewählt und Darstellt* (Springer, Berlin, 1930)
51. R. Rashed, *Thabit ibn Qurra, Science and Philosophy in Ninth-Century Baghdad* (Walter de Gruyter, Berlin, 2009)
52. K. Saito, 1985. Book II of Euclid's elements in the light of the theory of conic sections. Historia Scientiarum **28**, 31–60 (1985) (reprinted in J. Christianidis, 2004 editor, *Classics in the History of Greek Mathematics*, Kluwer Academic Publishers, 2004, pp. 139–168)
53. Simplicius, *On Aristotle Physics 1.3–4*, Translated by P. Huby and C.C.W. Taylor, General editor: Richard Sorabji, Bristol Classical Press, Ancient Commentators on Aristotle (2011)
54. Á. Szabó *The Beginnings of Greek Mathematics*, Synthese historical library 17, Reidel, 1978. Originally published as *Anfänge griechischen Mathematik* (Oldenbourg Wissenschaftsverlag, 1969)
55. P. Tannery, *De la solution géométrique des problèmes du second degré avant Euclide*, in his *Mémoires scientifiques*, ed. by J.L. Heiberg, H.G. Zeuthen (Toulouse and Paris, 1912) (1882), pp. 254–280
56. Theon of Smyrna, *Mathematics Useful for Understanding Plato*, Translated from the 1892 Greek/French edition of J. Dupuis by R. and D. Lawlor (Wizards Bookshelf, San Diego, 1979)
57. Thesaurus Linguae Graecae: a Digital Library of Greek Literature (TLG, Irvine, 2001)

58. O. Toeplitz, *Die Entwicklung der Infinitesimalrechnung. Eine Einleitung in die Infinitesimalrechnung nach der Genetischen Methode* (Springer, Berlin, 1949)
59. S. Unguru, On the need to rewrite the history of Greek mathematics. Arch. History Exact Sci. **15**, 67–114 (1975)
60. S. Unguru, Greek mathematics and mathematical induction. Physis **28**, 273–289 (1991)
61. B.L. van der Waerden, *Science Awakening* (Noordhoff, Groningen, 1954) (Originally published as *Ontwakende wetenschap*. Noordhoff, Groningen, 1950)
62. B.L. van der Waerden, Defence of a "shocking" point of view. Arch. History Exact Sci. **15**, 199–210 (1976)
63. B.L. van der Waerden, *Geometry and Algebra in Ancient Civilizations* (Springer, Berlin, 1983)
64. A. Wedberg, *Plato's Philosophy of Mathematics* (Almqvist and Wiksell, Stockholm, 1955)
65. A. Weil, Who betrayed Euclid? (Extract from a letter to the Editor). Arch. History Exact Sci. **19**, 91–93 (1978) (reprinted in J. Christianidis, 2004 editor, *Classics in the History of Greek Mathematics*, Kluwer Academic Publishers, 2004, pp. 447–450)
66. H.G. Zeuthen, *Die Lehre von den Kegelschnitten im Altertum* (Kopenhagen, 1886) (reprint, Hildesheim, 1966)
67. L. Zhmud, *The Origin of the History of Science in Classical Antiquity*. Translated from the Russian by A. Chernoglazov (W. de Gruyter, Berlin, 2006)

Index

A. M. Uludağ, A. Zeytin (eds.), *Essays on Geometry*,
https://doi.org/10.1007/978-3-031-76257-4

MIX
Papier aus verantwortungsvollen Quellen
Paper from responsible sources
FSC® C105338

If you have any concerns about our products,
you can contact us on
ProductSafety@springernature.com

In case Publisher is established outside the EU,
the EU authorized representative is:
Springer Nature Customer Service Center GmbH
Europaplatz 3, 69115 Heidelberg, Germany

Printed by Libri Plureos GmbH
in Hamburg, Germany